VITAMINS		MINERALS											
Vitamin E RDA (mg/day)[e]	Vitamin K AI (µg/day)	Calcium AI (mg/day)	Phosphorus RDA (mg/day)	Magnesium RDA (mg/day)	Iron RDA (mg/day)	Zinc RDA (mg/day)	Iodine RDA (µg/day)	Selenium RDA (µg/day)	Copper RDA (µg/day)	Manganese AI (mg/day)	Flu[illegible]/day)	[illegible]ium [illegible]y)	[illegible]denum [illegible]g/day)
4	2.0	210	100	30	0.27	2	110	15	200	0.003	0.01	[illegible]	[illegible]
5	2.5	270	275	75	11	3	130	20	220	0.6	0.5	[illegible]	[illegible]
6	30	500	460	80	7	3	90	20	340	1.2	0.7	[illegible]	[illegible]
7	55	800	500	130	10	5	90	30	440	1.5	1.0	15	22
11	60	1300	1250	240	8	8	120	40	700	1.9	2	25	34
15	75	1300	1250	410	11	11	150	55	890	2.2	3	35	43
15	120	1000	700	400	8	11	150	55	900	2.3	4	35	45
15	120	1000	700	420	8	11	150	55	900	2.3	4	35	45
15	120	1200	700	420	8	11	150	55	900	2.3	4	30	45
15	120	1200	700	420	8	11	150	55	900	2.3	4	30	45
11	60	1300	1250	240	8	8	120	40	700	1.6	2	21	34
15	75	1300	1250	360	15	9	150	55	890	1.6	3	24	43
15	90	1000	700	310	18	8	150	55	900	1.8	3	25	45
15	90	1000	700	320	18	8	150	55	900	1.8	3	25	45
15	90	1200	700	320	8	8	150	55	900	1.8	3	20	45
15	90	1200	700	320	8	8	150	55	900	1.8	3	20	45
15	75	1300	1250	400	27	13	220	60	1000	2.0	3	29	50
15	90	1000	700	350	27	11	220	60	1000	2.0	3	30	50
15	90	1000	700	360	27	11	220	60	1000	2.0	3	30	50
19	75	1300	1250	360	10	14	290	70	1300	2.6	3	44	50
19	90	1000	700	310	9	12	290	70	1300	2.6	3	45	50
19	90	1000	700	320	9	12	290	70	1300	2.6	3	45	50

[e] Vitamin E recommendations are expressed as α-tocopherol.

SOURCE: Adapted with permission from the *Dietary Reference Intakes* series, National Academy Press. Copyright 1997, 1998, 2000, 2001, by the National Academy of Sciences. Courtesy of the National Academy Press, Washington, D.C.

MINERALS									
Zinc (mg/day)	Iodine (µg/day)	Selenium (µg/day)	Copper (µg/day)	Manganese (mg/day)	Fluoride (mg/day)	Molybdenum (µg/day)	Boron (mg/day)	Nickel (mg/day)	Vanadium (mg/day)
4	—	45	—	—	0.7	—	—	—	—
5	—	60	—	—	0.9	—	—	—	—
7	200	90	1000	2	1.3	300	3	0.2	—
12	300	150	3000	3	2.2	600	6	0.3	—
23	600	280	5000	6	10	1100	11	0.6	—
34	900	400	8000	9	10	1700	17	1.0	—
40	1100	400	10,000	11	10	2000	20	1.0	1.8
40	1100	400	10,000	11	10	2000	20	1.0	1.8
34	900	400	8000	9	10	1700	17	1.0	—
40	1100	400	10,000	11	10	2000	20	1.0	—
34	900	400	8000	9	10	1700	17	1.0	—
40	1100	400	10,000	11	10	2000	20	1.0	—

NOTE: An Upper Limit was not established for vitamins and minerals not listed and for those age groups listed with a dash (—) because of a lack of data, not because these nutrients are safe to consume at any level of intake. All nutrients can have adverse effects when intakes are excessive.

SOURCE: Adapted with permission from the *Dietary Reference Intakes* series. National Academy Press. Copyright 1997, 1998, 2000, 2001, by the National Academy of Sciences. Courtesy of the National Academy Press, Washington, D.C.

GLOSSARY OF NUTRIENT MEASURES

kcal: kcalories; a unit by which energy is measured.

g: grams; a unit of weight equivalent to about 0.03 ounces.

mg: milligrams; one-thousandth of a gram.

µg: micrograms; one-millionth of a gram.

IU: international units; an old measure of vitamin activity determined by biological methods (as opposed to new measures that are determined by direct chemical analyses). Many fortified foods and supplements use IU on their labels.

- For vitamin A, 1 IU = 0.3 µg retinol, 3.6 µg β-carotene, or 7.2 µg other vitamin A carotenoids.
- For vitamin D, 1 IU = 0.025 µg cholecalciferol.
- For vitamin E, 1 IU = 0.67 natural α-tocopherol (other conversion factors are used for different forms of vitamin E).

mg NE: milligrams niacin equivalents; a measure of niacin activity.

- 1 NE = 1 mg niacin.
  = 60 mg tryptophan (an amino acid).

µg DFE: micrograms dietary folate equivalents; a measure of folate activity.

- 1 µg DFE = 1 µg food folate.
  = 0.6 µg fortified food or supplement folate.
  = 0.5 µg supplement folate taken on an empty stomach.

µg RAE: micrograms retinol activity equivalents; a measure of vitamin A activity.

- 1 µg RE = 1 µg retinol.
  = 12 µg β-carotene.
  = 24 µg other vitamin A carotenoids.

Basic Nutrition Counseling Skill Development

Basic Nutrition Counseling Skill Development

A Guideline for Lifestyle Management

KATHLEEN D. BAUER
CAROL A. SOKOLIK

MONTCLAIR STATE UNIVERSITY

Australia • Brazil • Japan • Korea • Mexico • Singapore • Spain • United Kingdom • United States

Kathleen D. Bauer, Ph.D., R.D., is the founder and has been the director of the Nutrition Counseling Clinic at Montclair State University for more than ten years. She teaches both undergraduate and graduate nutrition counseling courses. Her applied nutrition counseling experiences extend to wellness programs, fitness centers, hospitals, nursing homes, and private practice. A major area of interest has been the development of innovative nutrition education programs and materials for the general public and higher education. Dr. Bauer has also been director of the dietetics program at Montclair State University for the past fourteen years.

Carol A. Sokolik, M.S., R.D., is a founder and is currently the codirector of the Approved Pre-Professional Practice Program of Montclair State University and teaches undergraduate courses in the dietetics program. She is a dietetic consultant to several nursing homes in northern New Jersey. She has applied her nutrition counseling skills in a variety of settings, including physicians' offices and long-term care and acute care facilities.

Basic Nutrition Counseling Skill Development: A Guideline for Lifestyle Management
Kathleen D. Bauer, Carol A. Sokolik

Publisher: Peter Marshall

Development Editor: Elizabeth Howe

Assistant Editor: John Boyd

Editorial Assistant: Andrea Kesterke

Marketing Manager: Becky Tollerson

Advertising Project Manager: Brian Chaffee

Signing Representative: Melissa Lerner

Project Manager: Sandra Craig

Print/Media Buyer: Robert King

Permissions Editor: Stephanie Keough-Hedges

Production: Heidi E. Marschner

Text Designer: Brenda Duke

Copy Editor: Laura Larson

Cover Designer: Roger Knox

Cover Images: © PhotoDisc, Inc.

Compositor: Parkwood Composition Service, Inc.

All chapter opening photos:
Digital Imagery © copyright PhotoDisc, Inc.

Library of Congress Control Number: 2001026540
ISBN-13: 978-0-534-58977-6
ISBN-10: 0-534-58977-4

Wadsworth
10 Davis Drive
Belmont, CA 94002-3098
USA

Cengage Learning is a leading provider of customized learning solutions with office locations around the globe, including Singapore, the United Kingdom, Australia, Mexico, Brazil, and Japan. Locate your local office at: **www.cengage.com/global**

Cengage Learning products are represented in Canada by Nelson Education, Ltd.

To learn more about Wadsworth, visit **www.cengage.com/wadsworth**

Purchase any of our products at your local college store or at our preferred online store **www.ichapters.com**

Printed in the U.S.A
9 10 11 12 13 14 15 12 11 10

To my husband, Hank, and my children,
Emily so mee Rose and Kathryn sun hee Rose
KDB

To my husband, Van, and my children,
Katie and Kimberly
CAS

Thank you for patience, support, and love.

Contents

Chapter 5

PROMOTING CHANGE TO FACILITATE SELF-MANAGEMENT

Chapter 6

MAKING BEHAVIOR CHANGE LAST

Chapter 7

PHYSICAL ACTIVITY

Chapter 8

PROFESSIONALISM AND FINAL ISSUES

Chapter 9

GUIDED COUNSELING EXPERIENCE

Preface

Traditionally, nutrition counseling consisted of disseminating a list of facts with the expectation that behavior would change if clients simply had adequate information. However, changing lifelong behavior patterns is a complex process, and the results of this strategy were often disappointing. Poor counseling skills decrease the likelihood that nutrition messages have been clearly understood or that a behavior change will actually take place.

More recently, nutritionists have been exploring behavior change models and intervention strategies to find ways of having a greater impact on behavior change. An array of methods are available today. This book helps nutrition counseling students, entry-level practitioners, and seasoned professionals gain a fundamental understanding of the current knowledge and new developments in nutrition counseling.

Specifically, this interactive text supplies a step-by-step guided approach by providing (1) an exposure to a variety of counseling theories, behavior change models, and counseling approaches commonly used in lifestyle behavior counseling; (2) a guide to learning basic skills universal to all counseling strategies; and (3) a well-defined counseling protocol on which to build and alter according to a counselor's expertise and needs of the counseling situation.

DISTINGUISHING FEATURES OF THIS TEXT

1. **Practical examples:** This book is filled with concrete examples and first-person accounts in a variety of wellness, private practice, and institutional settings of how to implement various theoretical approaches, counseling strategies, and common nutritional assessment and documentation tools.
2. **Putting it all together-a four-week guided nutrition counseling program:** The text includes a step-by-step guide for a student working with a volunteer adult client for four sessions. The objective of this section is to demonstrate how the theoretical discussions, practice activities, and nutrition tools can be put together for an effective intervention.
3. **Integration of multicultural issues:** Multicultural issues have been integrated throughout the text. Case studies and first-person accounts of actual counseling experiences provide insight into the complexities of counseling across cultures. A cross-cultural algorithm is provided and elaborated on to guide cross-cultural interventions. For instructors who prefer to teach this concept as one unit, a cross-cultural outline that cuts across all chapters is supplied on Wadsworth's Web site for this text (www.nutrition.wadsworth.com).
4. **Action based:** Exercises are integrated into the chapters to give students ample opportunity and encouragement to interact with the concepts covered in each chapter. Instructors can choose to assign the activities to be done individually at home, or many of them can be used as classroom activities. Students are encouraged to journal their responses to the exercises to be used for classroom discussions, distance learning, or their own reflection. Instructors can assign journal entries and collect them for evaluation. This method provides a deep understanding of how students are grasping concepts. Each chapter has a culminating assignment and a case study that integrates all or most of the major topics covered throughout the chapter.
5. **Web site:** A Wadsworth Nutrition Counseling Web site supports the text; see www.nutrition.wadsworth.com. It contains copies of all the nutrition clinic forms, nutritional assessment questionnaires, and fact sheets.

BOOK DESIGN

The first eight chapters address fundamental nutrition counseling knowledge and skills. The final chapter outlines a step-by-step nutrition counseling intervention building on the learning experiences in the preceding chapters. The sequential flow of the chapters follows the needs of students to develop knowledge and skills during each step of the counseling process. Using the format of the book, broad student behavioral objectives flow in the following order:

1. Develop a foundation of knowledge about nutrition counseling goals and theories.
2. Understand the basics of an effective counseling relationship and broaden an understanding of their own traits and skills that can impact the relationship.
3. Develop interpersonal skills needed for a productive counseling intervention.
4. Utilize effective counseling responses.
5. Implement a counseling model and a motivational algorithm to direct the flow of a counseling interaction.
6. Follow a step-by-step protocol to implement a counseling intervention.
7. Develop a nutrition care plan.
8. Chart a counseling intervention using two methodologies.
9. Select appropriate food management tools according to a client's desire for structure.
10. Develop a client-friendly behavior tracking method.
11. Implement various behavior change strategies.
12. Utilize basic relapse prevention techniques.
13. Follow a protocol to incorporate physical activity into a nutrition counseling session.
14. Follow procedures for ending a counseling relationship and evaluating the counseling process.
15. Develop an understanding of various professional issues, such as ethics, standards, and client rights.
16. Explain selected topics useful to a beginning nutrition counselors, including procedures for handling difficult client behaviors and group counseling.

Chapter Content

Chapter 1 introduces students to the foundations of the counseling process, including an overview of selected counseling theories and behavior change models. The components of an effective counseling relationship are reviewed. The reader is encouraged to reflect on his or her own attributes and evaluate how they interface with counseling objectives.

Chapter 2 begins the process of skill development, including basic counseling responses. Emphasis is placed on six relationship-building responses, since development of a supportive relationship is the foundation of most counseling approaches and can be the instrument of change no matter what counseling method has been selected.

In Chapter 3, a nutrition counseling algorithm based on the stages of change model, motivational interviewing, solution-focused therapy, and the health behavior change method provides the framework for a systematic counseling protocol that is detailed throughout the chapter. This is integrated into cross-cultural nutrition counseling guidelines.

The basics of developing a nutrition care plan are attended to in Chapter 4, including goal setting, standard assessment procedures, and documentation. Chapters 5 and 6 review various types of educational and behavioral strategies for promoting a behavioral change and making the change last.

Research has shown that nutrition counselors find the need to address physical activity issues in their counseling sessions, and many do provide guidance for rudimentary exercise plans. Recently, the American College of Sports Medicine and the American Dietetic Association have been collaborating on ways to enhance crossover expertise. Therefore, a protocol for nutrition counselors counseling about physical activity has been supplied in Chapter 7. To assist clients in making changes in their physical activity patterns, counselors need a basic understanding of the interrelationship between physical movement and health. Since nutritionists have cited a need for greater expertise in this area, the basics of physical activity and wellness are addressed in this chapter.

Guidelines for ending and evaluating the counseling relationship have been outlined in Chapter 8. In addition, this chapter covers an array of professional concerns including professional standards, ethics, and the boundaries between psychotherapy and nutrition counseling. Finally, some special topics addressing the needs of a beginning counselor are covered in this chapter, including the basics of group counseling and a review of strategies for handling difficult client behavior.

The guided approach in Chapter 9 of the text has a step-by-step guide with a checklist of tasks to accomplish for four nutrition counseling sessions with a volunteer client. These also can be adapted to specific counseling needs.

Lifestyle Management Forms

Lifestyle Management forms are included in Appendix G and are found on the Wadsworth Nutrition Counseling Web site. These are tools that are utilized for completing chapter exercises and Chapter 9 intervention sessions.

Acknowledgments

We appreciate the helpful comments made by our reviewers. We would like to thank:

Francie Astrom, Ohio University
Katherine Cason, The Pennsylvania State University
Alana Cline, University of Northern Colorado
Carol Friesen, Ball State University
Patricia Hodges, University of Alabama
Tanya Horacek, Syracuse University
Marilyn Hughes, University of Georgia
Lisa Nicholson, University of Southern California
Linda Snetselaar, University of Iowa
Connie Vickery, University of Delaware

Chapter

PREPARING TO MEET YOUR CLIENT

We and They
Father, Mother, and Me
Sister and Auntie say
All the people like us are We,
And every one else is They
And They live over the sea,
While We live over the way,
But—would you believe it?—They look upon We
As only a sort of They!
We eat pork and beef
With cow-horn-handled knives.
They who gobble Their rice off a leaf,
Are horrified out of Their lives;
And They who live up a tree,
Feast on grubs and clay,
(Isn't it scandalous?) look upon We
As a simply disgusting They!
We eat kitcheny food.
We have doors that latch.
They drink milk or blood,
Under an open thatch.
We have Doctors to fee.
They have wizards to pay.
And (impudent heathen!) They look upon We
As a quite impossible They!
All good people agree,
And all good people say,
All nice people, like Us, are We
And every one else is They:
But if you cross over the sea,
Instead of over the way,
You may end by (think of it!) looking on We
As only a sort of They!

—Rudyard Kipling

BEHAVIORAL OBJECTIVES:

- Define nutrition counseling.
- Identify goals of a nutrition counselor.
- Identify major tenets of selected theoretical approaches to counseling.
- Identify major concepts of selected behavior change models.
- Describe characteristics of an effective counselor.
- Describe one's own cultural values.
- Identify factors affecting clients in a counseling relationship.
- Evaluate self for strengths and weaknesses in building a counseling relationship.
- Identify novice counselor issues.

KEY TERMS:

- **BEHAVIOR CHANGE MODELS:** a conceptual framework for analyzing and explaining behavior change
- **CULTURAL COMPETENCE:** a set of knowledge and interpersonal skills that allows individuals to increase their understanding and appreciation of cultural differences and similarities
- **CULTURAL GROUPS:** nonexclusive groups that have a set of values in common; an individual may be part of several cultural groups at the same time.
- **CULTURE:** learned patterns of thinking, feeling, and behaving that are shared by a group of people
- **ETHNOCENTRISM:** believing a particular cultural view is best
- **MODELS:** generalized descriptions used to analyze or explain something

- **MOTIVATION:** a state of readiness to change
- **MULTICULTURAL:** a description of a situation in which two or more distinct cultures interact
- **SELF-EFFICACY:** a behavior change model based on our personal belief of how capable we are of exercising control over events in our life
- **SELF-MOTIVATIONAL STATEMENTS:** arguments for making a behavior change made by the client
- **THEORETICAL TENETS:** major beliefs and assumptions held true by a particular theory
- **THEORY:** an explanation based on observation and reasoning of what happens in the world
- **WORLDVIEW:** the perception of the world as biased by culture and personal experience

COUNSELING DEFINED

N*utrition counseling* has been defined as the process of guiding a client toward a healthy nutrition lifestyle by meeting normal nutritional needs and solving problems that are barriers to change.[1] Haney and Leibsohn[2] have designed a **model** of counseling to enable the guiding to be effective, indicating that

> *counseling can be defined as an interaction in which the counselor focuses on client experience, client feeling, client thought, and client behavior with intentional responses to acknowledge, to explore, or to challenge. (p. 5)*

NUTRITION COUNSELING GOALS

Using the Haney and Leibsohn model of counseling, we may identify three specific goals in nutrition counseling. The first is to *facilitate lifestyle awareness,* which can be achieved by keeping the focus on your client; acknowledging feelings, experience, and behavior; and providing information. Exploring feelings, ambivalence, inner strengths, behavior, and alternative options can increase the likelihood of obtaining the second goal, *healthy lifestyle decision making.* The ultimate goal in nutrition counseling is for your client to *take appropriate action* to obtain a healthier lifestyle and become self-sufficient.* This is done by exploring issues and challenging your client to view his or her situation differently. In particular, nutrition counselors help clients take appropriate actions by encouraging them to take risks, tolerate incongruities, and give new behaviors and thoughts a chance before discounting them.[3] Nutrition counselors have a host of behavior change strategies that can be explored and tailored to meet the individual and cultural needs of their clients.

EXERCISE 1.1—DOVE Activity: Broadening Our Perspective (Awareness)

D—defer judgment
O—offbeat
V—vast
E—expand on other ideas

Divide into groups of three. Your instructor will select an object, such as a cup, and give you one minute to record all of the possible uses of the object. Draw a line under your list. Take about three minutes to share each other's ideas, and write the new ideas down below the line. Discuss other possibilities for using the object with your group and record these in your journal. Use the DOVE technique to guide your thinking and behavior during this activity. Do not pass judgment on thoughts that cross your mind or on the suggestions of others. Allow your mind to think of a vast number of possibilities that may even be offbeat. How many more ideas occurred with sharing? Did you see possibilities from another perspective? One of the goals of counseling is to help clients see things using different lenses. What does this mean? How does this activity relate to a counseling experience? Write your thoughts in your journal and share them with your colleagues.

Source: Dairy, Food, and Nutrition Council, *Facilitating Food Choices: Leaders Manual* (Cedar Knolls, NJ: Author; 1984).

***Although self-sufficiency is often stated as the ultimate goal of nutrition counseling, it should be noted that among a significant number of older diabetic patients, Anderson and colleagues (1998) found that these individuals do not desire an independent self-care role. Also, the guidelines developed by the expert panel for the National Institutes of Health (NIH) report, *Identification, Evaluation, and Treatment of Overweight and Obesity,* headed by Dr. F. Xavier Pi-Sunyer acknowledge that a weight maintenance program consisting of diet therapy, behavior therapy, and physical activity may need to be continued indefinitely for some individuals (NIH Publication 98-4083, 1998).**

THEORETICAL APPROACHES FOR LIFESTYLE AWARENESS AND MANAGEMENT

To alter food habits successfully, nutrition counselors work with clients to change behaviors influenced by a multitude of social, cultural, psychological, and physiological factors. Applying an ecological perspective to nutrition counseling, these factors can be numerated as they frequently apply to dietary behavior:[4]

- Awareness, **motivation,** and attitudes about the role of nutrition in both health and disease
- Types and amounts of foods eaten
- Food preparation methods
- Sensory perceptions
- Economic and time commitments
- Food availability
- Housing features
- Societal and family pressures

Historically, dietitians ignored many of these factors and attempted to change food behavior by simply dispensing facts and diets. The results were disappointing. Eventually, nutrition professionals recognized a need for a new approach to nutrition counseling. During the 1980s, the focus was on behavior modification, giving way to goal setting and client-centered counseling in the 1990s. More recently, the transtheoretical model and motivational interviewing have provided guides for instituting behavior change in the health arena. An array of counseling philosophies, theories, and behavior change strategies are currently available to deal with the complex process of changing behaviors influencing health. The following discussion summarizes the theoretical approaches most often identified as useful in guiding and appraising a nutrition counseling session. Theories provide the conceptual framework for helping practitioners understand the dynamics of health behaviors, including the effects of internal and external forces, and suggest processes for eliciting behavior changes. Some of the concepts overlap among the theories.

When I started working for the WIC Program, I worried that I might have trouble totally accepting an unmarried client who was pregnant or had a baby. However, my biggest problem was accepting the fact that the young women were very pleased with themselves and full of positive expectations about the upcoming births of their children.†

Person-Centered Therapy

Carl Rogers was the founder of person-centered therapy, also referred to as "nondirective" or "client centered."[5] The basic assumption in this **theory** of counseling is that humans are basically rational, socialized, and realistic and that there is an inherent tendency to strive toward growth, self-actualization, and self-direction. Clients actively participate in clarifying needs and exploring potential solutions.[6] They realize their potential for growth in an environment of unconditional positive self-regard. Counselors help develop this environment by totally accepting clients without passing judgments on their clients' thoughts, behavior, or physical self. This approach includes respecting clients, regardless of whether they have followed medical and counseling advice.

Total acceptance is extremely important for a level of trust to develop at which clients feel comfortable to express their thoughts freely. This portion of the theory has special meaning for nutrition counselors. A study of nutrition counselor perceptions and attitudes toward overweight clients indicates a need for training in sensitivity and empathy.[7]

Another important component of this theory for a nutrition counselor is the underlying assumption that simply listening to knowledge cannot help a client. In person-centered therapy, clients discover within themselves the capacity to use the relationship to change and grow, thereby promoting wellness and independence. Listening to a client's story has been compared to the role of a pharmacologic agent.[8] Nutrition counselors should not lose sight of the fact that the educational component of dietary therapy has been shown to be extremely valuable;[9] however, the person-centered theory of counseling can help guide nutrition counselors when facts should be kept to a minimum.

Behavioral Therapy

Behavioral counseling evolved from behavioral theories developed by Ivan Pavlov, B. F. Skinner, Joseph Wolpe, Edward Thorndike, and Albert Bandura. The

†Numerous first-person accounts from dietetic students or nutrition counselors working in the field are included throughout this book.

premise of this type of counseling is that many behaviors are learned, so it is possible to learn new ones. The focus is not on maintaining will power but on changing the environment so that it will be conducive to learning new behaviors. Williams[10] identifies three approaches to learning forming the basis for behavior modification:

1. **Classical conditioning** focuses on antecedents (stimuli, cues) that affect food behavior. For example, seeing or smelling food, watching television, studying, or experiencing boredom may be a stimulus to eat. In nutrition counseling, clients may be encouraged to identify and eliminate cues, such as removing the cookie jar from the kitchen counter.
2. **Operant conditioning** is based on the law of effect, which states that behaviors can be changed by their positive or negative effect. In nutrition counseling, generally a positive approach to conditioning is applied, such as a reward for obtaining a goal. The change in diet itself can be the reward, as in the alleviation of constipation by an increased intake of fluids and fiber.
3. **Modeling** is observational learning, such as learning by watching a video or demonstration, observing an associate, or hearing a success story.

I walked into the hospital room of an obese teenage boy to give a discharge calorie-controlled, weight reduction diet. As soon as I introduced myself and explained the purpose of my visit, the boy said he didn't want another diet. He said he tried them all before, and none of them worked. He said he was fat, his whole family was fat, and that is the way it would always be. Although I was sympathetic to his blight, I proceeded to explain the diet. During the whole explanation, he rolled his eyes, and the rest of his body language indicated that he was annoyed with me. Even at the time I knew that the encounter was not productive. I just transmitted a bunch of facts, even though he obviously was not listening. I felt it was my responsibility to go over the diet with him and chart in his record that the diet order was accomplished. Now that I have had a counseling course, I believe I would have spent the limited time I had with him dealing with his frustration and would have told him to come see me as an outpatient after discharge if he had a change of heart. Now I wouldn't even attempt to go over the diet.

Gestalt Therapy

Developed by Frederick Perls, Gestalt therapy emphasizes confronting problems. Counselors operate in the here and now instead of the past or the future.[11] A practical application of this theory in nutrition counseling is for clients to take responsibility for making dietary changes,[12] rather than blaming a spouse for not buying appropriate foods or a work supervisor for causing stress. Setting realistic goals with your client can aid the concern of disowning ownership of problems.[12]

Once I counseled a fifty-year-old man who had recently undergone bypass surgery. Our sessions did not appear to be very productive because he consistently blamed his wife for his inability to change his diet despite the fact that she was cooking low-fat meals. After we focused our discussions on who exactly was responsible for making changes, his ability to institute changes improved.

Cognitive Therapies

Leaders in this field include Albert Ellis, who developed *rational emotive behavior therapy* (REBT);[13,14] Aaron T. Beck, who developed *cognitive therapy* (CT);[15,16] and Donald Meichenbaum,[17] who developed *cognitive behavior modification.* The premise of this approach is that negative self-talk and irrational ideas are self-defeating learned behaviors and the most frequent source of people's emotional problems. Clients are taught that harmful self-monologues should be identified, eliminated, and replaced with productive self-talk. By influencing a person's pattern of thinking, feelings and actions are modified.[18] An example of an individual with a high cholesterol level using negative self-talk and creating an emotional turmoil for herself would be "I am a fool for eating that cheesecake. I have no self-control. I'll just die of a heart attack." This could be changed into better-coping self-talk: "I am learning how to handle these situations. Next time I will ask for a small taste. I am on the road to a healthier lifestyle."

Cognitive therapists have developed a number of techniques to improve positive feelings and help problem-solving ability.[1,19] These include relaxation training and therapy, mental imagery, thought stopping, meditation, biofeedback, cognitive restructuring and systematic desensitization. See Chapter 6 for elaboration on several of the strategies.

Family Therapy

The objective of family therapy is to assist individuals and families to change themselves and the system of relationships within the family.[20] Unless nutrition counselors have professional family counseling credentials, no attempt should be made to change complex family dynamics. However, an appreciation for the influences of significant relationships can help guide a nutrition counseling intervention. Hopefully there are friends or family members who are willing to actively support your clients in their attempts to make lifestyle changes. Including critical family members in counseling sessions is crucial among particular **cultures** for the success of an intervention. In some cases, significant others can impart a negative influence and ways to lessen the impact need to be explored. Family therapy has been shown to be a particularly successful mode of therapy in treating adolescent obesity.[3]

> *One of my clients was a forty-two-year-old overweight woman who lived in a group home. She had a lovely personality. Having no articulation ability, she could only communicate by facial expressions and body language. This client's diet instructions were hand-drawn pictures on a page kept on the refrigerator. As she consumed appropriate foods, she would put an x through the picture. A new diet page was posted everyday. This client did beautifully with the program, but her overweight sister was not happy. The sister took my client home on weekends and encouraged her to overeat. It wasn't until I included her sister in our counseling sessions that my client was truly successful.*

Solution-Focused Therapy

Insoo Kim Berg developed solution-focused therapy, and Steve de Shazer[21] brought the topic to international attention. Solution-focused therapists work with their clients to concentrate on solutions that have worked for them in the past and identify strengths to be expanded upon and used as resources. Focus of sessions is not on discovering and solving problems but may well be on the exception to the normal course of action—that is, the one time the client was able to positively cope. By investigating the accomplishment, no matter how small, adaptive strategies are likely to emerge. For example, a middle-aged executive who complains that business lunches and dinners are a frequent difficulty would be asked to think of an occasion when healthy food was consumed at one of these meals. After identifying the skills the executive used to make the meal a healthy experience, the nutrition counselor would focus on helping replicate and expand those skills. The aim is for clients to use solution-oriented language—to speak about what they can do differently, what resources they possess, and what they have done in the past that worked. Language (solution-talk) provides the guide in solution-focused therapy. Examples of questions a solution-focused therapist may ask include the following:

- What can I do that would be helpful to you?
- Was there a time when you ate a whole-grain food?
- When was the last time you ate fruit?
- Has a family member or friend ever encouraged you to eat low-sodium foods?

> *In the cardiac rehabilitation center where I worked, there was a client whose quality of life was severely affected by his weight. He was working as a security guard and had difficulty climbing steps or walking any reasonable distance due to his weight and his need to lug oxygen. After several months of trying a variety of intervention strategies, I asked him whether he had ever been on a diet that worked. He said the only time he lost weight was when he cut bread out of his diet. We set "no more bread" as a goal, and that was the beginning of a successful weight loss program that did allow grains in other forms, such as cereal, pasta, and rice.*

Multicultural Counseling

Conventional helping theories developed by predominately white males of Euro-American descent have gender and cultural limitations and are likely to disregard family issues.[3] Effective **multicultural** therapy respects unfamiliar perspectives and provides culturally appropriate interventions. Specifically, a counselor working with a culturally different client requires technique flexibility and training in **cultural sensitivity** regarding complaints and symptoms, cultural norms, and expectations and conceptions of the helping process.[22] Although multicultural issues are integrated throughout this book, an application of cross-cultural counseling techniques in nutrition counseling is detailed in Chapter 3.

Multicultural counseling and the other approaches we have discussed in the preceding sections are summarized in Table 1.1.

EXERCISE 1.2—Focus on Continuing

Think about what happens in your life (relationships, habits, activities, etc.) that you would like to continue to happen. Record two of these in your journal and identify what skills you have that facilitate these situations to exist.

Source: de Shazer S, *Keys to Solution in Brief Therapy* (New York: Norton; 1985).

TABLE 1.1—Summary of Counseling Approaches

COUNSELING APPROACH	KEY CONCEPTS
Person-centered therapy	• Counselors develop an environment of unconditional positive self-regard.
Behavioral therapy	• Focus is on changing the environment. • Behavior modification techniques address cues, substitutions, and consequences.
Gestalt therapy	• Clients take responsibility for making dietary changes.
Cognitive therapy	• Focus is on changing negative self-talk and irrational ideas.
Family therapy	• Significant relationships are explored as a means to provide support or to lessen the impact of negative influences.
Solution-focused therapy	• Clients focus on identifying strengths and expanding on past successes.
Multicultural counseling	• Counselors appreciate multiple perspectives and provide culturally appropriate interventions.

OVERVIEW OF BEHAVIOR CHANGE MODELS AND APPROACHES

Making a decision to change one's diet and implementing that decision is guided by a complex interaction of psychological factors.[23] Sigman-Grant[4] states that sixty different **behavior change models** have been developed to explain changes in health behavior and used to guide nutrition counseling interventions. A variety of counseling approaches has been developed based on these models. See Table 1.2 for an overview of the ones most often cited as a guide for explaining the nature and dynamics of food behavior. Those that provided the conceptual framework for the Motivational Nutrition Counseling Algorithm presented in Chapter 3 are discussed more fully below.

I once counseled a Hindu couple; the wife was about three months' pregnant and was fluent in English. Throughout the interview, the husband responded to every question I asked—no matter how detailed. The wife remained quiet during the whole session. At first I found myself getting angry; however, the wife did not seem to be bothered by this arrangement. After I realized there were obvious cultural behavior patterns that I should respect, the interview went more smoothly for me.

Self-Efficacy

The concept of **self-efficacy** as a basic component of behavior change was developed by Albert Bandura.[24] Although sometimes considered a separate model, self-efficacy has been widely accepted and incorporated into numerous behavior change models. Cormier and Cormier[25] define self-efficacy as "our personal belief of how capable we are of exercising control over events in our life" (p. 565), such as a belief in ability to change food patterns. Attainment of health behavior changes has been found to correlate solidly with a strong self-efficacy, [26] probably because self-perception of efficacy affects individual choices, the amount of effort put into a task, and willingness to pursue goals when faced with obstacles.[27] As a result, a person's confidence in his or her ability to accomplish a behavior change is more important than actual skill.[24]

In Chapter 5, several behavior change strategies to support self-efficacy are discussed, including reinforcement, behavioral contracting, encouragement, and tracking. However, giving your clients hope by letting them know that options and strategies are available to deal with their problems, pointing out strengths, setting achievable goals, relating success stories, and expressing optimism for their future

TABLE 1.2—Summary of Behavior Change Models and Approaches

BEHAVIOR CHANGE MODEL/APPROACH	FOCUS	KEY CONCEPTS
Self-efficacy	A component of numerous behavior change models; belief in ability to make a behavior change	• Positive self-efficacy increases probability of making a behavior change.
Health belief model	Perception of the health problem and appraisal of proposed behavioral changes are central to a decision to change.	• Perceived susceptibility • Perceived impact • Perceived advantages of change • Appraisal of barriers • Self-efficacy
Locus of control	Behavior change is more likely to occur if there is a feeling of personal control over life events.	• Social support • Choices
Social cognitive theory/ social learning theory	People and their environment interact continuously, each influencing the other.	• Self-efficacy • Knowledge and skills required • Learning occurs through taking action, observations of others taking action, and evaluation of the results of those actions.
Transtheoretical model	Behavior change is explained as a readiness to change.	• Behavior change is described as a series of changes. • Specific behavior change strategies are identified for each stage.
Motivational interviewing	Exploration of ambivalence	• Reduce resistance • Self-discovery is encouraged.
Health behavior change method	Guiding framework for method is patient centered and directive.	• Reduce resistance • Exchange information • Negotiate behavior change • Readiness to take action is based on importance and confidence.

counselor can promote self-efficacy during a counseling session.

Health Belief Model

The **health belief model** proposes that cognitive factors influence an individual's decision to make and maintain a specific health behavior change. Central to making this decision are the following beliefs:[23]

- The individual is susceptible to the health problem.
- The specific disease can severely impact quality of life.
- Changing health behavior will reduce the risk of the disease.
- Barriers to making the behavior change can be overcome with reasonable effort.
- The individual is capable of making the change (self-efficacy).

These beliefs interact with each other to determine a client's willingness to take action. For example, a women who loves to eat sweets may believe that she is susceptible to getting dental cavities, but if she perceives the adverse effect on her life to be minimal, then she will not have an impetus to change. Studies have shown that a person with few overt symptoms has lower dietary adherence.[28] Similarly, a man may believe that eating a plant-based diet will reduce his cholesterol levels, but he may not be willing or feel capable of taking the necessary steps to make the change.

Transtheoretical Model

This model, also known as the stages of change model, is often referred to as transtheoretical because it crosses over many behavior change models. It has been used effectively to design and evaluate successful dietary change interventions.[29,30,31,32] This model serves two purposes: to help understand (explain) the process of behavior change and to develop and select effective intervention strategies.

Motivational Stages This model describes behavior change as a process of passing through a sequence of distinct motivational stages (that is, five levels of readiness to take action). For an intended behavior change, an individual can begin at any one of the motivational levels or stages:

1. **Precontemplation:** A person in this stage has no intention of changing and in fact resists any efforts to modify the problem behavior.[31] The reasons for this include no awareness that a problem exists, denial of a problem, awareness of the problem but unwillingness to change, or feeling of hopelessness after attempting to change.[33,34]
2. **Contemplation:** Contemplators recognize a need to change but are in a state of ambivalence, alternating between reasons to change and reasons not to change. There is concern that the long-term health benefits of the change do not compensate for the short-term real or perceived costs.[35] Perceived barriers such as unacceptable tastes, economic constraints, or inconvenience are major obstacles. People can be stuck in this stage for years waiting for absolute certainty, the magic moment, or just wishing for different consequences without changing behavior.[36]
3. **Preparation:** Preparers believe the advantages outweigh the disadvantages of changing and are committed to take action in the near future (within the next thirty days). They may have taken small steps to prepare for a change, such as making an appointment with a nutrition counselor or inquiring about a walking club. A person in this stage would probably be willing to try a new recipe or to taste some new foods.
4. **Action:** Clients are considered to be in this stage if they have altered the target behavior to an acceptable degree for one day or up to six months and continue to work at it.[35] Although changes have been continuous in this stage, the new behaviors should not be viewed as permanent. The most common time for relapse to occur is the first three to six months of the action stage.[37]
5. **Maintenance:** A person in this stage has been engaging in the new behavior and is consolidating the gains attained during previous stages.[38] However, the individual needs to work actively to modify the environment to maintain the changed behavior and prevent a relapse.[37] As Prochaska and Norcross[39] explain, "Perhaps most important is the sense that one is becoming more of the kind of person one wants to be."

A review of the various stages would indicate that behavior change occurs in a linear order, in which people "graduate" from one stage to the next.[33] However, it is normal for individuals to slip back one or more stages,

EXERCISE 1.3—Determine Your Stage

The following is a list of health behaviors commonly accepted as desirable. Review the stages of change, and circle the corresponding number that indicates your stage.

Behavior					
• Floss teeth at least once a day.	1	2	3	4	5
• Exercise at least ninety minutes a week.	1	2	3	4	5
• Go to the dentist at least once a year.	1	2	3	4	5
• Eat at least five servings of fruits and vegetables a day.	1	2	3	4	5
• Always use a seat belt.	1	2	3	4	5
• Refrain from smoking.	1	2	3	4	5
• Consume at least one thousand milligrams of calcium every day.	1	2	3	4	5
• Eat at least two servings of whole grains every day.	1	2	3	4	5
• Consistently use sunscreens.	1	2	3	4	5

In your journal, write what you learned about yourself. Describe what you learned about the stages of change construct.

Source: This activity was adapted from one developed by Mary Finckenor, Adjunct Professor, Montclair State University, Upper Montclair, New Jersey. Used with permission.

even to have a relapse and then start to move forward again, progressing toward maintenance.[33] (See Lifestyle Management Form 6.4 in Appendix G.) Figure 1.1 depicts the concept that there is forward and backward movement in the various stages. Smoking research, for example, has shown that people commonly recycle four times through various stages before achieving long-term maintenance.[40] The fact that change is not perfectly maintained should not be viewed in a negative light.[38] By knowing from the onset that perfection is not realistic and lapses are to be expected, an intervention can be planned accordingly. Hopefully by understanding that relapses are a normal occurrence in the change process, clients and counselors can maintain a realistic perspective and not become demoralized when they occur.

Transtheoretical Model as a Counseling Guide
Besides helping understand and explain behavior change, the transtheoretical model also serves as a guide to identify potentially effective messages and intervention strategies to facilitate movement through the stages to reach and remain at the maintenance stage. Since the strategies clients find useful at each stage differ,[31] the treatment intervention needs to be tailored to a client's stage of change. Traditionally, nutrition interventions have not taken readiness into consideration and treated all people as if they were actively searching for ways to make behavior changes (giving information; offering advice; developing a diet plan). This approach has been counterproductive, because most individuals with dietary problems are in a preaction stage—precontemplation, contemplation, or preparation.[41] In fact, giving advice to individuals who do not believe they have a problem could make them feel beleaguered and defensive, making change even less likely to occur.[40] In some cases, nutrition counselors may have erroneously assumed that an individual enrolled in a program is ready to take action.[38] The person may in fact have decided to participate because of pressure from a loved one, or serious consideration may have been given to the problem but the person is not actually ready to make a behavior change. Authorities estimate that only 20 percent of the individuals who seek behavior change assistance are actually in the action stage.[42]

Prochaska and Norcross[39] have identified the most effective intervention strategies to assist clients' progress from one stage to another. In general, cognitive (thinking-related) and affective (feeling-related) strategies are more effective in the early stages, while behavioral (action-oriented) strategies in the latter stages are more likely to meet client needs.[38] Table 1.3 contains a summary of the stages and appropriate intervention strategies. As individuals move through stages, intervention strategies need to be adjusted; therefore, counselors need to reassess their client's stage periodically.

Using the Transtheoretical Model to Measure Outcomes By tracking movement through the various stages, the transtheoretical model has given nutrition counselors a new tool for measuring outcomes. For example, counselors should consider their intervention successful if a client has moved from "I do not need to make a change" to "Maybe I should give some thought to a change." This measure of success may provide encouragement to health professionals who become discouraged with the slow pace of change.[43]

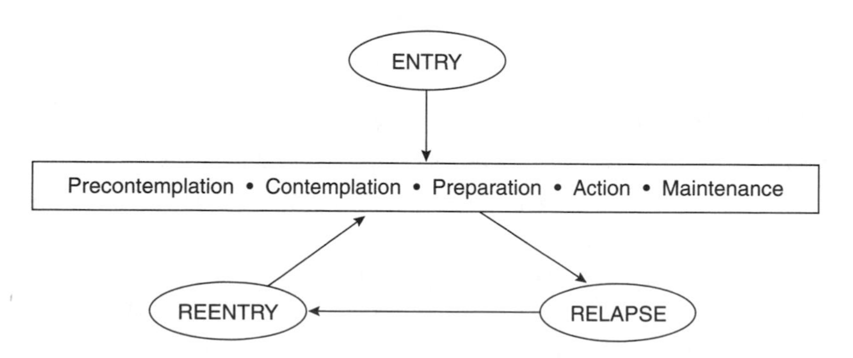

Figure 1.1 Prochaska and DiClemente's Stages of Change
Sources: Glanz K, Greene G, Shield JE, Understanding behavior change. In: American Dietetic Association, *Project Lean Resource Kit* (Chicago: American Dietetic Association, 1995). © 1995, The American Dietetic Association. "Project Lean Resource Kit." Used with permission. Also Sandoval WM, Heller KE, Wiese WH, Childs DA, Transtheoretical Model: a model of nutrition counseling. *Topics in Clinical Nutrition* 9:64–69; © 1994, Aspen Publishers, Inc.

TABLE 1.3—Stages of Change Summary

STAGE	KEY STRATEGIES	TREATMENT DO'S	TREATMENT DON'TS
Precontemplation	Increase in information and awareness; emotional acceptance.	• Provide personalized information. • Allow client to express emotions about the need to make dietary changes.	• Do not assume client has knowledge or expect that providing information will automatically lead to behavior change. • Do not ignore client's emotional adjustment to the need for dietary change, which overrides ability to process relevant information.
Contemplation	Increase in confidence in ability to adopt recommended behaviors.	• Discuss and resolve barriers to change. • Encourage support networks. • Give positive feedback about client's abilities. • Help clarify ambivalence about adopting behavior, and emphasize expected benefits.	• Do not ignore the potential impact of family members and others, on client's ability to comply. • Do not be alarmed or critical of a client's ambivalence.
Preparation	Initiation of change by resolving ambivalence, eliciting a firm commitment, and developing a specific action plan.	• Encourage client to set specific, achievable goals (e.g., use skim milk instead of whole milk). • Reinforce small changes that client may have already achieved. • Encourage client to make public the intended change.	• Do not recommend general behavior changes (e.g., "Eat less fat"). • Do not refer to small changes as "not good enough."

Motivational Interviewing

A major factor for backsliding on the readiness continuum is lack of motivation (that is, eagerness to change).[33] Motivational interviewing is an approach to counseling that complements the transtheoretical model because it entails a focus on strategies to help motivate clients to build commitment to make a behavior change. In this approach motivation is not viewed as a personality trait or a defense mechanism but considered a state of readiness to change that can fluctuate and be influenced by others.[40] Since counselors can impact motivation, to do so is considered an inherent part of their intervention responsibility. Counselors work at building motivation by using a directive, patient-centered counseling style that encourages clients to examine and resolve ambivalence about behavior change, thereby reducing resistance and encouraging action. Although motivational interviewing was originally developed to address addictive behavior, the main concepts of this process have been integrated into a brief intervention model and successfully

TABLE 1.3—Stages of Change Summary (Continued)

STAGE	KEY STRATEGIES	TREATMENT DO'S	TREATMENT DON'TS
Action	Behavioral skill training and social support.	• Develop or refer to education program to include self-management skills. • Provide self-help materials.	• Do not refer client to information-only classes. • Do not assume that initial action means permanent change.
Maintenance	Development of problem solving and encouragement of social and environmental support.	• Identify and plan for potential difficulties (e.g., maintaining dietary changes on vacation). • Collect information about local resources (e.g., support groups, shopping guides). • Provide relapse prevention counseling. • Recommend more challenging dietary changes if client is motivated.	• Do not be discouraged or judgmental about a lapse or relapse.

Source: Adapted from Kristal AR, Glanz K, Curry SJ, Patterson RE. How can stages of change be best used in dietary interventions? © 1999, The American Dietetic Association. Reprinted by permission from *Journal of the American Dietetic Association,* 99:683.

applied to promote dietary change.[44,45] These strategies are particularly useful in the early stages of behavior change when there is a great deal of ambivalence about making a decision to change.[44]

Basic Principles of Motivational Interviewing[46]

1. Express Empathy. Acceptance facilitates change. The underlying assumption of expressing empathy is an acceptance and understanding of a client's perspective. This does not mean that a counselor has the same perspective or would have made similar choices. However, basic acceptance ("You are OK") creates an environment for change.[46] A message of "You are not OK" creates resistance to change. Reluctance to change is viewed as a normal reaction, not a pathological condition.

2. Develop Discrepancy. Awareness of a discrepancy between present behavior and important goals facilitates change. The objective of motivational interviewing is to amplify discrepancy until it overrides the need to keep the present behavior. Encouraging clients to clarify important goals, vocalize reasons for change, and explore the consequences or potential consequences of their present behavior helps to create a sense of discomfort about the discrepancy and thereby assists clients in making a decision to change, particularly when acceptable alternatives have been identified.

3. Avoid Escalating Resistance. Defensiveness evokes resistance to change. Visible manifestations of resistance include engaging in denial, arguing, putting up objections, and showing reluctance to discuss a

Overview of What Is Motivational:

- Knowledge of consequences
- Self-efficacy
- A perception that a course of action has been chosen freely
- Self-analysis (giving arguments for change)
- Recognition of a discrepancy between present condition and desirable state of being
- Social support
- Feeling accepted

subject.[45] Some resistance to changing comfortable habits is expected, but counselors can escalate the condition by arguing, judging, persuading, discounting feelings, and interacting with clients as if they are looking for ways to implement a change when they simply want to consider it. If you find resistance is escalating during a counseling session, recognize it as a clue to change strategies.

4. Roll with Resistance. Ambivalence and reluctance are expected. When resistance is encountered, the behavior is acknowledged as understandable. The counseling environment should provide a comfortable atmosphere to allow clients to express fears about change without feeling judged or pressured into making a change.[45]

5. Support Self-Efficacy. Belief in the ability to change is an important motivator. Confidence in one's ability to successfully make a specific behavior change (self-efficacy) is an important component of motivation.[45]

Specific Motivational Interviewing Strategies[46]

1. Listen Reflectively. To listen reflectively means to use basic listening skills (which are covered in Chapter 2), make your best guess as to what was in the heart of your client's message, and reflect that guess back to your client. By acting as a mirror and reflecting back your understanding of the intent or your hypothesis of the underlying meaning, clients are encouraged to keep talking. This show of interest is an expression of empathy creating an environment for self-exploration about ambivalence to take action. The following dialogue illustrates a nutrition counselor listening reflectively and attempting to identify the underlying meaning of a client's statements:

Client: *Everyone is getting on my back about my cholesterol level—my wife, my doctor, my brother. I guess I have to do something about my diet.*

Counselor: *You're feeling harassed that other people are pushing you to change the way you eat.*

Client: *I suppose they're right, but I feel fine.*

Counselor: *You're worried about the future.*

Client: *Yeah. I have a lot of responsibilities. I have two children and I want to be around to take care of them, see them grow up and get married. But it doesn't thrill me to give up meatballs and pizza.*

Counselor: *You're wondering about what food habits you are willing to change.*

Client: *You know, I wouldn't mind eating more fish. I heard that was a good food to eat to lower cholesterol levels. What do you think about oatmeal?*

Note that the formulation of a response is an active process. You must decide what to reflect and what to ignore. In this dialogue example, the counselor chose to respond to the client's statement "I suppose they're right" rather than "I feel fine." The counselor guessed that if the client thought all those others were right, then he must be worried about his health. If the counselor had chosen to reflect on the feeling fine part of the client's second statement, what would have happened? Of course we can only "guess," but it doesn't seem likely that a client-initiated discussion of diet changes would have occurred so quickly. To respond reflectively is particularly useful after asking an open-ended question when you are trying to better understand your client's story.

The development of reflective listening skills can be a complex task for novice counselors.[45] If this is a skill you decide to develop, explore the motivational interviewing resources at the end of this chapter and consider attending motivational interviewing workshops.

EXHIBIT 1.1—General Motivational Interviewing Counseling Strategies

- Encourage clients to make their own appraisals of the benefits and losses of an intended change.
- Do not rush clients into decision making.
- Describe what other clients have done in similar situations.
- Give well-timed advice emphasizing that the client is the best judge of what can work.
- Provide information in a neutral, nonpersonal manner.
- Do not tell clients how they should feel about a medical or dietary assessment.
- Present choices.
- Clarify goals.
- Failure to reach a decision to change is not a failed consultation.
- Make sure clients understand that resolutions to change break down.
- Expect commitment to change to fluctuate, and empathize with the client's predicament.

Source: Rollnick S, Heather N, Bell A. Negotiating behavior change in medical settings: The development of brief motivational interviewing. *J Mental Health.* 1992;1:25–37.

2. Elicit Self-Motivational Statements. The objective of this strategy is to resolve ambivalence by providing opportunities and encouragement for the client, rather than the counselor, to make arguments for change. When clients express the need for change or the reasons why change is necessary, the balance of indecision begins to shift toward taking action. There are four categories of **self-motivational statements:**

1. Cognitive—problem recognition; for example, "I really have a serious problem here."
2. Cognitive—optimism for change; for example, "Lots of people have to take insulin. I can do it, too."
3. Affective—expression of concern; for example, "I'm so worried about my diabetes. I don't know how this could have happened to me."
4. Behavioral—intention to change; for example, "In the past I always enjoyed eating fruit. I just have to make a plan and start eating them again."

Several methods can be used to encourage clients to make these statements:

- Request clarification (how much, how many, give an example, etc.) on previous self-motivational statements.
- Formulate reflective listening statements that respond to previous self-motivational statements.
- Ask open-ended questions to explore the client's perceptions and concerns about his or her problems. (This is explained later.)
- Reinforce self-motivational statements both nonverbally (for example, a nod) and verbally with a statement such as "I can understand why this has been so difficult for you." Repeat any self-motivational comments made by your client in a summary.
- Tell the client that you are going to change roles, and ask the client to convince you to make the contemplated behavior change. Gradually allow the client to be persuaded.

3. Ask Open-Ended Questions to Elicit Self-Motivational Statements. In motivational interviewing, certain types of open-ended questions are used to evoke a self-motivational statement. To use these effectively, remember that your approach must communicate curiosity, concern, and respect. You should not appear to be conducting an inquisition to gather information against your client.

- Ask about the pros and the cons of the client's present eating pattern and the contemplated change.
- Ask about extremes related to the problem. For example, "What worries you the most?"
- Ask the client to envision the future after the change has been accomplished.
- Ask about priorities in life (that is, what is most important to the client). Then ask how this contemplated behavior change fits into the hierarchy.
- Ask your client to give a reaction to an assessment report.

Health Behavior Change Method

Rollnick et al.[45] describe a method of counseling for health care practitioners that approaches the "spirit" of motivational interviewing and addresses the needs of health professionals who are often involved in brief interventions. They provide some interesting insight based on their health care counseling experience and research of behavior change methodologies. After their analysis of various standard models of behavior change, two concepts—importance and confidence (self-efficacy)—were found to be central to the notion of readiness to make a change. These researchers maintain that both factors should be taken into consideration during an assessment of readiness to change as well as when selecting an intervention strategy. An individual may feel that a change is worthwhile and may even make self-motivational statements indicating the importance of change, but if that person has very little confidence in ability to make the change, then implementation of action strategies are not likely to be successful.

By assessing for both importance and confidence, a counselor will not waste time dealing with the importance of a change when the focus should be on increasing confidence in ability to make a change or vice versa. For example, a woman may feel confident in her ability to increase her calcium intake, but if she does not consider the project important enough, her degree of readiness is dampened. Likewise, a woman who feels an increase in calcium intake is important but does not feel confident in her ability to do so will be at a lower level of readiness to change. In general, lowest levels of readiness are often associated with low importance. Differences between the terms are illustrated in Table 1.4. The method described by Rollnick et al.[45] focuses

Garfield® **by Jim Davis**

on how to apply these questions constructively in a counseling intervention. Many of their suggestions for approaching health care counseling have been incorporated into the analysis and flow of a counseling session found in Chapter 3.

The art of nutrition counseling is an evolving process for both the profession and the professional. No one counseling orientation meets all the needs of a complex, fluid society, nor can one methodology be a perfect package for an individual nutrition counselor. Counselors must use their professional judgment regarding selection of an intervention. Addressing this issue, Gilliland et al.[46] suggest an eclectic approach merging the most useful ideas from various models. The motivational nutrition counseling algorithm presented in a step-by-step manner in Chapter 3 has taken this approach and provided a flexible client-centered solution orientation. In addition, a basic knowledge of a variety of counseling approaches is desirable to meet the needs of clients as they progress in a counseling program. Their readiness for particular interventions may change.[12]

TABLE 1.4—Three Topics in Talk about Behavior Change

IMPORTANCE **WHY?**	**CONFIDENCE** **HOW? WHAT?**	**READINESS** **WHEN?**
Is it worthwhile?	Can I?	Should I do it now?
Why should I?	How will I do it?	What about other priorities?
How will I benefit?	How will I cope with *x*, *y*, and *z*?	
What will change?	Will I succeed if . . . ?	
At what cost?	What change . . . ?	
Do I really want to?		
Will it make a difference?		

Source: Rollnick S, Mason P, Butler C. *Health Behavior Change: A Guide for Practitioners.* New York: Churchill Livingstone; © 1999; p. 21. Used with permission.

UNDERSTANDING AN EFFECTIVE COUNSELING RELATIONSHIP

No matter what theory or behavior change model is providing the greatest influence, the relationship between counselor and client is the guiding force for change. The effect of this relationship is most often cited as the reason for success or failure of a counseling interaction.[1] Helm and Klawitter[3] report that successful clients identify their personal interaction with their therapist as the single most important part of treatment. To set the stage for understanding the basics of an effective counseling relationship, you will investigate the characteristics of effective nutrition counselors, explore your own personality and culture, examine the special needs and issues of a person seeking nutrition counseling, and review the two phases of a helping relationship in the following sections.

Characteristics of Effective Nutrition Counselors

"Ideal helpers" have been described as possessing the following qualities:

> *They respect their clients and express that respect by being available to them, working with them, not judging them, trusting the constructive forces found in them, and ultimately placing the expectation on them to do whatever is necessary to handle their problems in living more effectively. They genuinely care for those who have come for help. They are nondefensive, spontaneous, and always willing to say what they think and feel, provided it is in the best interest of their clients. Good helpers are concrete in their expressions, dealing with actual feelings and actual behavior rather than vague formulations, obscure psychodynamics, or generalities.*[47] (p. 29)

After thoroughly reviewing the publications of the leading authorities in counseling, Okun[48] identified seven qualities of counselors considered to be the most influential in affecting the behaviors, attitudes, and feelings of helpees: knowledge, self-awareness, ethical integrity, congruence, honesty, ability to communicate, and gender and culture awareness. The following list describes these characteristics as well as those thought to be effective by nutrition counseling authorities:

- *Effective nutrition counselors are self-aware.* They are aware of their own beliefs, respond from an internal set of values and as a result have a clear sense of priorities. However, they are not afraid to reexamine their values and goals.[49] This awareness aids counselors with being honest with themselves as to why they want to be a counselor and avoid using the helping relationship to fulfill their own needs.[50]
- *They have a solid foundation of knowledge.* Nutrition counselors need to be knowledgeable in a vast array of subjects in the biological and social sciences as well as have an ability to apply the culinary arts. Since the science and art of nutrition is a dynamic field, the foundation of knowledge requires continuous updating. Clients particularly appreciate nutrition counselors who are experienced with the problems they face.[3]
- *They have ethical integrity.* Effective counselors value the dignity and worth of all people. Such clinicians work toward eliminating ways of thinking, speaking, and acting that reflect racism, sexism, ableism, ageism, homophobia, religious discrimination, and other negative ideologies.[51] Ethical integrity entails many facets that are addressed in the American Dietetic Association's Code of Ethics.[52]
- *They have congruence.* This means the counselor is unified. There are no contradictions between who the counselor is and what the counselor says, and there is consistency in verbal and nonverbal behaviors as well. (For example, if a client shared some unusual behavior such as eating a whole cake covered with French dressing, the counselor's behavior would not be congruent if the nonverbal behavior indicated surprise but the verbal response did not.)
- *They can communicate clearly.* Clinicians must be able to communicate factual information as well as a sincere regard for their clients. Effective nutrition counselors are able to make sensitive comments and communicate an understanding about fears concerning food and weight.[3]
- *They have a sense of gender and cultural awareness.* This requires that counselors be aware of how their own gender and culture influences them. Effective counselors have a respect for a diversity of values that arise from their clients' cultural orientations.

EXERCISE 1.4—Helper Assessment

Think of a time someone helped you, such as a friend, family member, teacher, or counselor. In your journal, write down the behaviors or characteristics the person possessed that made the interaction so effective. After reading over the characteristics of effective counselors, compare their qualities to those identified by the leading authorities. Do they differ? Share your thoughts with your colleagues.

EXERCISE 1.5—People Skills Inventory

- Do you expect the best from people? Do you assume that others will be conscientious, trustworthy, friendly, and easy to work with until they prove you wrong?
- Are you appreciative of other people's physical, mental, and emotional attributes—and do you point them out frequently?
- Are you approachable? Do you make an effort to be outgoing? Do you usually wear a pleasant expression on your face?
- Do you make the effort to remember people's names?
- Are you interested in other people—all kinds of people? Do you spend far less time talking about yourself than encouraging others to talk about themselves?
- Do you readily communicate to others your interest in their life stories?
- When someone is talking, do you give him or her 100 percent of your attention—without daydreaming, interruption, or planning what you are going to say next?
- Are you accepting and nonjudgmental of others' choices, decisions, and behavior?
- Do you wholeheartedly rejoice in other people's good fortune as easily as you sympathize with their troubles?
- Do you refuse to become childish, temperamental, moody, inconsistent, hostile, condescending, or aggressive in your dealings with other people—even if they do?
- Are you humble? Not to be confused with false modesty, being humble is the opposite of being arrogant and egotistical.
- Do you make it a rule never to resort to put-downs, sexist or ethnic jokes, sexual innuendoes, or ridicule for the sake of a laugh?
- Are you dependable? If you make commitments, do you keep them—no matter what? If you are entrusted with a secret, do you keep it—no matter what?
- Are you open-minded? Are you willing to listen to opposing points of view without becoming angry, impatient, or defensive?
- Are you able to hold onto the people and things in your life that cause you joy and let go of the people and things in your life that cause you sadness, anger, and resentment?
- Can you handle a reasonable amount of pressure and stress without losing control or falling apart?
- Are you reflective? Are you able to analyze your own feelings? If you make a mistake, are you willing to acknowledge and correct it without excuses or blaming others?
- Do you like and approve of yourself most of the time?

Affirmative answers indicate skills you possess that enhance your ability to relate to others.

Source: Adapted from Scott N, "Success Often Lies in Relating to Other People," *Dallas Morning News,* April 20, 1995, p. 14C.

- *They have a sense of humor.* Helping clients see the irony of their situation and laugh about their problems enriches relationships. Humor helps prevent clients from taking themselves and their problems too seriously.[49]
- *They are honest and genuine.* Such counselors appear authentic and sincere. They act human and do not live by pretenses hiding behind phony masks, defenses, and sterile roles.[49] Such counselors are honest and show spontaneity, congruence, openness, and willingness to disclose information about themselves when appropriate. Honest counselors are able to give effective feedback to their clients.
- *They are flexible.* This means not being a perfectionist. Such counselors do not have unrealistic expectations and are willing to work at a pace their clients can handle.[3]
- *They are optimistic and hopeful.* Clients want to believe that lifestyle changes are possible, and they appreciate reassurance that solutions will be found.[3]
- *They respect, value, care, and trust others.* This enables counselors to show warmth and caring authentically through nonjudgmental verbal and nonverbal behavior, listening attentively and behaving responsibly, such as returning phone calls and showing up on time. This behavior conveys the message that clients are valued and respected.
- *They can accurately understand what people feel from their frame of reference (empathy).* It is important for counselors to be aware of their own struggles and pain to have a frame of reference for identifying with others.[49]

Understanding Yourself—Personality and Culture

According to Brammer,[50] our personalities are one of the principal tools of the helping process. By taking an inventory of your personality characteristics, you can have a better understanding of the ones you wish to modify.

EXERCISE 1.6—How Do You Rate?

Ask a close friend or family member who you helped at one time to describe what it was about your behavior that was helpful. Write these reactions down in your journal. Review the desirable characteristics for an effective counselor described in the previous section, complete the personality inventory in Exercise 1.5, and then identify what characteristics you possess that will make you a good helper. What behaviors need improvement? Write in your journal specific ways that you need to change to improve your helper skills.

It is one of the most beautiful compensations of this life that no man can sincerely try to help another without helping himself.

—Ralph Waldo Emerson

Closely related to your personality evaluation is an examination of why you want to be a counselor. What you expect out of a counseling relationship, the way you view yourself, and the personal attitudes and values you possess can affect the direction of the counseling process. You should be aware that as a helper, your self-image is strengthened from the awareness that "I must be OK if I can help others in need." Also, since you are put into the perceptual world of others, you remove yourself from your own issues, diminishing concern for your own problems.[50]

Sometimes counselors seek to fulfill their own needs through the counseling relationship. Practitioners who have a need to express power and influence over others tend to be dictatorial, overly silent, and less likely to be open to listening to their clients. This type of counselor expects clients to obey suggestions without questions.[12] A counselor who is particularly needy for approval and acceptance will fear rejection. Belkin[53] warns that sometimes counselors try too hard to communicate the message "I want you to like me," rather than a more effective "I am here to help you." As a result, such counselors may be anxious to please their clients by trying to do everything for them, perhaps even doing favors.[12] The tendency will be to gloss over and hide difficult issues since the focus is on eliciting only positive feelings from their clients. Consequently, clients will not learn new management skills, and dietary changes will not take place.

EXERCISE 1.7—Why Do You Want to Be a Helper?

Describe in your journal what it means to you to be a helper and why you want to be a helper. How does it feel when you help someone? Is it possible that you have issues related to dominance or neediness that could overshadow interactions with your clients and hamper counseling relationships?

Another important component to understanding yourself so as to become a **culturally competent** nutrition counselor is to know what constitutes your **worldview** (cultural outlook). Each culture has a unique outlook on life, what people believe and value within their group. Our worldview has both conscious and unconscious influences. Kittler and Sucher[54] relate this unique outlook to its special meaning in the health community:

> *These standards typically express what is worth in a life well lived within a particular society and are the measures through which one assesses personal behavior and that of others. This cultural outlook, or worldview, influences individual perceptions about health and illness as well as the role each plays within the structure of society.* (p. 29)

Your worldview is determined by your life experiences. As a result, we share a commonality with those who are most like us. For example, many North Americans appreciate a friendly, open health care professional. People from other cultures, however, may feel uncomfortable interacting with a professional on such terms and may even view this behavior as a sign of incompetence.

Until you become aware that, as a member of a certain **cultural group,** you place a high degree of importance on certain beliefs, it is not possible to possess cultural sensitivity. This awareness can help you prevent personal biases, values, or problems from interfering with your ability to work with clients who are culturally different from you. This awareness can also help you appreciate the value that others of a culture different than yours put on their beliefs. When you can appreciate this point, then you are able truly to empathize with a client from a different culture who is trying to participate in the North American health care system. Your food habits are also a very important part of your worldview. For example, Hindus find eating beef to be abhorrent—much the way many Westerners feel about Asians consuming dog meat for a meal.

Conscious and unconscious prejudices unrelated to cultural issues that a counselor may possess could also

EXERCISE 1.8—What Is Your Worldview?

Indicate on the continuum the degree to which you share the following white North American cultural values; 1 indicates not at all, and 5 represents very much.

Not at All				Very Much	
1	2	3	4	5	Personal responsibility and self-help for preventing illness
1	2	3	4	5	Promptness, schedules, and rapid response-time dominates.
1	2	3	4	5	Future oriented—willing to make sacrifices to obtain future goals
1	2	3	4	5	Task oriented—desire direct participation in your own health care
1	2	3	4	5	Direct, honest, open dialogue is essential to effective communication.
1	2	3	4	5	Informal communication is a sign of friendliness.
1	2	3	4	5	Technology is of foremost importance in conquering illness.
1	2	3	4	5	Body and soul are separate.
1	2	3	4	5	Client confidentiality is of utmost importance; health care is for individuals, not families.
1	2	3	4	5	All patients deserve equal access to health care.
1	2	3	4	5	Youth, thin, fit
1	2	3	4	5	Competition and independence
1	2	3	4	5	Materialism

Can you think of a time that your values and beliefs were in conflict with a person you were trying to associate with? What were the circumstances and results of that conflict? Write your response in your journal, and share your stories with your colleagues.

Source: Adapted from Kittler P and Sucher K, *Food and Culture in America,* 2d ed. (Belmont, CA: West/Wadsworth; 1998); and Keenan, Debra P. In the face of diversity: Modifying nutrition education delivery to meet the needs of an increasingly multicultural consumer base, *J Nutr Ed.* 1996;28:86–91.

interfere with emotional objectivity in a counseling situation. Individuals could have exaggerated dislikes of personal characteristics such as obesity, lack of hair, color of hair, poorly dressed, odor, aggressiveness, meekness, and high-pitched voices. Awareness of these prejudices can help build tolerances and a commitment not to let them interfere with the counseling process through facial expressions and other nonverbal behavior.[27]

EXERCISE 1.9—What Are Your Food Habits?

Record answers to the following questions in your journal; share them with your colleagues.

- Who purchases and prepares most of the food consumed in your household?
- What is your ethnic background and religious affiliation?
- Are there foods you avoid eating for religious reasons?
- List two foods you believe are high-status items.
- What major holidays do you and your family celebrate?
- List two rules you follow when eating a meal (for example, "Don't sing at the table").
- Are there food habits that you find morally or ethically repugnant?
- Are you aware of any of your own food habits that others would consider repugnant?

Source: Adapted from Kittler P and Sucher K, *Food and Culture in America,* 2d ed. (Belmont, CA: West/Wadsworth; 1998), pp. 24–25.

Understanding Your Client

Just like counselors, clients come into nutrition counseling with unique personalities, cultural orientations, health care problems, and issues related to the counseling process. Each person's individual personality should be recognized and appreciated. Clients have their own set of needs, expectations, concerns, and prejudices that will have an impact on the counseling relationship. In the rushed atmosphere of some institutional settings, health care workers can lose sight of the need to show respect, especially if a client has lost some of his or her functions due to illness.

From a cultural perspective, clients are diverse in many ways: gender, sexual orientation, race, ethnicity, socioeconomic level, religious or spiritual affiliation, disability status, age, and so forth.[25] All of these factors plus a set of distinct life experiences contribute to each person having a unique view of the world. As a result, no two people can see the world in exactly the same way. This is one of the advantages of counseling: getting a fresh perspective. However, the farther away counselors are from their clients' cultural orientation, the more difficult it is to understand their worldview. If this is the case, then you will need to explore your clients' culture through books; newspapers; magazines; workshops; movies; and cultural encounters in markets, fairs, and restaurants.[55] Learning your clients' beliefs about illness and the various functions and meanings of food is particularly important.

My client, a robust man in youth, was a World War II veteran who took part in the invasion of Normandy. But at age seventy-five, he suffered a stroke and went into a veterans' hospital for treatment. During his hospital stay, he asked a health care worker to help him get into bed because he wanted to go to sleep. The worker told him he would be able to go to sleep after he finished his lunch. My client became very angry and threw his lunch tray at the health care worker.

The circumstances that bring clients to counseling can have a major impact on their readiness for nutrition counseling. Those who have been recently diagnosed with a serious illness may be too stunned or experiencing too much discomfort to deal effectively with complex dietary guidelines—or any guidelines at all. They may display a tendency toward rebelliousness, a denial of the existence of the problems, anxiety, anger, or depression.[1,56] When counseling an individual with a life-threatening illness, nutrition counselors need to take into account a client's position on the continuum of treatment and recovery.[57]

A classmate in my nutrition class was pregnant and an immigrant from Africa. One day she brought in clay pellets that had been sent to her from home. She said they tasted good and that all women in her country eat them when they are pregnant to be sure their children are born healthy. This surprised me. I had read about the practice, but I didn't think that an educated woman who was majoring in nutrition would eat clay. I guess I was being ethnocentric.

Anderson et al.[58] report of attitudinal studies of young and well-educated diabetic patients suggests that a significant number do not wish to be told what they should do to care for their diabetes. This response probably indicates a need for a collaborative relationship with their health care providers. On the other hand, this same study found that a significant number of elderly do not desire an independent self-care role in the process of controlling their diabetes. Promoting self-sufficiency is often stated as a goal of nutrition counseling[6]; however, for some clients, that goal may need to be modified. This issue has also been addressed by the expert panel for the NIH report, *Identification, Evaluation, and Treatment of Overweight and Obesity*,[9] which states that a weight maintenance program consisting of diet therapy, behavior therapy, and physical activity may need to be continued indefinitely for some individuals.

Some clients may regard the counseling process itself as an issue. The act of seeking and receiving help can create feelings of vulnerability and incompetence.[50] During counseling there is a presumed goal of doing something for the clients or changing them in some way. This implication of superiority can raise hostile feelings in the helpee because the act presumes that the helper is wiser, more competent, and more powerful than the helpee. This is illustrated in Helen Keller's account of her dreams about her teacher and lifelong friend, Annie Sullivan, who provided constant help for almost all aspects of Helen's existence:

> *[T]here are some unaccountable contradictions in my dreams. For instance, although I have the strongest, deepest affection for my teacher, yet when she appears to me in my sleep, we quarrel and fling the wildest reproaches at each other. She seizes me by the hand and drags me by main force towards I can never decide what—an abyss, a perilous mountain pass or a rushing torrent, whatever in my terror I may imagine.* (Herrman,[59] pp. 165–166)

To help alleviate the negative impact of such issues on the counseling process, the motive for help and the nature of the helping task as perceived by the counselor should be made clear to the receiver.[50]

EXERCISE 1.10—Exploring Food Habits of Others Assignment

Interview someone from a culture different than your own. Ask that person the questions in Exercise 1.9, and record his or her answers in your journal. What did you learn from this activity? How can you personally avoid **ethnocentric** judgments regarding food habits?

Relationship Between the Counselor and Client

The helping relationship is often divided into two phases: building a relationship and facilitating positive action.[50] Building a relationship requires the development of rapport, an ability to show empathy, and the formation of a trusting relationship.[60] The goals of this phase are to learn about the nature of the problems from the client's viewpoint, explore strengths, and promote self-exploration.

The first part of the counseling process lays the foundation for the second. The focus changes to helping clients identify specific behaviors to alter and to designing realistic behavior change strategies to facilitate positive action.[6] This means clients need to be open and honest about what they are willing and not willing to do. Lorenz et al.[61] (1996) state that in the successful Diabetes Control and Complications Trial, clients could better communicate their capabilities when health professionals articulated what problems could develop in attempting to improve glycemic control. They found honesty more likely to occur in an environment in which clients do not feel they will be criticized when difficulties occur but rather believe the caregivers will show understanding and work toward preparing for similar future circumstances. Nonjudgmental feedback was also an important component of the successful DASH (Dietary Approaches to Stop Hypertension) dietary trial for reducing hypertension.[62] Counselors must communicate their willingness to discover their clients' concerns and help them prioritize in a realistic manner.

In summary, it would be futile to start designing behavior change strategies when an effective relationship has not developed and you do not have a clear understanding of your clients' problems or an appreciation of their strengths. According to Laquatra and Danish:[60]

> *Attending to the second part of the counseling process without the strong foundation afforded by the first part results in dealing with the problem as being separate from the client, or worse yet, providing solutions to the wrong problems. Behavior-change strategies designed under these circumstances are not likely to succeed. (p. 352)*

The scenario in Exercise 1.11 illustrates a common mistake helpers make—indicating that everything will be fine. Because it has no basis for reality, the comment belittles the client's feelings. If the client actually feels reassured by the comment, the benefit is temporary since no solution to the problem has been sought. Patronizing a client is self-defeating: it indicates superiority and can automatically create negative feelings. Effective counselors provide reassurance through clarifying their role in the counseling process, identifying possible solutions, and explaining the counseling program.

EXERCISE 1.11—Starting a Relationship

Lilly is forty-two years old, has three children, and is about twenty pounds overweight. She sought the help of a fitness/nutrition counselor, Joe, because she wants to increase her energy level and endurance. She finds she tires quickly and feels that some exercise will help her stamina.

Joe: Hello, Lilly. It's great you came a little early. Let's get you right on the scale. OK, at 163 pounds, it looks to me like you need to shed about twenty pounds. You have a ways to go but worry not—we will get it off you.

Lilly: I really . . .

Joe: I am not kidding, Lilly—don't worry. We will start slowly. What you want to do is get your BMI down, your muscle tissue up, as well as get rid of the fat. If you follow me, I'll introduce you to everyone, sign you up for an aerobics class, and start you on your routine.

Lilly: Well, you see I only want . . .

Joe: Hey, Rick, this is Lilly. She is a newcomer.

Rick: Welcome, Lilly. Don't forget to take home some of our power bars—they are great for beginners who may not know how to eat right.

Joe: Yeah, and be sure to bring a sports drink in with you; you will get mighty thirsty. No pain, no gain!

In groups of three, brainstorm the concerns in this scenario. Why is this helping relationship off to a bad start? What questions or comments could Joe have made that may have been more helpful?

Novice Counselor Issues

New counselors typically have concerns about their competency. A counselor who feels inadequate may be reluctant to handle controversial nutrition issues, sometimes giving only partial answers and ignoring critical questions. Confidence in your ability will increase with experience.

Client: *Are high-protein diets a good way to lose weight?*

Counselor: *Some people say they lose weight on them.*

In this example, the counselor is talking like a politician—not taking a stand, trying not to offend anyone. If you are not clear about an issue, you may want to tell your client that it is a topic you have not thoroughly investigated and you will review the matter. If after investigating the issue, you still do not have a clear answer, you should provide your client with what you have found out regarding the positives and negatives of the topic. The American Dietetic Association Code of Ethics[52] states, "The dietetics practitioner presents substantiated information and interprets controversial information without personal bias, recognizing that legitimate differences of opinion exist."

Another issue for novice nutrition counselors is assuming the role of expert or empathizer.[12] Combining the two roles can contribute to an effective intervention, but a single approach is likely to hamper progress. An authority figure is impressive and appears to have all the answers. Clients blindly accept the direction of the "guru," but little work is done to determine how to make the lifestyle changes work for them. As a result, clients revert to old eating patterns. The empathizer puts so much effort into focusing on client problems that the client receives very little direction or information. With experience and determination, the two roles can be effectively combined.

EXERCISE 1.12—Helping Relationships

After reading Case Study 1, record in your journal five behaviors, characteristics, or physical concerns that a nutrition counselor needs to consider prior to any interventions, and indicate how they will impact the relationship. Also, take into consideration that this is an institution, and a nutrition counselor has a limited amount of time to spend with each client.

- What is preventing the professional staff from building a helping relationship?
- What issues regarding John could be brought up at a medical staff meeting?
- What steps could the nutritionist take to develop an effective helping relationship?

CASE STUDY 1—Helping Relationships

John is a seventy-year-old white veteran of the Korean War who was admitted to the nursing home because he was no longer able to care for himself. His diagnoses on admission included cerebrovascular accident (stroke), angina, cancer of the prostate, and major depression. He is generally confined to a wheelchair, but he can ambulate eight to ten steps with the assistance of two people. He is unable to dress himself due to right-sided hemiparesis (paralysis); he is continent of bowel but at times is incontinent of bladder. All of his laboratory values are within normal limits. John is mentally alert and not at all confused but has clinical depression. He has no family or visitors.

John is able to feed himself and has an excellent appetite. He consistently consumes 100 percent of his meals. His weight on admission can be indicated only as over 300 pounds, as the scale cannot measure over 299 pounds. It is estimated that he weights about 320 pounds, which is approximately 100 pounds overweight. He has no difficulty chewing or swallowing and receives a regular diet of regular consistency. He loves to eat, and some staff members bring him food items from home, especially on the 3–11 shift. This helps calm him down during the evening hours, allowing the nurses to do their work.

At times John has outbursts of anger at the staff, particularly when given instructions on what he should do or when he is awakened from sleep. He calls the nurses "Babe" or "Sweetie" and can often be heard telling staff that they "look good today" and "you have a great set of gams." Most of the nurses, recreation staff, and social service staff are relatively young and find his comments to be offensive. Their attitude toward him is tolerant at best, and they do very little for him beyond his basic care. The staff openly talk at the nurses' station about his repulsive attitude toward women. The nurses' aids complain about his weight because it is very difficult to get him in and out of bed. Allowing him to ambulate as per doctor's orders is also a challenge because it takes two to three people to assist. John once fell, and the fire department had to come to get him off the floor because he was too heavy for staff members to lift.

In November, John had surgery for the removal of a cancerous prostate, and his prognosis continues to remain relatively poor considering his cancer and his heart disease. He will frequently comment on how he wants to lose weight; however, he will also say things like, "I could sure go for another one of those eclairs," or "That cook sure can make a great meat loaf—I could have eaten another whole lunch." At times he has even gotten angry if the staff does not meet his requests for seconds.

John spends his day sitting in the hallway watching the activities at the nurses' station and chain smoking. He enjoys some game shows and listens to country music. He does not attend recreational activities.

EXERCISE 1.13—Applying Theoretical Approaches for the Helping Relationships Case Study

COUNSELING APPROACH	APPLICATION OF THE COUNSELING APPROACH
Person centered	A person-centered therapist would have unconditional positive regard for John and clearly communicate understanding of his concerns. Nutrition facts would be kept to a minimum, and the nutritionist would take direction from John as to what nutrition goals should be formulated.
Behavioral	A behavioral therapist would work on changing the environment to improve John's food management. This may involve offering rewards, making certain foods available, or showing a video.
Gestalt	A Gestalt therapist would focus on John taking ownership of his problems. An intervention could include working with John to set realistic goals.
Cognitive	A cognitive therapist would be concerned with John's irrational thought pattern. Intervention could focus on cognitive restructuring (changing thought patterns) or relaxation techniques.
Family	A family therapist would be concerned with John's lack of family and may seek a surrogate family for him.
Solution focused	A solution-oriented therapist could ask John whether he ever did exercises in his wheelchair or when he believes he is eating healthy foods at the nursing home. After identifying the resources that have worked for him in the past, opportunities to expand on those resources would be sought.
Multicultural	A multicultural therapist would explore John's cultural orientation and provide assistance to complement his worldview of food and service.

Review the theoretical approaches for interacting with John. Record in your journal which theoretical approach or approaches you believe would provide success in dealing with this client. As a nutrition counselor, what do you believe would be your goals for working with John? Identify some specific actions you would take to achieve these goals.

REVIEW QUESTIONS

1. Define *nutrition counseling.*

2. Using the Haney and Leibson model of counseling, identify three specific goals in nutrition counseling.

3. This chapter contains seven theoretical approaches for nutrition counseling. Explain the major tenets of each.

4. Explain self-efficacy, the health belief model, the transtheoretical model, motivational interviewing, and the health behavior change method.

5. Name and explain the seven qualities of counselors considered to be the most influential by leading authorities as identified by Okun.

6. Explain how taking on the role of helper improves the self-image of the helper.

7. Identify and explain two needs of counselors that, if they seek to fulfill them through a counseling relationship, will be detrimental to the relationship.

8. Why is it important for counselors to understand their worldviews to achieve cultural sensitivity?

9. What can a counselor do to alleviate the negative impact of the counseling process itself?

10. Name and explain the two phases of the helping relationship.

11. Why is indicating to a client that everything will be fine unlikely to be productive? What is a more useful approach?

12. Identify three issues for novice counselors.

ASSIGNMENT—Observation of a Nutrition Counselor

Observe a nutrition counselor in an inpatient, outpatient clinic, or private office setting for two hours.‡ Answer the following questions in your journal or in a typed, formal paper to be handed in to your instructor. Use the corresponding number or letter for each answer.

1. Identify the name of the setting, address, starting and ending time of the observation, date, and name of the counselor you observed.

2. Describe the physical setting where the nutrition counseling sessions took place.

3. Describe the counselor's attire and its appropriateness.

4. Select a client you observed, and give the following information to the best of your ability:

- **a.** What was the client's gender, age, and cultural/ethnic orientation?
- **b.** Was a helping relationship established? If not, why not? If yes, what did the counselor specifically do and/or say to encourage an effective relationship?
- **c.** Explain the nature of the client's problem.
- **d.** Was there evidence of collaboration between the counselor and the client to define dietary objectives? Explain.
- **e.** Were short- or long-term goals established? If yes, what were they?
- **f.** Describe any teaching or visual aids.
- **g.** Was there evidence of tailoring dietary objectives to address the client's lifestyle issues? Explain.
- **h.** Give your impression of the client's educational level and needs.
- **i.** Give your impression of the client's health belief and self-efficacy regarding their dietary objectives.
- **j.** What were the client's barriers to meeting their dietary objectives?
- **k.** Was there evidence of social support for the client to meet the dietary objectives?
- **l.** Complete a counseling observation checklist (see p. 24).

5. Review the seven theoretical approaches to counseling covered in this chapter. List each one, and indicate whether any components of the approaches were demonstrated in your observations. If yes, explain. If no, how could they have been incorporated?

6. Describe your general impressions of the counseling session. What did you learn from this experience?

‡If a counselor is not available, an alternative would be to use a video of a counseling session that could be critiqued individually or in groups.

COUNSELING OBSERVATION CHECKLIST

	RARELY	OCCASIONALLY	UNDECIDED	OFTEN	ALMOST ALWAYS
Did the nutrition counselor appear to be comfortable with the client and with the subject areas discussed?					
Did the counselor avoid imposing values on the client?					
Did the counselor remain objective?					
Did the counselor focus on the client, not just on the procedure of providing a diet instruction?					
Were the counselor's skills spontaneous and nonmechanical?					
How would you describe the likelihood that the client would return to this nutrition counselor again?					

Comments:

Source: Adapted from Cormier WH and Cormier LS, *Interviewing Strategies for Helpers: Fundamental Skills and Cognitive Behavioral Interviews* (Pacific Grove, CA: Brooks/Cole; 1991).

SUGGESTED READINGS, MATERIALS, AND INTERNET RESOURCES

General Counseling

Corey G. *Theory and Practice of Counseling and Psychotherapy.* Pacific Grove, CA: Brooks/Cole; 1996. An overview of counseling theories and strategies.

Cormier S, Cormier B. *Interviewing Strategies for Helpers: Fundamental Skills and Cognitive Behavioral Interventions.* Pacific Grove, CA: Brooks/Cole; 1998. An excellent resource for the development of fundamental helping skills.

Application of Psychological Theories

Curry KR, Jaffe A. *Nutrition Counseling and Communication Skills.* Philadelphia: Saunders; 1998. An analysis of how selected psychological theories have application in nutrition counseling can be found in Chapter 5.

Glanz K, Barbara R. *Theory at a Glance: A Guide for Health Promotion Practice.* Washington DC: National Cancer Institute; 1997. A readable monograph describing theories of health-related behaviors. Order it from the Cancer Information Service at (800) 4-CANCER; ask for NIH publication 95-3896.

Transtheoretical Model

Prochaska JO, Norcross JC, DiClemente CC. *Changing for Good.* New York: Avon Books; 1994. Definitive guidelines written in a readable format to implement the stages of change model. Nutrition counselors will find the coping skills for each stage and the assessments useful.

Kristal AR, Glanz K, Curry SJ, Patterson RE. How can stages of change be best used in dietary interventions? *J Am Diet Assoc.* 1999;99:679–684. Practical advice given for identifying a client's specific stage of change and selecting effective strategies.

Miller WR, Heather N. eds. *Treating Addictive Behaviors,* 2d ed. New York: Plenum; 1998. A critique of the stages of change model suggesting that in some cases there has been a tendency to jump too quickly regarding the usability of the model.

Motivational Interviewing

Miller WR, Rollnick S. *Motivational Interviewing: Preparing People to Change Addictive Behavior.* New York: Guilford; 1991. The landmark book on the motivational interviewing approach, clearly written, an enjoyable read, and full of practical advice.

Miller WR, Rollnick S. *Motivational Interviewing.* 1998. These six clinical demonstration videotapes to assist in training can be ordered from Delilah Yao, Department of Psychology, University of New Mexico, Albuquerque, New Mexico 87131–1161; Dyao@unm.edu.

Brief Motivational Intervention Resources

Berg-Smith SM, Stevens VJ, Brown KM, Van Horn L, Gernhofer N, Peters E, Greenberg R, Snetselaar L, Ahrens L, Smith K for the Dietary Intervention Study in Children (DISC) Research Group. A brief motivational intervention to improve dietary adherence in adolescents. *Health Education Research.* 1999; 14:101–112. Guidelines for applying a brief motivational intervention to dietary change with adolescents. Much of what was discussed can be applied to older age groups.

Rollnick S, Mason P, Butler C. *Health Behavior Change: A Guide for Practitioners.* New York: Churchill Livingstone; 1999. The authors present a collection of useful strategies and readiness assessments referred to as a "method" based on attempts to develop a brief form of motivational interviewing for health counselors.

Solution-Focused Brief Therapy

Hawkes D, Marsh TI, Wilgosh R. *Solution Focused Therapy: A Handbook for Health Care Professionals.* Hanover, NH: Butterworth-Heinemann Medical; 1998. This step-by-step guide supplies many case studies and strategies.

REFERENCES

[1]Curry KR., Jaffe A. *Nutrition Counseling & Communication Skills.* Philadelphia: Saunders; 1998.

[2]Haney JH, Leibsohn J. *Basic Counseling Responses.* Pacific Grove, CA: Brooks/Cole; 1999.

[3]Helm KK, Klawitter B. *Nutrition Therapy: Advanced Counseling Skills.* Lake Dallas, TX: Helm Seminars; 1995.

[4]Sigman-Grant M. Change strategies for dietary behaviors in pregnancy and lactation. *University of Minnesota Educational Videos from the 1999 National Maternal Nutrition Intensive Course.* Minneapolis: University of Minnesota School of Public Health; 1999.

[5]Rogers CR. *Client-Centered Therapy.* Boston: Houghton Mifflin; 1951.

[6]Berry M, Krummel D. Promoting dietary adherence. In: Kris-Etherton P, Burns JH, eds. *Cardiovascular Nutrition-Strategies and Tools for Disease Management and Prevention.* Chicago: American Dietetic Association; 1998:203–215.

[7]McArthur LH, Ross JK. Attitudes of registered dietitians toward personal overweight and overweight clients. *J Am Diet Assoc.* 1997;97:63–66.

[8]Butler C, Rollnick S, Stott N. The practitioner, the patient and resistance to change: Recent ideas on compliance. *Can Med Assoc J.* 1996;154:1357–1362.

[9]National Institutes of Health (NIH) Obesity Health Initiative. *Clinical Guidelines on the Identification, Evaluation, and Treatment of Overweight and Obesity in Adults,* NIH Publication No. 98-4083. Washington DC: US Department of Health and Human Services; 1998.

[10]Williams AB. Behavior modification. In: Holli BB, Calabrese R. *Communication and Education Skills for Dietetics.* 3d ed. Philadelphia: Williams & Wilkins; 1998.

[11]Perls FS. *The Gestalt Approach and Eyewitness to Therapy.* Palo Alto, CA: Science and Behavior Books; 1973.

[12]Snetselaar LG. *Nutrition Counseling Skills for Medical Nutrition Therapy.* Gaithersburg, MD: Aspen; 1997.

[13]Ellis A, Harper R. *A Guide to Rational Living.* North Hollywood, CA: Wilshire; 1997.

[14]Ellis A, Dryden W. Rational-emotive therapy. *Nurse Pract.* 1984;12:16.

[15]Beck AT. *Cognitive Therapy and Emotional Disorders.* New York: International Universities Press; 1976.

[16]Beck AT. Cognitive therapy: past, present, and future. *J Consult Clinl Psychol.* 1993;61:194–198.

[17]Meichenbaum D. *Cognitive Behavior Modification: An Integrative Approach.* New York: Plenum; 1977.

[18]Baldwin TT, Falcigia GA. Application of cognitive behavioral theories to dietary change in clients. *J Am Diet Assoc.* 1995;95:1315–1317.

[19]Parham, Ellen S. Promoting body size acceptance in weight management counseling. *J Am Diet Assoc.* 1999;99:920–925.

[20]Bowen M. *Family Therapy in Clinical Practice.* New York: Aronson; 1978.

[21]de Shazer S. *Keys to Solutions in Brief Therapy.* New York: Norton; 1985.

[22]Baruth LG, Manning ML. *Multicultural Counseling and Psychotherapy: A Lifespan Perspective.* 2d ed. Upper Saddle River, NJ: Prentice Hall; 1999.

[23]Brownell KD, Cohen LR. Adherence to dietary regimens: An overview of research. *Behav Med.* 1995;20:149–155.

[24]Bandura A. Self-efficacy: Toward a unifying theory of behavioral change. *Psychosoc Rev.* 1977;191–215.

[25]Cormier S, Cormier B. *Interviewing Strategies for Helpers: Fundamental Skills and Cognitive Behavioral Interventions.* 4th ed. Pacific Grove, CA: Brooks/Cole; 1998.

[26]Strecher VJ, DeVellis BM, Becker MH, Rosenstock IM. The role of self-efficacy in achieving health behavior change. *Health Educ Q.* 1986;13:73–91.

[27]Holli BB, Calabrese RJ. *Communication and Education Skills for Dietetics Professionals.* 3d ed. Baltimore, MD: Williams & Wilkins; 1998.

[28]Meichenbaum D, Turk DC. *Facilitating Treatment Adherence: A Practitioner's Guidebook.* New York: Plenum; 1987.

[29]Campbell MK, DeVellis BM, Strecher NJ, Ammerman AS, DeVellis RF, Sandler RS. Improving dietary behavior: The effectiveness of tailored messages in primary care settings. *Am J Public Health.* 1994;84:783–787.

[30]Glanz K, Patterson RE, Kristal AR, DiClemente CC, Heimendinger J, Linnan L, McLerran DF. Stages of change in adopting health diets: Fat, fiber, and correlates of nutrient intake. *Health Educ Q.* 1994;21:499–519.

[31]Greene GW, Rossi SR, Reed GR, Willey C, Prochaska JO. Transtheoretical model for reducing dietary fat to 30% of energy or less. *J Am Diet Assoc.* 1994;94:1105–1110.

[32]Brug J, Van Assema P. Factors differentiating between stages of change for fat consumption reduction. *Appetite.* 1995;24:295. Abstract.

[33]Glanz K, Greene G, Shield JE. Understanding behavior. In: American Dietetic Association. *Project Lean Resource Kit.* Chicago: American Dietetic Association; 1995:142–189.

[34]Brownell, KD, Cohen R. Adherence to dietary regimens: Components of effective intervention. *Behav Med.* 1995;20:155–165.

[35]Ruggiero L, Prochaska JO. Introduction. *Diabetes Spectrum.* 1993;6:22–24.

[36]The 6 stages of change. *Tufts University Diet & Nutrition Letter.* 1996;14(7):5.

[37]Sandoval WM, Heller KE, Wiese WH, Childs DA. Stages of change: A model for nutrition counseling. *Top Clin Nutr.* 1994;9:64–69.

[38]Grommet JK. Weight management; framework for changing behavior. In: Dalton S. ed. *Overweight and Weight Management: The Health Professional's Guide to Understanding and Practice.* Gaithersburg, MD: Aspen; 1997:332–347.

[39]Prochaska JO, Norcross JC. *Systems of Psychotherapy: A Transtheoretical Analysis.* 3d ed. Pacific Grove, CA: Brooks/Cole; 1994.

[40]Maher L. Motivational interviewing: what, when, and why. *Patient Care.* 1998;32:55–64.

[41]Snetselaar L. Counseling for change. In: Mahan LK, Escott-Stump S, eds. *Krause's Food, Nutrition, & Diet Therapy,* 10th ed. Philadelphia: Saunders; 2000:451–462.

[42]DiClemente CC, Prochaska J. Toward a comprehensive, transtheoretical model of change: Stages of change and addictive behaviors. In: Miller WR, Heather N. eds. *Treating Addictive Behaviors,* 2d ed. New York: Phenum; 1998.

[43]Sigman-Grant M. Stages of change: A framework for nutrition interventions. *Nutr Today.* 1996;31:162–170.

[44]Berg-Smith SM, Stevens VJ, Brown KM, Van Horn L, Gernhofer N, Peters E, Greenberg R, Snetselaar L, Ahrens L, Smith K for the Dietary Intervention Study in Children (DISC) Research Group. A brief motivational intervention to improve dietary adherence in adolescents. *Health Educ Res.* 1999;14:101–112.

[45]Rollnick S, Mason P, Butler C. *Health Behavior Change: A Guide for Practitioners.* New York, NY: Churchill Livingstone; 1999.

[46]Miller WR, Rollnick S. *Motivational Interviewing—Preparing People to Change Addictive Behavior.* New York: Guilford; 1991; Gilliland BE, James, RK, Roberts GR, Bowman JT. *Theories and Strategies in Counseling and Psychotherapy.* Englewood Cliffs, NJ: Prentice Hall; 1984.

[47]Egan G. *The Skilled Helper.* 5th ed. Pacific Grove, CA: Brooks/Cole; 1994.

[48]Okun B. *Effective Helping: Interviewing and Counseling Techniques.* Pacific Grove, CA: Brooks/Cole; 1997.

[49]Corey G. *Theory and Practice of Counseling and Psychotherapy.* 5th ed. Pacific Grove, CA: Brooks/Cole; 1996.

[50]Brammer LM. *The Helping Relationship Process and Skills.* 4th ed. Englewood Cliffs, NJ: Prentice Hall; 1988.

[51]Murphy BC, Dillon C. *Interviewing in Action: Process and Practice.* Pacific Grove, CA: Brooks/Cole; 1998.

[52]American Dietetic Association. Code of ethics for the profession of dietetics. *J Am Diet Assoc.* 1999;99:109–113.

[53]Belkin GS. *Introduction to Counseling.* Dubuque, IA: Brown; 1984.

[54]Kittler PG, Sucher KP. *Food and Culture in America: A Nutrition Handbook.* 2d ed. Belmont, CA: West/Wadsworth; 1998.

[55]Terry RD. Needed: A new appreciation of culture and food behavior. *J Am Diet Assoc.* 1999;99:501–503.

[56]Cohen-Cole SA. *The Medical Interview: The Three-Function Approach.* St. Louis, MO: Mosby Year-Book; 1991.

[57]Individualizing nutrition counseling for patients with cancer. *J Am Diet Assoc.* 1999;99:1221.

[58]Anderson RM., Donnelly MB, Dedrick RF. Diabetes attitude scale. In: Redman BK, ed. *Measurement Tools in Patient Education.* New York: Springer; 1998:59–66.

[59]Herrman D. *Helen Keller A Life.* New York: Knopf; 1998.

[60]Laquatra I, Danish SJ. Practitioner counseling skill in weight management. In: Dalton S ed. *Overweight and Weight Management: The Health Professional's Guide to Understanding and Practice.* Gaithersburg, MD: Aspen; 1997:348–371.

[61]Lorenz RA., Bubb J, Davis D, Jacobson A, Jannasch K, Kramer J, Lipps J, Schlundt D. Changing behavior: Practical lessons from the Diabetes Control and Complications Trial. *Diabetes Care.* 1996;19:648–655.

[62]Windhauser MM., Evans MA, McCullough ML, Swain JF, Lin PH, Hobe KP, Plaisted CS, Karanja NM, Vollmer WM. Dietary adherence in the dietary approaches to stop hypertension trial. *J Am Diet Assoc.* 1999;99:S76–S83.

Chapter

2

BUILDING A RELATIONSHIP: BASIC COUNSELING RESPONSES

An ounce of dialogue is worth a pound of monologue.
—Anonymous

BEHAVIORAL OBJECTIVES:

- Identify stages of skill development.
- Describe the impact of communication dynamics on counseling.
- Explain intercultural influence on communication.
- Identify three intents for formulating counseling responses.
- Evaluate effectiveness of counselor's nonverbal communication.
- Identify common North American body messages.
- Identify communication roadblocks.
- Demonstrate skills for building a relationship.
- Utilize basic counseling responses.

KEY TERMS:

- **COMMUNICATION ROADBLOCKS:** obstacles that hamper self-exploration
- **COUNSELING FOCUS:** placement of emphasis in a counseling response
- **COUNSELING INTENT:** rationale for selecting a particular intent
- **EMPATHY:** true understanding of another's perspective
- **INTERCULTURAL COMMUNICATION:** face-to-face interactions among people of diverse cultures
- **SKILL:** an acquired ability to perform a given task
- **SYNCHRONY:** harmony of body language
- **TRAIT:** an inherent quality of mind or a personality characteristic

STAGES OF SKILL DEVELOPMENT

A student of nutrition counseling has the task of learning many new technical, social, and conceptual skills. For some individuals the job will be easier than for others. This is because we are all born with **traits,** a quality of mind or a personality characteristic. The special traits that are part of our essence may or may not easily interface with the interpersonal skills needed for counseling. However, no matter what special abilities we possess, we can all learn counseling skills through patience and practice.

As with learning any new skill, we must pass through a sequence of steps before mastering the skill. While reviewing the following sequence, keep in mind a skill that took you some time to develop, such as learning to drive a car.

1. **Motivation.** The first important step for developing a skill is having a desire to learn. A motivated student will progress to a much higher level of expertise than an unmotivated one, no matter what special traits a person possesses. Motivation can be enhanced by learning in a supportive environment that encourages success by mastering skills in a sequential, stepwise manner.
2. **Learning.** Acquiring knowledge, skills, and attitudes necessary to become an effective nutrition counselor comes from reading, participating in learning activities, making observations, engaging in discussions, and listening to presentations.
3. **Awkwardness.** If possible, the initial attempts at using a new counseling skill should be undertaken with volunteers under supervision, such as a role-playing situation. A novice counselor must be willing to go through a period of discomfort in order to acquire effective counseling skills. A degree of awkwardness should also be expected when conducting your first counseling session.
4. **Conscious awareness.** As ability is gained, a counselor is likely to feel more comfortable using the specific skill but will still be consciously aware of the process.
5. **Automatic response.** Eventually the skill will become an automatic reaction with little or no forethought or discomfort.
6. **Proficiency.** A high level of expertise will be obtained when a counselor can perform and modify the skill under varying conditions. As nutrition counselors gain proficiency, they are likely to feel free to experiment with new approaches and to modify and expand their skills.

MODEL OF COMMUNICATION

Understanding the dynamics of communication is essential for developing good counseling skills. Counselors in particular need to realize that a speaker's statements can be interpreted in several ways, as illustrated in Figure 2.1. In this model a speaker's intended meaning can be distorted at three main junctures: (1) The talker may not communicate clearly because of a faulty *encoding process,* or the ability to express a thought. This happens when language skills are not adequate or a person uses abstractions or generalizations as a way of dealing with denial or anxiety. (2) Distortions also occur when words are not

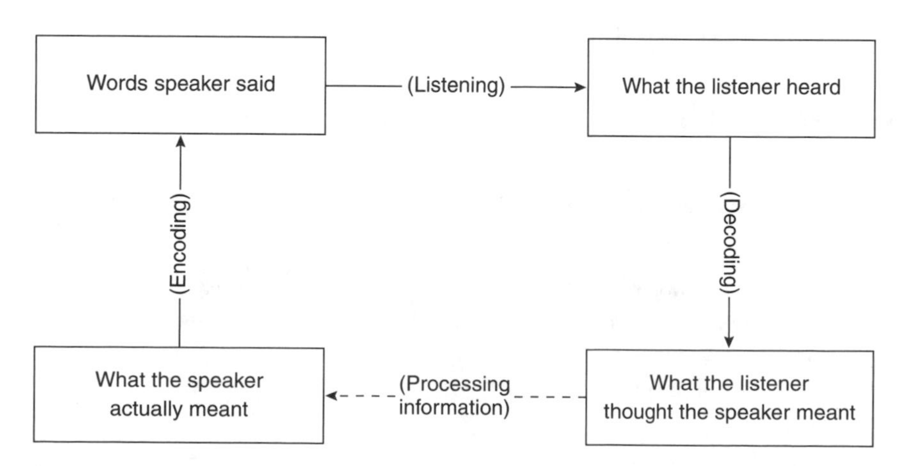

Figure 2.1 A Model of Communication
Source: Adapted from Gordon T, *Parent Effectiveness Training* (New York: Three Rivers; 2000).

heard properly. (3) Lastly, a listener can distort a message during the *decoding process,* which is analysis of the thoughts expressed by others. We all interpret statements through mental filters created by past experiences. Since no two people have precisely the same life experiences, their filters will differ, and interpretations simply become best guesses. Any remark contains multiple meanings. For example, the statement "I binged last Saturday" could mean eating two brownies, consuming half the food in the refrigerator, or getting drunk. If counselors silently equate their interpretations with exact meaning, communication will break down.

INTERCULTURAL INFLUENCE ON COMMUNICATION

Cultural orientation has a major impact on the process of communication. The closer two individuals share a common culture, the greater the likelihood that distortions will be minimal and conversation will flow smoothly. Each society has a conscious and an unconscious series of expected reciprocal responses.[1] For example, in the United States a "You're welcome" is expected to follow a "Thank you." When expected responses do not happen, a feeling of discomfort ensues. Race, gender, age, and nationality have the greatest influence on **intercultural communication**, although equally influential can be degree of acculturation or assimilation, socioeconomic status, health condition, religion, educational background, group membership, sexual orientation, or political affiliation.[1]

The following is a summary of key differences among cultural groups.[1,2,3] While reviewing this list, keep in mind that considerable individual variation exists within any particular cultural group.

Body language The same hand wave gesture that means "Come here" in Nigeria signals "Hello" in the United States. The expressive body language of African Americans may be considered excessive and intense among European North Americans. Friendly behavior for Caribbeans

EXERCISE 2.1—Generating Alternative Meanings

The purpose of this activity is to encourage you to practice generating alternative meanings. Work with colleagues in triads. Each person should write an answer to the question "One thing that I like about myself is that I . . ." The answer should be relatively abstract and have a degree of ambiguity. Concrete statements such as physical attributes should be avoided, such as "One thing I like about myself is that I have blue eyes." Each person should take a turn making one of the statements. The two listeners consider various meanings and respond five times with "Do you mean that you . . . ?" The volunteer can only answer yes or no. Here is an example:

Speaker: One thing I like about myself is that I am strong.

Listener: Do you mean that you can lift a lot of weight?

Speaker: No.

Listener: Do you mean that you have a strong odor?

Speaker: No.

Listener: Do you mean that you are there to help people if there is a problem?

Speaker: Yes.

Listener: Do you mean that you can handle a lot of problems at one time?

Speaker: No.

Listener: Do you mean that you don't fall apart when a problem occurs?

Speaker: Yes.

In a counseling situation, you would not interrogate a client with a series of "What do you mean?" questions but rather would listen closely and consider alternative meanings. In the following sections of this chapter, various counseling responses will be covered to help clients clarify their meanings to you and to themselves.

When you were the speaker in this activity, were you frustrated by the limitation of only being able to answer yes or no? Generally as attempts are made to clarify meanings, a person undergoes a deeper self-evaluation and will feel the need to elaborate. How does this activity relate to the counseling process?

Source: Miller WR, Rollnick S. *Motivational Interviewing.* New York: Guilford Press; 1991:168.

often involves touching that European North Americans and Asians may find uncomfortable. Asian Americans or Latinos are not likely to appreciate a slap on the back.

Vocal style Latinos often use expressive language and engage in lengthy pleasant talk before getting down to business. European North Americans prefer a quiet, controlled style that other groups may consider manipulative and cold.

Verbal following Asians are more likely to use an indirect and subtle form of communication and may find the direct styles of African Americans and European Americans too confrontational. Native Americans find direct personal questions particularly offensive.

Eye contact Making eye contact for European North Americans is a sign of respect, but among Latinos, Native Americans, and Asian Americans, avoidance of eye contact is often considered proper behavior. Latinos and Filipinos use sustained eye contact to challenge authority. In addition, Filipinos express sexual interest with eye contact. Many African Americans use more eye contact during talking than when listening, while the opposite is true of most European North Americans.

Physical space Conversational space in Arab and Middle Eastern cultures is commonly six to twelve inches; among European North Americans distance is ordinarily arm's length. Latinos prefer closer proximity than European North Americans, but Asians desire even a greater distance.

Silence Duration of silence considered acceptable differs among cultures. Native Americans may take ninety seconds to formulate a response to a question, but that amount of silence can seem intolerable to others.

GUIDELINES FOR ENHANCING COUNSELING COMMUNICATION EFFECTIVENESS

This section reviews selective skills for enhancing communication in a counseling session. These include an introduction into the use of focuses and intents for the formulation of responses, an overview of effective nonverbal behavior, an explanation of the value of harmonizing verbal and nonverbal behaviors, a review of analysis of a client's nonverbal behavior, guidelines for interacting in a culturally sensitive manner, methods for utilizing translators appropriately, analysis of communication roadblocks, and procedures for developing relationship-building skills.

Use Focuses and Intents When Formulating Responses

Flow of communication in a counseling setting for the most part is not like having a conversation with a friend. Counselors need to modify some previously learned behaviors such as talking about oneself, asking a lot of questions, and avoiding lulls and silences.[4] Counselors use verbal and nonverbal counseling responses with a specific intent and focus to address counseling objectives. **Counseling intent** is a rationale for selecting a particular response, and **counseling focus** is the placement of the emphasis in a response. The focus of a response could be placed on information about a client or a client's general experiences. The experience response can be subdivided further into feelings, thoughts, or behaviors.

The counseling model developed by Haney and Leibsohn[4] (see Figure 2.2) has three intents, or rationales for selecting a particular response. By recognizing the hoped-for outcome of a response, a counselor is

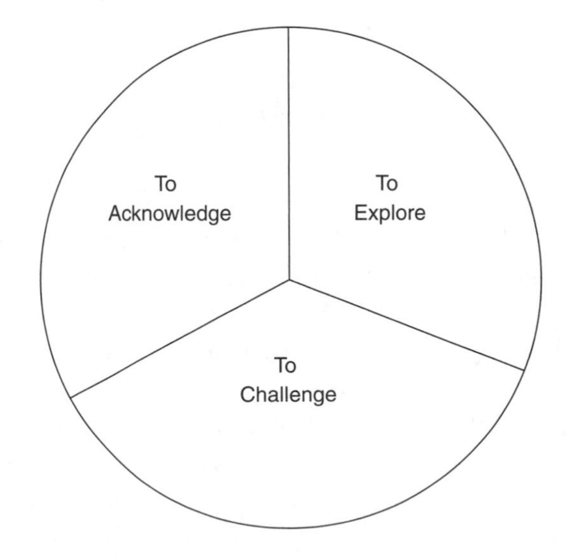

Figure 2.2 Counseling Response Intents
Source: Haney JH, Leibsohn J. *Basic Counseling Responses.* Pacific Grove, CA: Brooks/Cole, Wadsworth © 1999. Reprinted with permission.

better able to formulate an effective response. The intents include the following:

1. **To acknowledge.** If a counselor's intent is to acknowledge, then responses would be selected that identify observations, affirm, show respect, or recognize the worthiness of a client. Relationship-building responses would fall into this category.
2. **To explore.** The objective of a response could be for a client to explore ambivalence, consider new information, or gain insight. If this is the case, counselors might ask questions, provide information, or make clarifying responses.
3. **To challenge.** If the intent is to help clients see their situation differently or to take a different course of action, then a response that notes a discrepancy could be selected.

The following dialogue illustrates responses utilizing various focuses and intents:

Counselor: *Your blood evaluation indicates that you have a cholesterol level of 330. Your dietary evaluation shows a high saturated fat intake and a low intake of fiber, fruits, and vegetables.* (information focus, intent to explore)

How did you handle the party last week? (experience focus*, intent to explore)

You have a right to feel angry about having to handle another dietary modification. (feeling focus, intent to acknowledge)

I am getting the impression that you are thinking that you are a bad person because you ate a lot of cheese at the party. (thought focus, intent to explore)

You set a goal to limit your intake of cheese to one ounce a day, but at the party you ate more much more. (behavior focus, intent to challenge)

The communication analysis in the case study at the end of this chapter presents an examination of the focus and intent of responses made to a particular client. Note again that a particular interpretation of an intent or a focus can be debated since communication is influenced by a multitude of factors—cultural orientation, body language, and voice inflection to name a few. However, by studying intent and focus evaluations, counseling students can enhance their abilities in formulating counseling responses. See Table 2.1 for a list of possible counseling intents for common counseling responses.

EXERCISE 2.2—Identifying Effective Nonverbal Behavior

Work with a colleague and do this exercise twice, exchanging roles as speaker and listener. The speaker engages in a monologue for five minutes on what it was like growing up in his or her home. The listener facilitates the discussion using only body language. No verbal sounds, such as "Mm-hmm," are permitted. After completing this exercise, exchange information with your associate as to what was done that communicated listening and encouraged you to keep talking. What would you have said if you could have talked while you were the listener?

Source: Miller WR, Rollnick S. *Motivational Interviewing.* New York: Guilford; 1991, pp. 164–165.

Use Effective Nonverbal Behavior

A great deal of our communication, up to 85 percent, is based on body language.[2] Generally people learn to trust perceptions of nonverbal behavior over verbal remarks as a truer indication of the meaning of messages. In other words, people inherently believe the adage "Actions speak louder than words." This tendency probably occurs because much of our body language is under unconscious control, while verbal statements are more likely to be deliberate and subject to censorship.[5]

Developing good nonverbal behavior is an extremely important skill for counselors to create an environment conducive to the development of a trusting relationship. Facilitative body behaviors have been shown to result in positive client ratings, even in the presence of ineffective and/or detracting verbal messages.[6] A counselor's joyful expression or attentive silence can communicate an understanding of a client's emotional state. Match the intensity of your own verbal and nonverbal messages with each other to create congruence. Communication will be hampered by unproductive nonverbal behavior such as frequently looking at a watch, yawning, slouching, tapping or swinging feet, or playing with hair or a pencil.[2] These distracting behaviors indicate that the listener is not very interested in what the speaker has to say.

Table 2.2 lists effective and less effective nonverbal behaviors; however, before condemning any behaviors, the context of an encounter needs to be taken into consideration, including type of client, verbal content, timing in session, and the client's perceptual style.[3] The list should be considered as a guide, not as steadfast rules.

*Note that this is a general question allowing the client to choose a more specific focus (that is, feeling, thought, or behavior).

TABLE 2.1—Summary of Possible Counseling Intents

	POSSIBLE COUNSELING INTENTS				
	ACKNOWLEDGE			EXPLORE	CHALLENGE
RESPONSES	RELATIONSHIP BUILDING	CONTINUE TALKING	COUNSELOR IS LISTENING	CLARIFY CONCERN	
1. Attending	X	X	X		
2. Empathizing	X	X	X	X	
3. Legitimation	X				
4. Respect	X				
5. Personal support	X				
6. Partnership	X				
7. Mirroring			X		
8. Paraphrasing		X	X	X	
9. Giving feedback				X	X
10. Questioning				X	
11. Clarifying				X	
12. Noting discrepancy					X
13. Directing				X	X
14. Advice				X	X
15. Silence				X	
16. Self-disclosing	X	X			

Harmonize Verbal and Nonverbal Behaviors

Your behaviors should also harmonize with your clients' expressive state. For example, a client who is animated and loud will have more trouble getting in synch with a counselor who has reserved body movements and a quiet voice. Body language harmony between two people is referred to as **synchrony.**

Mirroring and matching a client's body language have been advocated for business and sales personnel as a way to increase sensitivity and establish rapport.[7] Similarly, Magnus[8] suggests that a counselor mirror a client's silence

TABLE 2.2—Effective and Ineffective Counselor Nonverbal Behavior

NONVERBAL MODE OF COMMUNICATION	INEFFECTIVE NONVERBAL COUNSELOR BEHAVIOR	EFFECTIVE NONVERBAL COUNSELOR BEHAVIOR
Space	Distant or very close	Approximately arm's length
Posture	Slouching; rigid; seated leaning away	Relaxed but attentive; seated leaning slightly toward
Eye contact	Absent; defiant; jittery	Regular
Time	You continue with what you are doing before responding; in a hurry	Respond at first opportunity; share time with client
Feet and legs (in sitting)	Used to keep distance between the persons	Unobtrusive
Furniture	Used as a barrier	Used to draw persons together
Facial expression	Does not match feelings; scowl; blank look	Match your own or other's feelings; smile
Gestures	Compete for attention with your words	Highlight your words; unobtrusive; smooth
Mannerisms	Obvious; distracting	None or unobtrusive
Voice: volume	Very loud or very soft	Clearly audible
Voice: rate	Impatient or staccato; very slow or hesitant	Average or a bit slower
Energy level	Apathetic; sleepy; jumpy; pushy	Alert; stay alert throughout a long conversation

Source: Cormier S, Cormier B. *Interviewing Strategies for Helpers Fundamental Skills and Cognitive Behavioral Interventions.* 4th ed. Pacific Grove, CA: Brooks/Cole; 1998 as reprinted from: *Amity: Friendship in Action. Part 1: Basic Friendship Skill,* by Richard P. Walters, Copyright © 1980 by Richard P. Walters, Christian Helpers, Inc., Boulder, CO. Reprinted by permission.

EXERCISE 2.3—Videotaping and Analyzing Nonverbal Behavior

Videotape a five-minute conversation with a colleague or friend, during class or out of class. After the discussion, write down your feelings during the dialogue. Play the videotape twice—with the sound on and again with the sound off. Describe your nonverbal behavior each time. Was your nonverbal behavior congruent with your recorded feelings? Analyze your behavior. Do you have any distracting habits that communicate inattention (for example, biting lips, playing with hair, and so forth)? What did you learn from this experience?

behavior for those who are culturally accustomed to long periods of silence. In an investigation reported by Curry and Jaffe,[7] students who were able to calibrate their behavior to match their clients had more successful counseling interventions. These students matched behaviors, such as cocking of the head, or made responses incorporating words used by their clients. Care should be taken not to use this method to an extreme; otherwise, clients will feel as if they are being mocked. In order not to feel overwhelmed when learning this strategy, try to select only one aspect of your client's behavior to mimic.

Analyze Nonverbal Behavior of Your Client

Besides paying attention to your own nonverbal behavior, care should be taken to observe and interpret your client's body language. Clues regarding a client's feelings can come from body language, including expressions of autonomic nervous system reactivity (sweaty palms, flushed face, and so forth).[9] A client who is nodding in agreement while listening to diet instructions offers cues as to what he or she is thinking and feeling.

The first client I ever counseled was in a nutrition counseling class. My first impression was that this person was not communicative, as she sat with her arms crossed in front of her. I thought she was putting up a barrier. Later she put her head in her hand and rested her elbow on the table. For some reason I instinctively followed both behaviors, even though they would not be found on a counseling etiquette list. This client gradually opened up, and I felt mimicking her behavior contributed to the harmony that developed between us.

Habit and culture complicate the overall task of interpreting nonverbal behavior, so counselors need to be wary of jumping to conclusions. Studies indicate that no single aspect of nonverbal communication can be universally translated across all cultural groups.[10] For example, nodding of the head usually means yes in North American, but a single nod in the Middle East means no. More than seven thousand different gestures have been recognized,[11] thereby creating many opportunities for misunderstanding of particular cultural meanings.

EXERCISE 2.4—Interpreting Common Nonverbal Cues Among North Americans

Select a person to act out the following behaviors. Write down your interpretation of the message portrayed by the behavior. The end of the chapter has a list of the common meanings of these behaviors for North Americans.

Behaviors

1. Hand over mouth
2. Finger wagging
3. Crossed arms
4. Clenched fists
5. Tugging at the collar
6. Hand over eyes
7. Hands on hips
8. Eyes wide, eyebrows raised
9. Smile
10. Shaking head
11. Scratching the head
12. Making eye contact
13. Avoiding eye contact
14. Wringing hands
15. Biting the lip
16. Tapping feet
17. Hunching over
18. Erect posture
19. Slouching in seat
20. Shifting in seat
21. Sitting on edge of seat

How did you do? Do not feel too bad if you weren't able to correctly identify all behaviors. Generally nonverbal behaviors are expressed in clusters, and we generally do not focus on one aspect of the cluster. We interpret nonverbal behavior based on a general impression.

Magnus[8] suggests if you are unsure of a particular behavior, you should ask for clarification. For example, you could ask, "I notice that you are mostly looking down. Would you tell me what that means for you?"

Interact in a Culturally Sensitive Manner

Because of individual variability in body language and differences in meanings of some behaviors among cultural groups, it is possible to insult a client inadvertently. Magnus[12] relates a story of offending individuals in Thailand by pointing her toes at them. Since it is not possible to know the rules of proper etiquette for every culture, Magnus[12] suggests that you follow your intuition if you believe something you are doing is causing a problem and ask, "There seems to be a problem—is something I am doing offending you?" Once informed of a difficulty, immediately apologize and admit, "I am sorry. I didn't mean to offend you." However, clients seen in a host country are likely to have experienced some degree of acculturation and do not expect to be treated exactly as they would in their homeland.[1]

The specific terms used to identify an individual's ethnicity can be a touchy issue. Depending on the setting, this information may not have been supplied to the nutrition counselor before meeting a client. By making an assumption regarding ethnicity, a nutrition counselor could inadvertently offend a person. For example, *Asian, Oriental, Chinese,* and *Chinese American* have been used to describe individuals of a similar background, but the terms are not acceptable to all. Magnus[8] suggests that to avoid alienation, a counselor should directly inquire about heritage with questions such as "How do you describe your ethnicity?"

Use Translators Appropriately

Too often health care providers resort to using nonprofessional translators, such as friends or relatives of clients or housekeeping staff.[1] This approach has been shown to present numerous problems.[13] Sometimes patients are reluctant or embarrassed to discuss certain problems in front of close relations, or the nonprofessional translator may decide that certain information is irrelevant or unnecessary and does not do a complete translation. Such a translator may be unfamiliar with medical terminology and unknowingly make mistakes. One study of untrained translators showed that 23 to 52 percent of phrases were mistranslated; for example, *laxative* was used to describe diarrhea.[1] Professional translators are bilingual and can recognize cultural contexts of health issues as well.[13]

When working with a translator, be sure to look at and speak directly to the client, and watch the client rather than the translator during the translation. Speak in short units of speech. To check on the client's understanding and the accuracy of the translation, ask the client to back-translate whatever has been communicated, with the translator facilitating. This technique may also open the conversation to questions by the client.[1] The difficulties of communication across cultures is illustrated in the story of an epileptic Hmong child, Lia:[14]

> *Lia developed an infection and severely seized ("like something out of the* Exorcist*") continuously for nearly two hours. Doctors in the local community hospital had a very difficult time stopping her seizures, and when they finally did, Lia was unconscious. Because she was very sick, arrangements were made to transport her to a children's hospital with an intensive care unit. With the help of an interpreter, the situation was explained to Lia's non-English speaking parents. The attending physician charted, "Parents spoken to and understood critical condition." Later investigation revealed that the parents thought their child had to go to another hospital because the doctors at the community hospital were going on vacation.*

Be Aware of Communication Roadblocks and Use Them Only When Justified

Communication roadblocks, are obstacles that counselors inadvertently put up that block self-exploration. They happen when counselors impose their own views, feelings, opinions, prejudices, and judgments. Statements that are roadblocks can be effectively used in the counseling process; however, too often they are employed too soon or too often. They are frequently made with good intentions and not meant to block

> *For a nutrition counseling assignment, I visited an Indian Hindu temple for a ceremony, followed by a meal. The temple was extremely crowded, and in the beginning of my visit, people were busy preparing for the ceremony and meal. Several times I was physically pushed aside with no apology. I found myself getting angry about the whole experience until I discussed the situation with one of the women. I asked if maybe this was some kind of cultural thing, and the woman told me that no offense was meant. She explained that gentle pushing was common when their temple was crowded and there was a lot of work to be done. I still do not know if this is an Indian Hindu practice or just what happens at that particular temple, but I did feel much better after the conversation. In fact, I thoroughly enjoyed the experience, and I found the people at the temple to be very warm and anxious to share and explain their culture to me.*

communication. Twelve kinds of responses that create roadblocks have been identified by Gordon[15] and elaborated on by Miller and Jackson[16] (see Table 2.3).

The following dialogue illustrates a nutrition counselor using roadblocks (the numbers in parentheses correspond to a roadblock response in Table 2.3). Compare this dialogue with the reflective listening example in Chapter 1.

Roadblock	Speaker	Response
	Client	*Everyone is getting on my back about my cholesterol levels—my wife, my doctor, my brother.*
Disagreeing	**Counselor**	*They have your best interests in mind. (6)*
	Client	*I suppose they're right, but I feel fine.*
Warning	**Counselor**	*You feel fine now but a cholesterol level of 300 is nothing to take lightly. (2)*
	Client	*I guess I have to do something about my diet.*
Agreeing; giving advice	**Counselor**	*I think so, too. (7). Eating the proper foods can really help bring down cholesterol levels. (3)*
	Client	*But it doesn't thrill me to give up meatballs and pizza.*
Reassuring; making suggestions	**Counselor**	*It really isn't that bad. (10) It's true that you probably can't have them as often as you have been eating them, but they could be worked into a meal plan. (3) Now, have you ever tried eating soy foods, like tofu? (3)*

Roadblock responses are not necessarily bad, and they do have a place in nutrition counseling. However, a

TABLE 2.3—Examples of Roadblocks

RESPONSE	EXAMPLES
1. Ordering, directing, or commanding	"Don't say that." "Go right back and tell her . . ."
2. Warning or threatening	"You're really asking for trouble when you eat like that." "You better get your blood pressure down." "It is risky to carry around so much weight."
3. Giving advice, making suggestions, or providing solutions	"What you need to do . . ."
4. Persuading with logic, arguing, or lecturing	"Yes, but . . ." "Let's reason this through. . ."
5. Moralizing, preaching, or telling them their duty	"You should . . ." "You really ought to . . ."
6. Judging, criticizing, disagreeing, or blaming	"You're wrong. It is too bad that you can't . . ." "You did this to yourself." "You have only yourself to blame for this condition."
7. Agreeing, approving, or praising	"You did the right thing." "You're doing good at . . ." "You are absolutely right."
8. Shaming, ridiculing, or name-calling	"How foolish can you be!" "You are acting like a child." "You should be ashamed of yourself."
9. Interpreting or analyzing	"You know what your real problem is?" "I know what's troubling you." "You didn't really mean to do that."
10. Reassuring, sympathizing, or consoling	"Everything is going to be all right." "You will have your cholesterol down in no time." "Before you know it, it will all be over."
11. Questioning or probing	"Why did you say that?" "How did you come to that conclusion?"
12. Withdrawing, distracting, humoring, or changing the subject	"We can talk about that next week." "Let me tell you about what happened to me." "Look at how hard the rain is falling."

Source: Miller WR, Jackson KA. *Practical Psychology for Pastors.* 2d ed. Englewood Cliffs, NJ: Prentice Hall; 1995.

counselor needs to be aware of their affect of blocking, stopping, diverting, or changing direction of communication. According to Miller and Jackson,[16] the underlying message of the counselor is, "Listen to me because I know better, I'm more important, or there is something wrong with you." There are times in a counseling session that you do want to take a new direction and one of the responses in Table 2.3 would be appropriate. Generally, this would be after you believe you have listened carefully and understood your client's "story."

Develop Skills for Building a Helping Relationship

Emphasis for developing an effective relationship is often placed on the first meeting and at the beginning of each session. However, a productive relationship needs to be continuously nurtured, for this process can in itself be an instrument of change. The core of such a relationship is the counselor's ability to experience and show empathy. Without this quality a therapeutic relationship cannot move forward.[17] The degree of empathy demonstrated by counselors has been shown to significantly affect outcomes in a study on drinking behavior.[18]

Empathy is a true understanding of another's unique perspective and experience without judging, criticizing, or blaming.[18,19] To allow empathic insight to enhance counseling, the counselor becomes immersed in another's experience without losing one's own sense of self. By maintaining a separate perspective, a counselor can gain insight for designing worthwhile interventions. To help explain the process of empathy, Murphy and Dillon[17] offer the following clarifications:

> 1. *Empathy is not sympathy. Sympathy is what I feel* toward *you; empathy is what I feel* as *you.*
>
> 2. *Empathy is much more than just putting oneself in the other person's shoes. Empathy requires a shift of perspective. It's not what I would experience* as me *in your shoes; empathy is what I experience* as you *in your shoes.*
>
> 3. *Empathy requires a constant shifting between my experiencing* as you *what you feel, and my being able to think* as me *about your experience. (p. 88)*

Empathy will not have a meaningful impact on a counseling relationship unless effectively communicated both verbally and nonverbally. Once a client perceives that she or he has been accurately seen and heard by another, a supportive environment is created conducive to growth and finding solutions. The skills reviewed in this section will aid in communicating this message. However, the greatest challenge of learning empathic skills lies in the integration of these skills into an interpersonal style that feels genuine to the counselor and is perceived as such by the client.[9]

To be able to empathize with another individual requires the ability to hear and sense the experiential world of that person.[2] This can be a challenge when a client's experiences are totally different from your own. However, empathy is a developmental process that can be consciously fostered and strengthened by expanding life experiences with people who are different from ourselves.[17] This could also happen via indirect encounters through watching a movie, play, or interview or reading a biography or novel. For example, movies such as *Philadelphia Story* or *Hoop Dreams* and books such as *Angela's Ashes* by Frank McCourt, *My Own Country: A Doctor's Story* by Abraham Verghese, and or *When I Was Puerto Rican* by Susan Sheehan could be the conduit for such an experience. In this way the range of reactions that people have to various situations can be learned.

Many skills can be employed to foster a helpful empathetic counseling interaction. The next section describes some of these in detail.

EXERCISE 2.5—Entering the World of Others

Think of a time in your life when you were able to enter into the emotional world of a person whose life experiences were alien to you. Describe the experience in your journal, explain whether in any way the experience influenced your ability to empathize with others.

BASIC COUNSELING RESPONSES

Nutrition counselors need a fundamental knowledge of counseling responses. The following list of basic responses are geared to accomplish three objectives: (1) to develop productive relationships, (2) to enhance listening and exploring to understand clients' messages—their needs and concerns, and (3) to provide the tools to utilize motivational strategies covered throughout this text. The following responses have been identified as being particularly useful in the health care arena (since a particular response may be known by several names, alternative terms are given in parentheses):[9]

1. Attending (listening)
2. Reflection (empathizing)
3. Legitimation (affirmation, normalization)
4. Respect
5. Personal support
6. Partnership
7. Mirroring (parroting)
8. Paraphrasing (summarizing)
9. Giving feedback (immediacy)
10. Questioning
11. Clarifying (probing, prompting)
12. Noting a discrepancy (confrontation, challenging)
13. Directing (instructions)
14. Advice
15. Allowing silence
16. Self-referent (self-disclosing and self-involving

To Build a Relationship, Use the Following Responses:

- Attending
- Reflection
- Legitimation
- Respect
- Partnership
- Personal support

Attending (Listening)

The most basic skill on which all other counseling skills build is *attending*. This involves giving undivided attention to your clients, listening for verbal messages, and observing nonverbal behavior. Your focus is on what you see and hear, not on what you know. This allows you to understand your clients' needs and concerns and how they view the world. Many attending behaviors are nonverbal but some nondescript verbal sounds need to be used in order to convey an impression of being engaged. Ivey et al.[2] identify four key components of attending behavior:

1. **Eye contact.** Look at your client during dialogs. Refrain from staring, and permit natural breaks.
2. **Attentive body language.** In North American culture, this generally means a slight forward trunk lean with a calm, flexible posture and an empathetic facial expression. Gestures should be relaxed but kept to a minimum.
3. **Vocal style.** Speech rate, volume, and tone should indicate concern.
4. **Verbal following.** Give brief verbal and nonverbal responses, such as nods or an occasional "Hmm-hmm" or "Yes, I see," to indicate that a client's message has been received. Responses should relate to the topic.

Listening

Nature has given men one tongue and two ears,
that we may hear twice as much as we speak
—Epictetus

Sometimes listening coupled with attending behavior is referred to as *active listening*. Actually, listening and attending are interrelated skills. One does not come without the other. Both are essential for the development of rapport and to communicate empathy. Murphy and Dillon[18] explain the interrelationship of listening, attending, and empathy:

> *It is important for clinicians to create an ambiance of focused attention in which meaningful communication can occur. Clinicians attend in order to listen; they listen in order to understand. Understanding contributes to empathy, and empathy engenders a readiness to respond. Thus, focused attending is an essential component of the therapeutic process.* (pp. 55–56)

Nutrition counselors have been criticized for controlling too much counseling time by doing most of the talking and spending too much time giving diet instructions and advice.[20] In one study, counselors who talked a great deal during sessions were described as unhelpful, inattentive, nonunderstanding, and disliked by the client.[21] A counselor who is doing most of the talking may have "missed the boat." Inattentive counselors solve problems and address issues important to themselves, which may not be of concern to their clients. Only by attending and listening can counselors accurately hear and understand their clients, respond appropriately, and find effective interventions. Good listening skills and attentive behavior indicate caring and concern, creating the impression that the counselor is capable and effective.[5]

The problem for nutrition counselors, as for most people, is that listening skills are not well developed. Active listening is not simply a matter of hearing words but rather hard work requiring focused attention and concentration. Curry-Bartley[22] has identified three essential components of effective listening:

1. **Openness.** Good listeners are willing to allow others to influence their perception of the world. Personal biases are put aside to hear viewpoints that could be in conflict with one's own belief system.
2. **Concentration.** Conscious attention needs to be focused on the speaker while tuning out everything else including fears, rational and irrational thoughts, and peripheral noises or activities.[22] If you have not been fully listening, your body language is likely to portray the fact unconsciously. This creates a barrier indicating to your client that you are not particularly interested in what he or she has to say. The most common reasons for interference with attention include the following:
   - **a. Lag time self-talk.** An individual with an average intelligence can process information at speeds approximately five times faster than human speech,[5]

EXERCISE 2.6—Attending Success

Think of a time you were telling a story to someone and it was obvious that the person was engaged. In your journal describe this experience and explain what you felt during the encounter. Did this attentive behavior surprise you? What affect did this encounter have on your relationship?

creating time for unproductive mental dialogue such as "Where did she buy those lovely earrings? They would go perfect with the dress I bought for my cousin's wedding. Oh, why is my cousin marrying such a dingbat?"

b. Rehearsing rebuttals. Using the extra mental capacity to rehearse rebuttals or questions will also break concentration.

c. Assumptions. Assuming you know a solution and deciding that what the client has to say is uninteresting or irrelevant can interfere with communication as well.

3 Comprehension. By attending to the first two skills of listening—openness and concentration—the counselor increases the likelihood of comprehending the meaning and importance of what was said.

Guidelines for Improving Counseling Listening Skills

1. Remind yourself to focus solely on your client before each session.
2. Remind yourself to listen with an open mind and a spirit of inquiry.
3. Watch for nonverbal clues to help understand meaning and identify what is important.
4. Use thinking-speaking lag time to analyze and understand the message.
5. Determine whether nonverbal behavior is congruent with the verbal message.
6. Provide feedback with nonverbal reactions or verbal responses such as "Uh-hum" or paraphrasing.

Source: Holli BB, Calabrese, RJ. 1998. *Communication and Education Skills for Dietetics Professionals,* 3d ed. Baltimore, MD: Williams & Wilkins.

Reflection (Empathizing)

Reflection is labeling a client's expressed verbal and/or nonverbal emotion. When a counselor has accurately sensed an emotional state and has effectively employed reflection responses, clients feel understood, thereby facilitating self-acceptance and self-understanding. Teyber[23] explains the importance of understanding in a counseling relationship: "Clients begin to feel that they have been seen and are no longer invisible, alone, strange, or unimportant. At that moment, the client begins to perceive the therapist as someone who is different from most other people and possibly as someone who can help" (p. 49). No matter how empathic you feel, your client will not know if you do not verbally acknowledge your client's feelings. Although your body language may convey empathy, nonverbal signals could be missed or misinterpreted, especially if a cultural difference exists.

Steps in Reflecting The following steps will help you reflect more effectively as you communicate with clients.

1. Correctly Identify the Feeling Being Expressed. There are five major feeling categories: anger, fear, conflict, sadness, and happiness. Table 2.4 presents a list of commonly used feeling words at three levels of intensity.

This step requires careful listening, closely observing nonverbal behavior and voice quality. Sometimes you

EXERCISE 2.7—Listening Awareness

Over the next two days, choose three distinctly different listening encounters and represent each as an *X, O,* and *R.* For example, listening to your mother during dinner could be represented as an *X,* listening to your friend on the telephone could be an *O,* and listening to your psychology teacher could be an *R.* These are illustrated on the first line. Put the symbols in the following continuum categories on the place that best fits your style of listening for each situation. I could describe myself as:

am alert . . .X . . .R . . . . .O . . . . am bored
feel nonjudgmental . . . . feel judgmental
feel calm . . . . feel volatile
listen to emotional messages . . . . listen only to facts
listen attentively . . . . fake attention
think, then respond . . . . react before thinking
am in the here and now . . . . am occupied with past or future

In your journal, describe the context and the participants of the listening encounter, and identify the corresponding symbol on the continuum. Compare and contrast the experiences. What did you learn from this experience?

Source: Adapted with permission from Curry-Bartley KR. The art of science and listening. *Topics in Clinical Nutrition.* 1:18–19 © 1986 Aspen Publishers.

TABLE 2.4—Feeling Words

	FEELING CATEGORY				
RELATIVE INTENSITY OF WORDS	**ANGER**	**CONFLICT**	**FEAR**	**HAPPINESS**	**SADNESS**
Mild feeling	Annoyed Bothered Bugged Irked Irritated Peeved Ticked	Blocked Bound Caught Caught in a bind Pulled	Apprehensive Concerned Tense Tight Uneasy	Amused Anticipating Comfortable Confident Contented Glad Pleased Relieved	Apathetic Bored Confused Disappointed Discontented Mixed up Resigned Unsure
Moderate feeling	Disgusted Hacked Harassed Mad Provoked Put upon Resentful Set up Spiteful Used	Locked Pressured Torn	Afraid Alarmed Anxious Fearful Frightened Shook Threatened Worried	Delighted Eager Happy Hopeful Joyful Surprised Up	Abandoned Burdened Discouraged Distressed Down Drained Empty Hurt Lonely Lost Sad Unhappy Weighted
Intense feeling	Angry Boiled Burned Contemptful Enraged Fuming Furious Hateful Hot Infuriated Pissed Smoldering Steamed	Ripped Wrenched	Desperate Overwhelmed Panicky Petrified Scared Terrified Terror-stricken Tortured	Bursting Ecstatic Elated Enthusiastic Enthralled Excited Free Fulfilled Moved Proud Terrific Thrilled Turned on	Anguished Crushed Deadened Depressed Despairing Helpless Hopeless Humiliated Miserable Overwhelmed Smothered Tortured

Source: From *Helping Relationships and Strategies*, second edition by D. Hutchins and C. Cole. Copyright © Brooks/Cole. Reprinted by permission of Wadsworth.

will need to rely on your intuition.[24] You can also imagine how you would feel in a similar situation. For example, consider this scenario:

Client: *Now that I have diabetes, I have to think about what I eat all the time. I don't know if I will ever learn to cope with this. There are so many things I have to do in life already. The children need constant attention, and I have a stressful job. It just doesn't seem possible to think constantly about my blood sugar and insulin all day.*

Feeling: *Overwhelmed*

Emotions are easy to identify for those who are demonstrative by nature and vividly display their feelings. However, people who present themselves in a straightforward and businesslike manner are not so easily understood. Dubé et al.[25] suggest using the following question as a way to opening the door to a discussion of feelings: "How has this whole illness (problem) been for you and your family—I mean, emotionally?"

2. Reflect the Feeling You Have Identified to the Client. Drop the tone of your voice at the end of the statement; do not bring it up as if you are asking a question. Questions give a slight indication that you think the client should not feel that way and should reconsider his or her feelings.[18] A reflection statement rather than a question communicates understanding and acceptance. As an illustration say the following to yourself:

Less effective: *You're really angry that the burden of your father's care has been put on your shoulders?* (voice turns up at the end)

More effective: *You're really angry that the burden of your father's care has been put on your shoulders.* (voice turns down at the end)

You may also want to begin your sentence with a *stem tentative phrase.* However, care should be taken not to overdo such phrases since they can become annoying, especially if the same one is used repeatedly. Here are some examples:

"Perhaps you are feeling . . ."
"I imagine that you're feeling . . ."
"It appears that you are feeling . . ."
"It sounds like . . ."
"It seems that . . ."

In response to the client who seems overwhelmed in the prior example, you might say, "It seems that you are feeling overwhelmed with trying to fit the care of diabetes into your life."

3. Match the Intensity of Your Response to the Level of Feeling Expressed by the Client. This can be done by choosing an appropriate word in Table 2.4 or by using modifying words such as *a little, sort of,* or *somewhat* to soften the response or *really, very,* or *quite* to make the feeling response stronger. Consider the following exchange:

Client: *I hate myself! I am such a jerk! I sit in front of the television eating junk food all night!*

Counselor (less effective): *It sounds like you are slightly annoyed with yourself.*

Counselor (effective): *It sounds like you are very angry with yourself.*

When in doubt, it is better to undershoot rather than overshoot your response. The effect of overstating a feeling can be a denial of the feeling and backing away from a feeling discussion.[16] Understating does not tend to have that effect; a client is likely to clarify the level of feeling—for example, "A 'little' happy! I'm elated!"

4. You Should Respond to the Feelings of Your Client, Not to the Feelings of Others.[24] Take a look at this sample dialogue:

Client: *I was upset last night at dinner when my sister kept talking about my weight.*

Counselor (ineffective): *Your sister is feeling uneasy about your weight.*

Counselor (effective): *You feel annoyed when someone nags you about your weight.*

Reflection responses have been presented here as a relationship-building skill; however, this response has other advantages, too. For example, Laquatra and Danish[20] emphasize the use of this response as a technique to encourage your clients to continue talking and to help clarify their problems. This clarification also helps counselors understand problems from the viewpoint of their clients. Finally, although these responses are very useful, they should not be used too frequently, or else they can make your clients feel uncomfortable and lead to a denial of their feelings.[3]

Legitimation (Affirmation, Normalization)

Reflection responses involve identification and acknowledgment of a client's feelings; *legitimation* communicates the acceptance and validation of the

EXERCISE 2.8—Practice Reflection Responses

Over the next two days, make it a point to practice acknowledging feelings with friends, family members, coworkers, supermarket clerks, and others. Record in your journal three of the experiences describing when, where, and what happened. What is your impression of the effect of reflection statements?

client's emotional experience.[9] The counselor acknowledges that it is normal to have such feelings and reactions. Usually it is a good idea to receive verification that you have correctly identified your client's feelings before making a statement that the feelings are legitimate and make sense to you. For example:

Counselor: *It seems to me that you are feeling overwhelmed with the whole ordeal of this illness.* (empathizing response)

Client: *Yeah, it sucks.*

Counselor: *I can understand why you would feel like this. Anyone would under the circumstances.* (legitimation statement)

This statement could also be made without first identifying the feeling if your client is especially communicative about his or her feelings. For example:

Client: *This is terrific! I am so happy! Exercise and eating all that rabbit food have really paid off. I actually enjoy all those fruits, vegetables, and whole grains. My blood pressure is so good my doctor is taking me off medication. This is wonderful!*

Counselor: *You deserve to feel so happy after getting such good results and working so hard to make changes in your life.*

Respect

Respect for your client and his or her coping ability is implied by attentive listening and nonverbal behavior. However, explicit statements of respect show genuine appreciation for the worth of the client and can help build rapport, improve the client-counselor relationship, and help your client cope with difficult situations.[9] Respect responses include words of appreciation on the ability to overcome adversity and adjust to difficult situations. The fact that the client has come to a nutrition counseling session can in itself show positive coping behavior. The person can be complimented on a willingness to search for nutrition interventions to deal with the problem. Examples of statements a nutrition counselor could make include these:

> "I am impressed that you are here searching for ways to lower your cholesterol levels through diet."

> "Despite the fact that you have so many responsibilities, you have done a great job of making exercise a priority in your life."

> "You have done such a terrific job of keeping a food journal."

Personal Support

You should make clear to your clients that strategies for solving their problems are available, and you are there to help them implement those strategies. Your clients should know you want to help. For statements of support to have a positive effect on building a relationship, they need to be honest. The following is an example of a supportive statement:

> "There a number of dietary options and strategies available to get your diabetes under control. I look forward to working with you to make that happen."

Partnership

Successful interventions begin with establishing a collaborative relationship with your client. This means that the client and counselor respect each other and work together to find solutions. The following is an example of a partnership statement:

> "I want us to work together to find and implement strategies that will work for you. After we talk about your problems and strengths, we will look at some options for finding a solution."

Mirroring (Parroting)

Parroting or *mirroring* responses repeat back to a client exactly what was said or with very few words changed. This response lets a client know you are listening and encourages the person to keep talking and exploring. Care should be taken not to overdo this response, or else your client is likely to talk less. Here's an example of a mirroring exchange:

Client: *I had chocolate hidden under my bed.*

Counselor: *You had chocolate hidden under your bed.*

Paraphrasing (Summarizing)

Paraphrasing responses are a rephrasing of the content of what the client said and meant. They can summarize prior statements or several statements of a conversation. Remember that the model of communication illustrated in Figure 2.1 indicates that alternative meanings are possible for any statement. These responses are a counselor's best guess as to what a client actually means. Paraphrasing responses let a client know you are listening, encourage clients to continue talking, and assist a client in clarifying concerns to him- or herself and the counselor. You should not be concerned if you have missed the actual meaning because the client's typical response will be to clarify back to the counselor the intended meaning. The following is an example of a counselor using paraphrasing responses:

Client: *I was really surprised to find that my cholesterol jumped to 300. It had always been around 190. I guess now that I am well into menopause it is going to be harder to control. I thought my diet was pretty good, so this is really annoying.*

Counselor: *It must seem unfair to you to have this happen.*

Client: *Yeah. When we go out to eat with my brother-in-law, he always orders an expensive steak, loads the butter on the baked potato, and never asks for salad dressing on the side. I don't know his cholesterol level, but he is alive and kicking. Well, these are the cards I was dealt in life, and I guess I can live with it.*

Counselor: *Even though you have some negative feelings about what has happened, you think that you can cope with what has to be done to get your cholesterol under control.*

Client: *That's right. I've been reading about good foods to eat to lower cholesterol levels. I am eating oatmeal and drinking soy milk just about every day. One problem I have is with the soy nuts. I buy double chocolate chips to mix with the nuts to make them tastier, but I find that when I feel stressed, I am going for the chips and eating too many of them. I have got to stop that.*

Counselor: *You're looking for a way to put an end to eating the chips.*

Client: *Not really. I am looking for a way to end eating them out of control, but I guess if I am not successful, I should stop buying them.*

Note that in the last statement the counselor did not fully pick up on the client's intended meaning, but that did not present a problem. The client simply clarified her meaning and even continued with a deeper self-exploration.

Paraphrasing an extended interaction is referred to as *summarizing* (Chapters 3 and 7 provide examples of counseling summaries). Periodic summaries of what has transpired in a counseling session can be used to transition to a new topic; integrate client behavior, thoughts, and feelings; provide closure at the end of a session; furnish a vehicle to elicit self-motivational statements; and allow checking for any misunderstandings.[18] They communicate a sense that the counselor is listening and trying to understand. If needed, this technique can provide a "therapeutic breathing space" for counselors to make a decision on what the next step should be in the session.[9] The following are possible lead-ins to summaries:

"Since our session is about to come to a close, I would like to review what we covered today so we can agree on where we are and where we are going."

"Let me summarize what we have covered so far and see whether we are in agreement."

Giving Feedback (Immediacy)

Giving feedback is telling clients what you have directly observed about their verbal and nonverbal behavior. Often this is not new information to your client, but by pointing out the behavior, you are inviting the client to examine the implications and increase self-awareness. Haney and Leibsohn[4] provide the following guidelines for giving feedback: be positive and specific; note behavior, not traits; and do not put the client on the defensive. Here are a couple examples:

"When you said you wanted to give up drinking so much coffee, you looked sad."

"I noticed that you started to wring your hands when you started to talk about your mother."

Questioning

Questions are effective responses for gathering information, encouraging exploration, and changing the focus of a discussion. See Exhibit 2.1 for a list of questions historically found effective for clinicians.[17] Although the use of questions appears to be an easily learned response, the challenge is asking appropriate questions and timing them well. New as well as seasoned nutrition counselors have been criticized for asking too many questions to fill silence or to satisfy curiosity.[26] Questions should only be asked if there is a particular therapeutic purpose in mind. They interrupt concentration and lead to a discussion of concerns that interest the counselor but not necessarily the client.

Closed-ended questions elicit yes or no or very short answers. They commonly begin with *is, are, was, were, have, had, do, does,* or *did.* For example: "Do you eat breakfast?" or "Have you ever counted calories?" These types of questions do not allow for expansion of ideas so their usefulness is limited. Additionally, they tend to influence answers because the questions hint at an expected response. For example, the question "How many fruits do you eat each day?" assumes fruits are eaten each day, and it would not be easy to give the answer "Zero." Closed-ended questions are useful for ending a lengthy discourse or for soliciting a specific answer.

Open-ended questions give a person a great deal of freedom to answer and encourage elaboration. They commonly begin with *what* or *how.* For example: "How did your goals for the week work out?" or "What problems do you have with the diet plan?" Clients are not likely to feel threatened by open-ended questions, and they communicate interest and trust. This type of question is generally preferable when possible. However, open-ended questions can lead to rambling, lengthy answers.

Funneling questions are a sequence of questions beginning with a broad topic and narrowing down to a specific item. For example: "Can you tell me what a typical day looks like regarding your food intake?" "Do you generally drink anything with your lunch?" "What kind of milk do you put in your coffee?"

Problematic Questions *Why questions* are generally to be avoided because they often sound judgmental, put clients on the defensive, and seem to require an excuse. For example: "Why didn't you follow your plan and eat the orange when you came home from work?" Clients are likely to respond with an evasive answer, which provides no useful information, such as "I don't know." The client is likely to feel ashamed or unnatural. If you believe an investigation into motives for behaviors or feelings is warranted, Murphy and Dillon[17] recommend the following questions: "As you look back, what do you think was going on?" or "What do you think caused that to happen?"

Double questions ask clients to respond to more than one question at a time. For example, "How did your goals to increase calcium work out? What did your family think of the changes?" Clients will become confused trying to decide which question to answer first.

Question-answer traps, termed by Miller and Rollnick,[18] are a series of questions causing clients to feel as if they are under interrogation. See the following example:

Counselor: *You're here to talk about a diet to lower your blood pressure. Is that right?*

Client: *Yes, that's right.*

Counselor: *Did your doctor give you any information about diet and blood pressure?*

Client: *No.*

Counselor: *Did your doctor talk to you about your weight?*

EXHIBIT 2.1—Tried-and-True Questions

What brings you here to see me? (reasons for coming)

Why are you seeking help now? (timing of request)

How did you think I might help? (anticipation of the experience)

What would you like to get done today? (client as the driver)

Where would you like to begin? (client as the driver)

Can you tell me more about your situation? (elaboration of person/situation)

Who else is available as a support or to help in this? (situation dynamics)

Who, if anyone, is making things more complicated just now? (situation dynamics)

Have you ever spoken to a nutrition counselor before? If so, how did it go? (vision of the work)

Are there other things you haven't mentioned yet that would be important for me to know? (elaboration)

What will we look for, to know that the changes you want have actually taken place? (concretizing desired outcome)

What is it like for you to be talking about these things with me? (relationship building, checking in)

How does the work we're doing compare with what you thought it would be like? (checking in)

Are there any other things that should be on our list of things to talk about? (double-checking)

Does what I'm saying make sense? (clarifying)

Could you put that in other words so I can understand it better? (not knowing)

Can you say more about that? (elaboration)

Source: Murphy BC, Dillon C. *Interviewing in Action.* Pacific Grove, CA: Brooks/Cole Publishing © 1998.

EXERCISE 2.9—Practice Using Questions

Practice with a partner, each taking turns discussing an individual concern (for example, getting the children to eat breakfast, switching from whole milk to skim milk, and so forth). One person should take on the role of client and the other counselor. First the client should ask five closed questions and then five open questions. Explain what happened. How did you feel during each set of questions?

Client: *Yes, she said I should lose about fifteen pounds.*

Counselor: *Do you want to work on losing weight?*

Client: *Yes.*

Counselor: *Have you ever been on a diet before?*

Client: *Yes, once or twice.*

Counselor: *Did you lose weight on the diets?*

Client: *Some, but I gained it back.*

Counselor: *Did you use a selection of foods from food groups on the diets you followed?*

Client: *Yes.*

The question-answer trap is not effective for several reasons. It encourages a client to give short answers, does not allow for much in the way of self-exploration, does not elicit much information, and sets up the counselor as the expert who will provide the magical solution after enough questions have been asked. Even a series of open-ended questions can lead to a less obvious trap. Miller and Rollnick[18] suggest a general rule of no more than three questions in a row.

Clarifying (Probing, Prompting)

Clarifying responses encourage clients to continue talking about their concerns in order to be clear about their feelings and experiences. Stories that are not clear to you may also not be clear to your client. Clarifying responses can take the following forms:

- Communicate "Tell me more" through body language such as nods of the head or short comments such as "Uh-huh" or "Go on."
- Use trailing words such as "and . . ." or "and then . . ." or the last few words spoken by your client.
- Ask clarifying questions, such as the following:[27]
  "Can you explain that in a slightly different way?"
  "Can you think of an example in your life where that happens as well?"
  "Let me make sure I understand what you have said because it seems to me that you are sharing something very, very important."
  "Anything else?"
  "Could you please clarify something you said earlier regarding . . ."
  "I am very interested in something you said before about . . . . Could we talk a little more about it?"

Noting a Discrepancy (Confrontation, Challenging)

Individuals often experience a great deal of resistance to making lifestyle changes and giving up comfortable behavior patterns. This is likely to lead to denial or distortions. Commonly observed discrepancies occur between two statements, verbal and nonverbal communication, stated feelings and the way most others would feel, and what the client is saying and his or her actual behavior.[26] A counselor who chooses to ignore observed discrepancies that lead to self-defeating or unreasonable behavior actually becomes an enabler.[26] Contradictions and inconsistencies are often not obvious, but once they are brought to our attention, they can be illuminating. By pointing out discrepancies, clients are better able to examine their ambivalent thoughts, feelings, and behaviors, becoming able to make decisions and move forward in making lifestyle changes.

There are several ways of noting a discrepancy, some less intimidating than others. No matter what way is used, the intent should never be to criticize or attack. The counselor should come across as caring and nonjudgmental. The following illustrate some ways to note a discrepancy:

- State observation without *but* (a softer approach).

Counselor: *You say you want to increase your level of physical activity, and you feel that you do not have time to exercise.*

- State observation with *but* (a harder approach).

Counselor: *When we talked earlier you said that you do not eat fruits, but you ate some of the oranges served after the Chinese meal.*

- Use "on one hand . . . ; on the other hand, . . ." expression.

Counselor: *On one hand, you say that you would cut off your right arm to lose weight; on the other hand, you say that you do not want to exercise.*

- Tentatively name it (with *seems, appears, could there be, have a feeling*).

Counselor: *I know you said you understood, but I get the feeling that there are aspects of these instructions you would like to go over again.*

- Directly name it ("I see an inconsistency"; "I'm hearing two things at once").

Counselor: *I see an inconsistency between what you say about your concern for yourself because of your father's heart attack and your willingness to cut back on your fat intake.*

Directing (Instructions)

The *directing* response is telling a client exactly what needs to be done. In nutrition counseling, directives are often important components of the educational portion of a session. For example, a counselor may explain to a client with a liver or kidney condition how to calculate fluid intake, or a diabetes educator may need to instruct an athlete in a new way of balancing food intake and insulin before a workout.

When giving directives, it is important to be clear and concise and to determine whether the instructions were completely understood. To be sure the message has been accurately communicated, Snetselaar[28] suggests that clients repeat back the instructions. (Additional suggestions for ensuring that educational objectives have been accomplished are addressed in Chapter 5.)

Advice

Advice is providing possible solutions for problems. Nutrition counselors have sometimes been criticized for giving too much advice; however, going to the other extreme and not giving any clear advice may leave your client confused and floundering.

Advice from physicians has been shown to be successful for changing smoking, exercise, and alcohol behavior.[29,30] To be effective, advice should (1) be given in a nonjudgmental manner, (2) identify the problem, (3) explain the need to change, (4) advocate an explicit plan of action, and (5) end with an open-ended question to elicit a response from the client.[18,31] Advice should only be given when there is a clear understanding of the problem, previous attempts to deal with the difficulty have been investigated, and the counselor has definite ideas for possible solutions.[20] If these criteria have not been met, advice giving is likely to lead to a "Yes, but . . ." scenario.[26] For example:

Counselor: *You might try walking to increase your physical activity.*

Client: *I have tried walking, but there just isn't enough time in the day.*

Counselor: *How about walking at lunchtime?*

Client: *There just isn't anywhere for me to walk at work.*

Counselor: *You could walk in the parking lot.*

Client: *Yes, but I wear high heels to work.*

Counselor: *You could bring sneakers to work and put them on for walking.*

Client: *Yes, but then I would get all sweaty, and I would not be comfortable at work for the rest of the day.*

This is an example of a counseling session going down the tubes. Obviously the criteria for giving advice have not been met. If you find yourself in a "yes, but" scenario, stop giving advice, back up, and spend more time exploring your client's issues.

Since giving advice is a roadblock to self-exploration, the timing of recommendations is very important. Generally advice should be avoided early in a session when the goal is to understand and clarify issues. Ideally a client makes a request for advice; research indicates, however, that clients do not frequently make the request.[32]

Allowing Silence

At times in a counseling interaction, silence is a valuable tool. Sometimes clients need space for internal refection and self-analysis. For a novice counselor, the thought of allowing silence may seem intimidating when there may already be a fear that silence will occur because of not knowing what to say. Also, previous social conditioning can lead one to feel that talking is preferable to silence, no matter what the content.[33] Using silence effectively is an art and a skill that comes with practice. However, by attending to your client and using good listening skills, the point at which to use silence is likely to naturally flow.

In nutrition counseling, the need for silent contemplation could occur after a client has been given the results of an evaluation, during instructions of a complex dietary regimen, or after an emotional outburst due to the demands of coping with a newly diagnosed illness. If it appears that your client needs some space to process information, then it would be appropriate to divert your eyes for the moment and not try to maintain eye contact. Effective silent periods could be about thirty to sixty seconds.[5]

One way to break the silence is to repeat the last sentence or phrase spoken by your client. For example, if your client's last words were "This all means making some big changes in my life," then your response could be "Some big changes in your life . . ."

Another possibility is asking your client what she or he was thinking about during the silence. A counselor should not always feel compelled to break a silence. There are times when you may want to challenge your client to formulate a response.

When I was working as a hospital dietitian, I was assigned to give dietary guidance to a middle-aged man newly diagnosed with diabetes. When I walked into his room and explained the purpose of my visit, he exploded with anger, berating just about everything concerning the hospital and diabetes. When he finished, we were silent for a while, and then I acknowledged that he was obviously very upset and had a right to be after all he had gone through. I asked if he would like me to come back at a later time. To my surprise, he wanted to go over the diet. We actually had a very good session. I believe both the silence as well as the legitimation statement changed his emotional state and allowed us to explore dietary modifications.

Self-Referent (Self-Disclosing and Self-Involving)

In a counseling relationship, there are likely to be times when shifting the focus of attention to yourself can be advantageous. The benefits of self-referent responses include increasing openness, building trust, providing a model to increase client level of self-disclosure, developing new perspectives, and creating a more impersonal atmosphere.[3,26] If these responses occur too frequently or too soon in the counseling relationship, the result could be a "chatty" session or a perspective that the counselor is self-absorbed. In addition, a client, possibly due to a cultural perspective, may perceive that the counselor lacks discretion. Before using a self-referent response, a counselor should always assess whether the intended sharing will be in the best interest of the client.

Two types of self-referent responses are self-disclosing and self-involving. A *self-disclosing* response involves providing information about oneself; generally this is related to coping experiences. For example, a nutrition counselor who is also a kidney dialysis patient might disclose that fact and commiserate that sometimes the diet requirements are frustrating. In another case, a counselor could explain how she herself incorporates exercise into her busy schedule.

Self-involving responses actively incorporate a counselor's feelings and emotions into a session. Laquatra and Danish[26] provide the following format for a self-involving response: I (the counselor) feel (name feeling) about what you (the client) said or did. These responses can be used to provide feedback or to sensitively confront.

Here are examples of both self-involving and self-disclosing responses:

Client: *My diet has worked so well that my doctor lowered my blood pressure medication.*

Counselor: *I am delighted that you are doing so well.* (self-involving)

Client: *You are going to be so disappointed with me. I ate a ton of chocolate and didn't eat enough vegetables this week. You must wonder what I am doing here.*

Counselor: *I'm concerned that you appear to be making changes for me rather than for yourself.* (self-involving)

Client: *I've always been afraid of shots, and now I need to give myself insulin injections.*

Counselor: *One of the reasons I became a diabetic counselor was because I also have diabetes. I was just like you. I wondered how I could possibly give myself shots.* (self-disclosing)

• • •

As described in Case Study 1, many nursing home employees were offended by John's remarks. This greatly impacted on their relationship with him and the quality of care John received. John never had visitors, and it appeared that he often made scenes to receive attention in order to counteract his loneliness.

Keeping these factors in mind, Table 2.5 lists various types of responses that could have been made by a nurse to address the relationship issue and hopefully to improve the situation. A focus and an intent are identified for each response. In some cases, your interpretation of the underlying focus and intent may be different from those listed here. Also, actual reasons that a particular response is formulated will be influenced by the total context of the counseling intervention, including personalities of the counselor and counselee and the history of the relationship. In addition, your decoding of a response will be based on your life experiences, which gives you a unique perspective. As Haney and Leibsohn[4] emphasize, counseling is an art, and as a result a certain amount of ambiguity is to be expected. However, the process of evaluating why and how a certain response was phrased helps counselors in training grasp the basics of the counseling process.

CASE STUDY 2—Communication Analysis of John's Interactions

This case study delves deeper into the communication difficulties described in the Chapter 1 case study between John and nursing home staff. To prepare for this investigation, reread the "Helping Relationships" case study in Chapter 1, and review the model of communication illustrated in Figure 2.1. The following diagram illustrates distortions that occurred during the encoding and decoding process associated with a typical comment made by John to a member of the nursing staff.

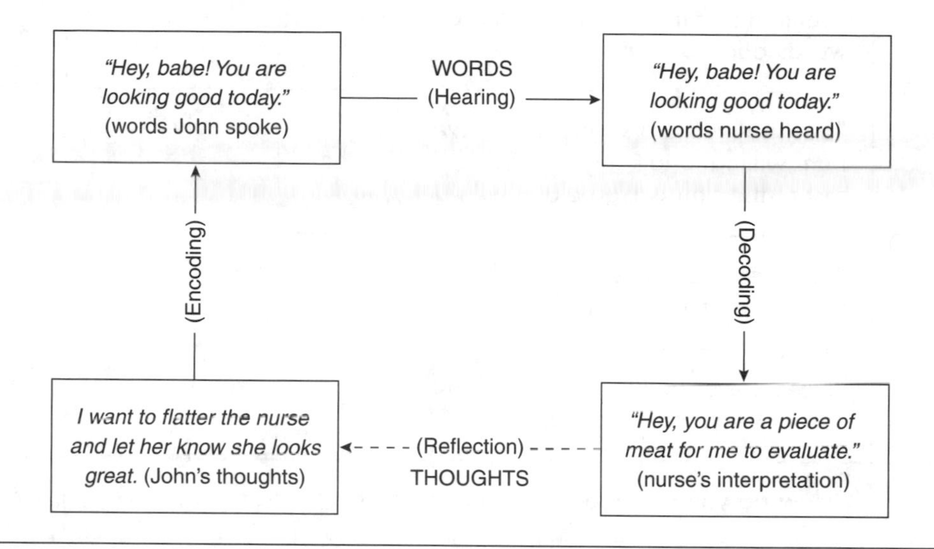

TABLE 2.5—Helping Relationships Case Study Response Analysis

TYPE OF RESPONSE	EXAMPLE OF NURSE RESPONSE	FOCUS OF RESPONSE[1]	INTENT FOR SELECTING A RESPONSE[2]
Attending	The nurse leans forward and shows interest when John commented on something that appeared on television.	Experience	Acknowledge
Empathizing	"You must feel confused about what kinds of comments women find flattering today as compared to what you knew to be true in the 1950s."	Feeling	Acknowledge
Legitimation	"You have a right to feel angry after all you have gone through."	Feeling	Acknowledge
Respect	"It must have been some experience to have been part of the Korean War. You must be proud of yourself."	Feeling	Acknowledge
Personal support	"John, I want you to know that I want to help you get along better with the staff here."	Behavior	Acknowledge
Partnership	"I hope we can work together to make your experience here better."	Experience	Acknowledge

(continued)

TABLE 2.5—Helping Relationships Case Study Response Analysis (Continued)

TYPE OF RESPONSE	EXAMPLE OF NURSE RESPONSE	FOCUS OF RESPONSE[1]	INTENT FOR SELECTING A RESPONSE[2]
Mirroring	John says the nursing staff doesn't pay enough attention to him, and the nurse repeats the words back to John.	Experience	Acknowledge
Paraphrasing	"So what you are saying is that even though I am willing to talk to you, you don't think the rest of the staff will go along with our plan."	Thought	Explore
Feedback	"I noticed that you used the word *babe* four times this afternoon when talking to the staff."	Behavior	Explore
Question—closed	"Does it ever bother you that the nurses find some of your comments offensive?"	Feeling	Challenge
Question—open	"How do you feel about the nursing staff here?"	Feeling	Explore
Clarifying	"John, tell me more about what you want to say when you think someone is annoyed with you."	Thought	Explore
Noting a discrepancy	"On one hand, you say the staff doesn't care about you; on the other hand, you say they bring you special treats from home."	Experience	Challenge
Directing	"Before I see you tomorrow I want you to think of two ways to complement a staff member that you do not believe will be offensive."	Thought	Challenge
Advice	"To every action there is a reaction. If you act in a respectful manner to the staff, they will be respectful in return."	Behavior	Challenge
Silence	The nurse remains silent for thirty seconds after John bitterly complains about the staff following a question regarding how he feels about the staff at the nursing home.	Experience	Acknowledge
Self-referent—self-disclosing	"We all have times when we say things that others perceive differently. It happens with my husband and me, but we talk it through and it works out."	Experience	Challenge

[1]Focus can be information or experience. Experience can be subdivided into feeling, thought, or behavior.
[2]Intent can be to acknowledge, to explore, or to challenge.

EXERCISE 2.10—Evaluating Focuses and Intentions

Continuing with the Helping Relationships Case Study communication analysis, you will explore a comment made by John about food and identify the focus and intent of responses made to him by a dietitian. Keep in mind John's complicated medical condition and a weight problem that is seriously affecting the quality of his life. Refer to your readings and the communication analysis illustrated in Table 2.5 to provide guidance for completing the response table in this exercise.

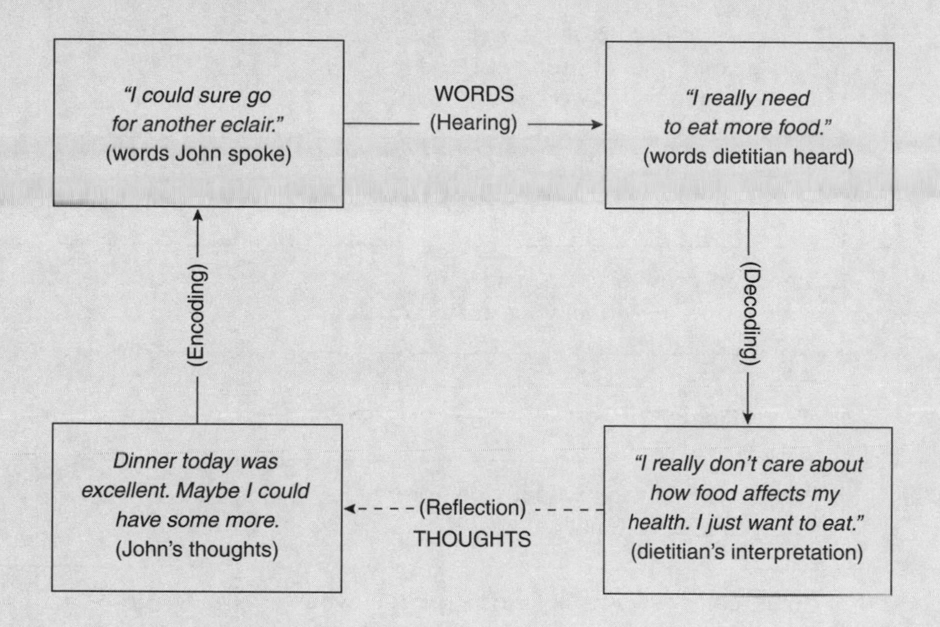

TYPE OF RESPONSE	EXAMPLE OF DIETITIAN'S RESPONSE	FOCUS OF RESPONSE[1]	INTENT FOR SELECTING A RESPONSE[2]
Attending	The dietitian gently touches his arm, sits in a chair next to his wheelchair, and looks directly at him as he is speaking.		
Empathizing	"It's frustrating for you when you have food that you really enjoy and the staff tells you that you can't have more of it."		
Legitimation	"You have a right to feel upset. Eating is the highlight of your day."		
Respect	"I think you have made some very difficult decisions in the past, and I will respect how you want to handle this serious weight issue."		
Personal support	"I want you to know that I am here to help you, even if that means making only baby steps toward your goals."		
Partnership	"If your weight is something you want to work on, you and I will work together to find a solution."		

(Continued)

EXERCISE 2.10—Evaluating Focuses and Intentions (Continued)

TYPE OF RESPONSE	EXAMPLE OF DIETITIAN'S RESPONSE	FOCUS OF RESPONSE[1]	INTENT FOR SELECTING A RESPONSE[2]
Mirroring	John complains that all his weight is causing him a lot of problems, and the dietitian repeats his words.		
Paraphrasing	"So you are saying that you are willing to diet but that I shouldn't expect miracles."		
Feedback	"I noticed your voice became very soft when we talked about your dessert goal."		
Question—closed	"Are you willing to eat only one snack after dinner?"		
Question—open	"What do you think would work for you to help you lose weight?"		
Clarifying	"John, can we go back to talking about what happened at lunch yesterday. It seemed to go well and maybe you could tell me more about why it worked."		
Noting a discrepancy	"You say you would probably feel better if you lost some weight, but eating appears to be the best part of your day."		
Directing	"Before we meet for lunch tomorrow, I want you to tell me two things you think you can do to help you lose weight."		
Advice	"If you want to walk again and feel better, you will need to lose weight."		
Silence	The dietitian asked John what he thought she could do to help him to lose weight. She waited during the silence until he was ready to answer.		
Self-referent—self-disclosing	"I've had times when I've had to lose weight and for me getting started is the hardest part of the process."		

[1]Focus can be information or experience. Experience can be subdivided into feeling, thought, or behavior.
[2]Intent can be to acknowledge, to explore, or to challenge.

REVIEW QUESTIONS

1. List six stages of skill development. Define *trait* and *skill*.

2. Explain three reasons why there could be distortions of a speaker's intended meaning.

3. Explain why intercultural communication can be a challenge.

4. Explain the use of focuses and intents for formulating counseling responses.

5. What is the value of harmonizing verbal and nonverbal behaviors with a client?

6. Give three examples of effective counseling nonverbal behavior.

7. Explain why a counselor needs to be wary of jumping to conclusions regarding interpretation of a client's nonverbal behavior.

8. Why should professional translators be used to communicate with non-English-speaking clients?

9. Describe some guidelines for talking with a client with the aid of a translator.

10. Why do roadblocks impede self-exploration?

11. Describe empathy.

12. Identify the six relationship-building responses.

13. Describe how nonverbal skills, empathy, attending, and listening can be used to enhance a counseling relationship.

14. List Ivey's four key components of attending behavior.

15. Explain the three essential components of effective listening.

16. Give one example of each of the three relative intensity levels for each feeling category: anger, conflict, fear, happiness, and sadness.

17. Give an example of each counseling response described in this chapter.

EXERCISE 2.4—ANSWERS

	Nonverbal behavior	Meaning
1.	Hand over mouth	should not have spoken, regret
2.	Finger wagging	judging
3.	Crossed arms	angry, disapproving, disagreeing, defensive, aggressive
4.	Clenched fists	anger, hostility
5.	Tugging at the collar	discomfort, cornered
6.	Hand over eyes	wish to hide, often from self
7.	Hands on hips	anger, superiority
8.	Eyes wide, eyebrows raised	surprise, guilt
9.	Smile	happiness
10.	Shaking head	disagreeing, shocked, disbelieving
11.	Scratching the head	bewildered, disbelieving
12.	Making eye contact	friendly, sincere, self-confident, assertive
13.	Avoiding eye contact	cold, evasive, indifferent, insecure, passive, frightened, nervous, concealment
14.	Wringing hands	nervous, anxious, fearful
15.	Biting the lip	nervous, anxious, fearful
16.	Tapping feet	nervous
17.	Hunching over	insecure, passive
18.	Erect posture	self-confident, assertive
19.	Slouching in seat	bored, relaxed
20.	Shifting in seat	restless, bored, nervous, apprehensive
21.	Sitting on edge of seat	anxious, nervous, apprehensive

Sources: Arthur D. The importance of body language. *HR Focus.* 1995; 72:22–23 and Curry KR, Himburg SP. © 1988, The American Dietetic Association. *Establishing an Effective Nutrition Education/Counseling Program.* Used with permission.

ASSIGNMENT—Observation and Analysis of a Television Interview

Observe a one-hour television interview. Record the interview on videotape so that the program can be reviewed for analysis. Your report should be typed. Use complete sentences to answer the following questions, and number each of your answers.

1. Record the name, date, and time of show observed; note who did the interviewing and who was interviewed.

2. Identify the purpose of the interview.

3. Explain how the interviewer handled the opening part of the interview: How did the interviewer address the interviewee (that is, Mr., Miss, first name, and so forth)? Was a rapport established? What statements were made or questions asked by the interviewer, and/or what body language of the interviewer facilitated or hampered the development of a rapport? Did the interviewee appear comfortable and willing to disclose information about him- or herself?

4. Explain how the interviewer handled the exploration phase. Did it appear that the interviewer had preplanned and prepared an "interview guide"? Did the interviewees talk 60 to 70 percent of the time in response to questions?

5. The following list contains names of responses that the interviewer could have made. Give an example for each of the following, identify whether they were effective, and give an evaluation as to the effect of the response on the course of the interview. State whether there were no examples of a specific response in the interview.

- Attending
- Reflection
- Legitimation
- Respect
- Mirroring
- Paraphrasing
- Summarizing
- Giving feedback
- Open questions
- Closed questions
- Why questions
- Clarifying
- Noting a discrepancy
- Directing
- Advice
- Self-disclosing
- Self-involving

6. Explain how the closing was handled. In your opinion was this an effective way to end the interview? Why?

7. Play the tape for ten minutes without sound. Describe the body language of the interviewee and the interviewer. Was their body language congruent with what you heard verbally?

8. Identify three things you learned from this activity.

9. Are there things you observed regarding the manner in which the interviewer handled the session that you would definitely not do or would like to emulate in your work as a nutrition counselor?

SUGGESTED READINGS, MATERIALS, AND INTERNET RESOURCES

Counseling Responses

Haney JH, Leibsohn J. *Basic Counseling Responses: A Multimedia Learning System for the Helping Professions.* Pacific Grove, CA: Brooks/Cole Wadsworth; 1999. This book contains an effective interactive learning CD-ROM to promote understanding of basic counseling responses.

Cultural and Ethnic Food Behavior, Intercultural Communication, and/or Multicultural Values

American Dietetic Association and American Diabetes Association, Inc. (1995–1998) *Ethnic and Regional Food Practices: A Series.* American Dietetic Association, 216 W. Jackson Boulevard, Chicago, IL 60606-6995; ADA Customer Service, (800) 877-1600, ext. 5000.
http://www.eatright.org/bibethnic.html

Fadiman, Anne. *The Spirit Catches You and You Fall Down.* New York: Noonday Press; 1997. Winner of the National Book Critics Circle Award. Insight regarding the multicultural issues in the American medical care giving community can be gained by reading this compelling narrative of a very sick Hmong child.

Haffner L. Translation is not enough: Interpreting in a medical setting. *Cross-cultural Medicine—A Decade Later* (Special Issue). *Western Journal of Medicine,* 1992;157:255–259. Emphasizes the need to use professional translators.

Keenan, Debra P. In the face of diversity: Modifying nutrition education delivery to meet the needs of an increasingly multicultural consumer base. *Journal of Nutrition Education,* 1996;28:86–91.

Kittler PG, Sucher KP. *Food and Culture in America: A Nutrition Handbook.* 2d ed. Pacific Grove, CA: West/Wadsworth; 1998.

REFERENCES

[1]Kittler PG, Sucher KP. *Food and Culture in America: A Nutrition Handbook.* 2d ed. Pacific Grove, CA: West/Wadsworth; 1998.

[2]Ivey AE, Gluckstern N, Ivey MB. *Basic Attending Skills.* 3d ed. North Amherst, MA: Microtraining Associates; 1997.

[3]Cormier S, Cormier B. *Interviewing Strategies for Helpers: Fundamental Skills and Cognitive Behavioral Interventions.* 4th ed. Pacific Grove, CA: Brooks/Cole; 1998.

[4]Haney JH, Leibsohn J. *Basic Counseling Responses: A Multimedia Learning System for the Helping Professions.* Pacific Grove, CA: Brooks/Cole Wadsworth; 1999.

[5]Holli BB, Calabrese RJ. *Communication and Education Skills for Dietetics Professionals.* 3d ed. Baltimore, MD: Williams & Wilkins; 1998.

[6]Fretz BR, Corn R, Tuemmier JM, Bellet W. Counselor nonverbal behaviors and client evaluations. *J Counsel Psych.* 1979;26:304–311.

[7]Curry KR, Jaffe A. *Nutrition Counseling and Communication Skills.* Philadelphia: Saunders; 1998.

[8]Magnus M. What's your IQ on cross-cultural nutrition counseling? *The Diabetes Educator.* 1996;96:57–62.

[9]Cohen-Cole SA. *The Medical Interview: The Three-Function Approach.* St Louis, MO: Mosby–Year Book; 1991.

[10]Arthur D. The importance of body language. *HR Focus.* 1995;72:22–23.

[11]Axtell R.E. *Gestures.* New York: Wiley; 1991.

[12]Magnus M. *Cross-cultural Counseling: Research and Practice.* ADA's 1994 Annual Meeting, Palm Desert, CA: Convention Cassettes Unlimited; 1994.

[13]Haffner, L. Translation is not enough: Interpreting in a medical setting. *West J Med.* 1992;157:255–259.

[14]Fadiman A. *The Spirit Catches You and You Fall Down.* New York: Noonday; 1997.

[15]Gordon T. *Parent Effectiveness Training.* New York: Three Rivers Press; 2000.

[16]Miller WR, Jackson KA. *Practical Psychology for Pastors.* 2d ed. Englewood Cliffs, NJ: Prentice Hall; 1995.

[17]Murphy BC, Dillon C. *Interviewing in Action: Process and Practice.* Pacific Grove, CA: Brooks/Cole; 1998.

[18]Miller WR, Rollnick S. *Motivational Interviewing Preparing People to Change Addictive Behavior.* New York: Guilford; 1991.

[19]Maher L. Motivational interviewing: what, when, and why. *Patient Care.* 1998;32:55–64.

[20]Laquatra I, Danish SJ. Practitioner counseling skill in weight management. In: Dalton S, ed. *Overweight and Weight Management: The Health Professional's Guide to Understanding and Practice.* Gaithersburg, MD: Aspen; 1997:348–371.

[21]Kleinke DL, Tully TB. Influence of talking level on perceptions of counselors. *J Counsel Psych.* 1979;26:23–29.

[22]Curry-Bartley K. The art and science of listening. *Top Clin Nutr.* 1986;1:14–24.

[23]Teyber E. *Interpersonal Processes in Psychotherapy.* 3d ed. Pacific Grove, CA: Brooks/Cole; 1997.

[24]Danish S, D'Augelli AR., Hauer AL. *Helping Skills: A Basic Training Program.* 2d ed. New York: Human Sciences Press, Inc.; 1980.

[25]Dubé C, Novack D, Goldstein M. *Faculty Syllabus & Guide: Medical Interviewing.* Providence, RI: Brown University School of Medicine; 1999.

[26]Laquatra I, Danish SJ. Counseling skills for behavior change. In: Helm KK, Klawitter B, eds. *Nutrition Therapy Advanced Counseling Skills.* Lake Dallas, TX: Helm Seminars; 1995.

[27]King NL. *Counseling for Health & Fitness.* Eureka, CA: Nutrition Dimension; 1999.

[28]Snetselaar LG. *Nutrition Counseling Skills for Medical Nutrition Therapy.* Gaithersburg, MD: Aspen; 1997.

[29]Long B, Woolen W, Patrick K, Calfas K, Sharpe D, Sallis J. *Project PACE Physician Manual.* Atlanta, GA: Centers for Disease Control; 1992.

[30]Russell MA, Wilson C, Taylor C, Baker CD. Effect of general practitioners' advice against smoking. *Br Med J.* 1979;2:231–235.

[31]Berg-Smith SM, Stevens VJ, Brown KM, Van Horn L, Gernhofer N, Peters E, Greenberg R, Snetselaar L, Ahrens L, Smith K for the Dietary Intervention Study in Children (DISC) Research Group. A brief motivational intervention to improve dietary adherence in adolescents. *Health Educ Res.* 1999;14:101–112.

[32]Rollnick S, Mason P, Butler C. *Health Behavior Change: A Guide for Practitioners.* New York: Churchill Livingstone; 1999.

[33]Dyer WW, Vriend J. *Counseling Techniques That Work.* Alexandria, VA: American Counseling Association; 1988.

Chapter

3

MEETING YOUR CLIENT: THE COUNSELING INTERVIEW

There are two ways of spreading light;
To be the candle or the mirror that reflects it.
—Edith Wharton

BEHAVIORAL OBJECTIVES:

- Explain the usefulness of counseling models.
- Describe the motivational nutrition counseling algorithm.
- Utilize a variety of readiness-to-change assessment tools.
- Demonstrate selected counseling strategies.
- Depict parts of a counseling interview.
- Describe the changing demographic trends in North America.
- Demonstrate intercultural counseling skills.

KEY TERMS:

- **ALGORITHM:** a step-by-step procedure for accomplishing a particular end
- **INTERCULTURAL COUNSELING:** when the counselor and the client are from significantly different cultures
- **LEARN COUNSELING MODEL:** a cross-cultural counseling model applicable to health care interventions
- **MODELS:** generalized descriptions used to analyze or explain something
- **MOTIVATION:** a state of readiness to change
- **MOTIVATIONAL NUTRITION COUNSELING ALGORITHM:** a step-by-step procedure to direct the flow of a nutrition counseling session
- **RESPONDENT-DRIVEN INTERVIEW:** a culturally sensitive approach to gathering information by asking simple, open-ended questions

NUTRITION COUNSELING MODELS

Models and **algorithms** can provide structure for conceptualizing the counseling process. These are aids for planning, implementing, and evaluating a counseling intervention since the session can be broken down into component parts and addressed individually. The actual flow of a counseling session will adjust to the skills of the counselor and to the needs of the client. However, some structure can help a counselor visualize the direction of the counseling experience as well as the expected end point.

Several models have been developed to guide nutrition counseling.[1,2,3] Snetselaar's[4] model for nutrition counseling (Figure 3.1) can be used to review the major components of the total interaction a counselor has with a client. This model addresses the need for counselors to assume several roles to accomplish counseling tasks. As a diagnostician, a nutrition counselor reviews medical records, food patterns, medication intake, health history, socioeconomic conditions, and other factors. This *preparation* can occur to some degree before the first counseling session, during the first session, and periodically thereafter to better understand problems, skills, and resources related to food intake and readiness to take action. Nutrition counseling also requires an educational component that entails an *explanation* of the counseling process. In addition, nutrition counselors repeatedly assume the role of educator when communicating pertinent nutrition information or providing hands-on educational experiences. During the *intervention* or treatment component of counseling, nutrition counselors take on the role of problem solver and expert using a variety of intervention strategies to help implement dietary goals.[4] Counselors resume the role of diagnostician to assess intervention strategies and evaluate client progress. The *assessment, intervention,* and *evaluation* components are part of each counseling session until the decision is made to conclude the program. At the *conclusion* of each session as well as the total program, the nutrition counselor again assumes the role of expert when reviewing major issues and goals. The counselor becomes a learner in the last component involving *self-evaluation.* Here the objective is to learn from specific counseling experiences for the purpose of improving helping skills.

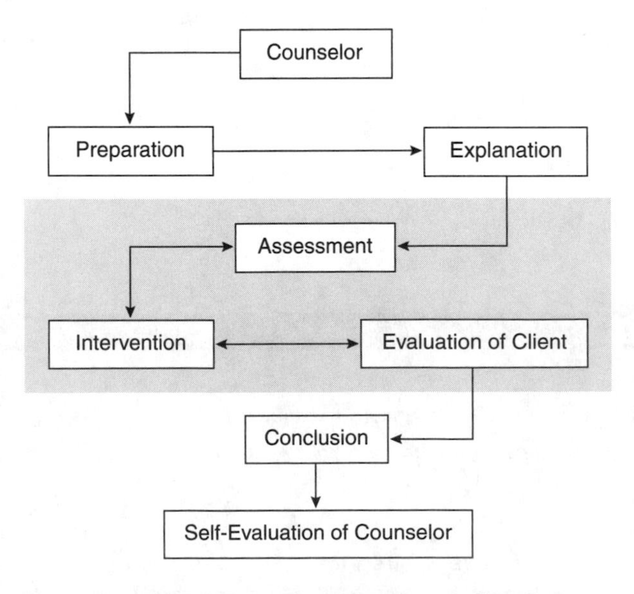

Figure 3.1 Model for a Nutrition Counseling Program
Source: Modified from *Nutrition Counseling Skills for Medical Nutrition Therapy* by L. G. Snetselaar. Copyright © 1997 by Aspen Publishers.

EXERCISE 3.1—Explore Counseling Models
Review the counseling model presented in Figure 3.1. List two activities you might do to address the function depicted in each box.

You cannot teach a man anything;
you can only help him find it within himself.

—Galileo

MOTIVATIONAL ALGORITHM FOR A NUTRITION COUNSELING INTERVENTION

A **motivational nutrition counseling algorithm** is presented in Figure 3.2 to direct the flow of a nutrition counseling session. It is based on one described by Snetselaar[5] and developed by Berg-Smith et al.[6] It takes into consideration that motivation is the underlying force for behavior change and that clients come into counseling at varying levels of readiness to take action. This algorithm incorporates concepts of several of the intervention models and behavior change theories covered in Chapter 1 to provide the guiding force to influence motivation and change lifestyle behavior, including

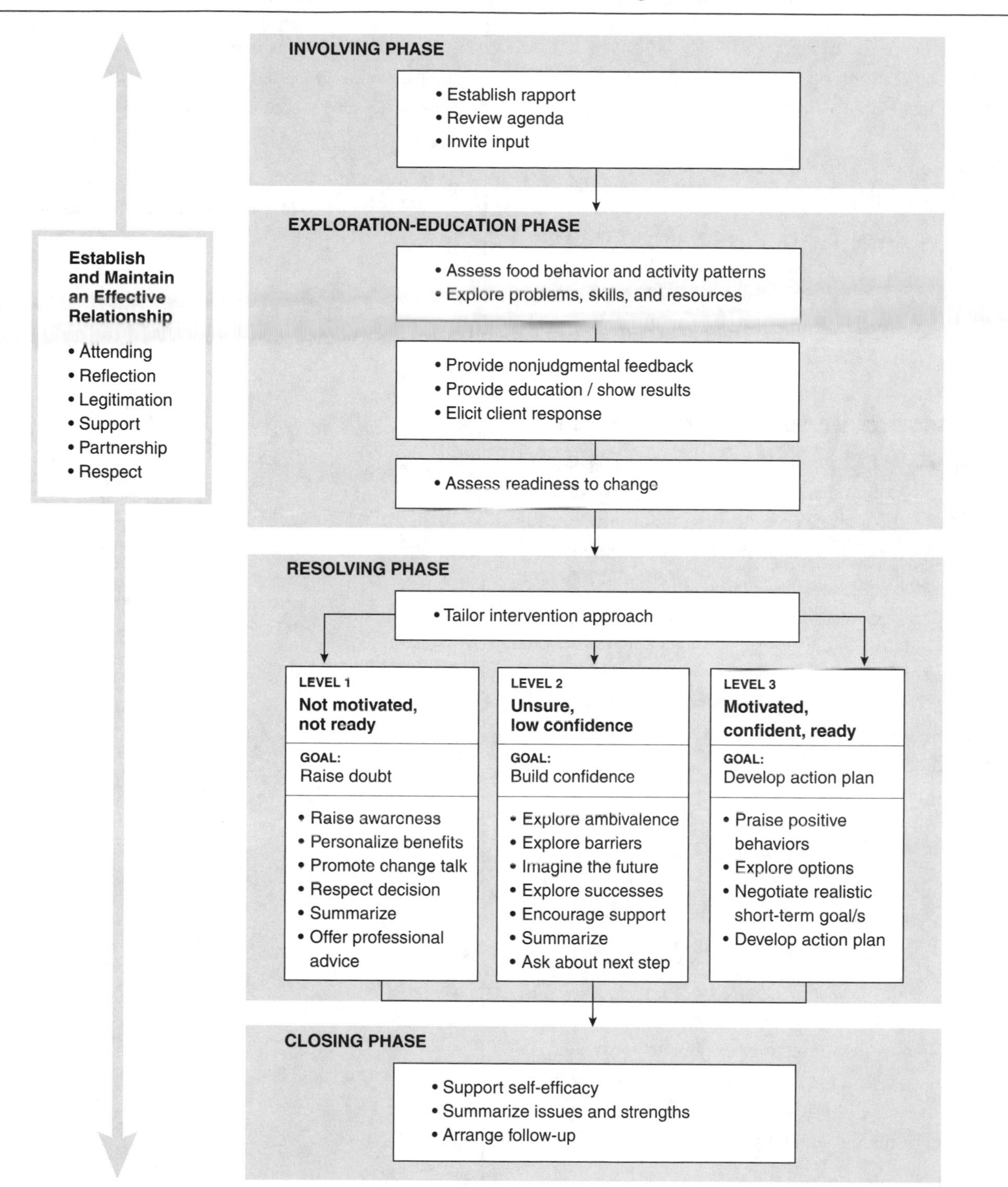

Figure 3.2 Motivational Nutrition Counseling Algorithm
Source: Adapted from Snetselaar L, *Counseling for Change*. In: Mahan LK, Escott-Stump S, eds. *Krause's Food, Nutrition, & Diet Therapy*, 10th ed. (Philadelphia: Saunders; 2000), pp. 451–462; and Berg-Smith SM, Stevens VJ, Brown KM, Van Horn L, Gernhofer N, Peters E, Greenberg R, Snetselaar L, Ahrens L, Smith K for the Dietary Intervention Study in Children (DISC) Research Group, A brief motivational intervention to improve dietary adherence in adolescents. *Health Educ Res.* 1999;14:101–112.

- transtheoretical model,[7,8]
- motivational interviewing,[9]
- brief negotiation—health behavior change method,[10,11]
- solution-focused brief therapy,[12] and
- self-efficacy.[13]

This algorithm provides the direction for the counseling interview/counseling protocol found in this chapter and for each session of the step-by-step guided counseling program outlined in Chapter 9 of this text.

ASSESSING READINESS TO CHANGE

The motivational nutrition counseling algorithm presented in Figure 3.2 provides for three levels of motivation requiring a need to make an assessment of readiness to change. Professionals have offered a variety of strategies to assess motivational level. The strategies generally attempt to identify a particular stage of the transtheoretical model or a person identifies a position on a scale. Figure 3.3 on page 61 shows questions and an algorithm that has been used in several nutritional studies to determine stage of change for adopting a low-fat diet. It can be easily modified to assess stages for other dietary factors.

Not all counseling situations lend themselves to a formal type of assessment, particularly when monitoring on a regular basis. A debate has emerged over whether to use a looser continuum of readiness to change in a therapeutic setting as an alternative to strictly defined stages of change.[11] Several looser assessments have been developed:

Readiness-to-change questions. Pastors et al.[2] provide simple readiness-to-change statements for two lifestyle behaviors, working on changing weight and on becoming more physically active (p. 3). Here is an adaptation of these statements:

- ❑ Yes, right now.
- ❑ Yes, but I can't right now.
- ❑ No, but I will think it over.
- ❑ No, not now.
- ❑ No, I'm not interested.

Readiness-to-change ruler: Snetselaar[5] advocates showing a ruler or a picture of a ruler and asking a question such as, "If number 1 corresponds to not thinking about a change and the top number corresponds to highly motivated to change, how interested are you in lowering the salt in your diet?" Numbers indicate the following: 0–4 = not ready (precontemplation), 5–8 = unsure

ASSESSMENT RULER

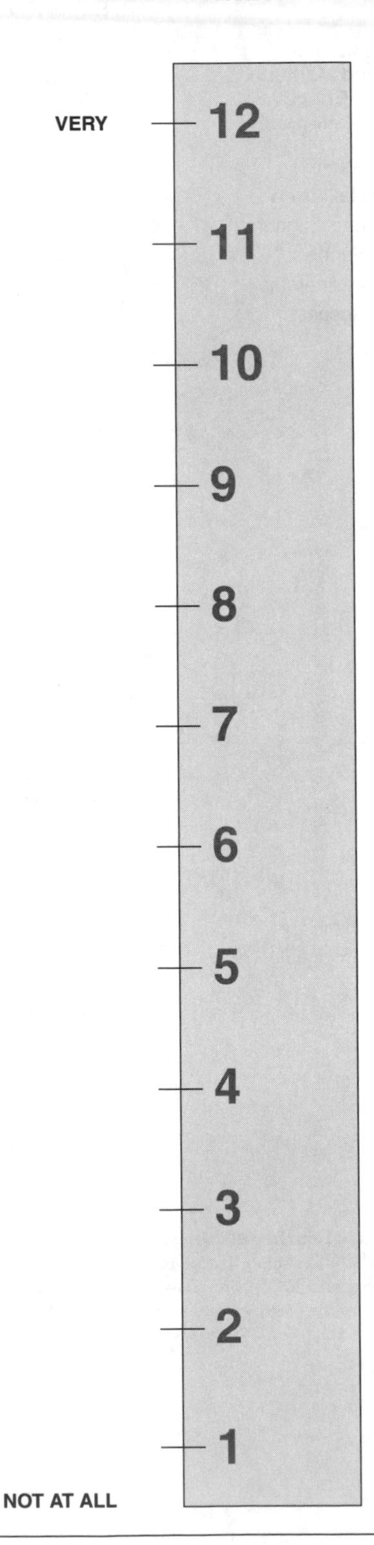

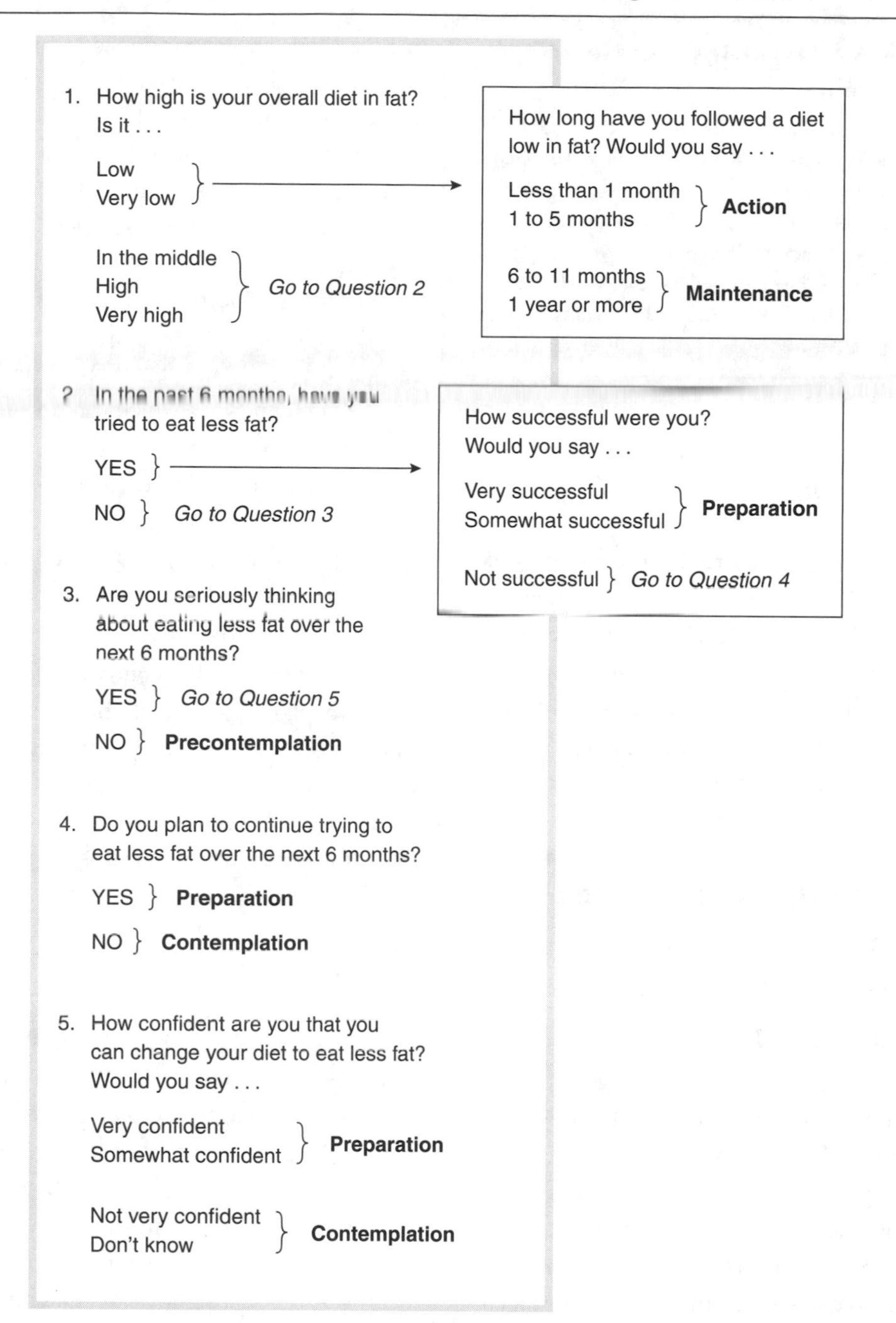

Figure 3.3 Questions and Algorithm Used to Assign Stages of Change for a Low-Fat Diet
Source: Kristal AR, Glanz K, Curry SJ, Patterson RE. How can stages of change be best used in dietary interventions? © 1999, The American Dietetic Association. Reprinted by permission from *Journal of the American Dietetic Association*, 99:680.

(contemplation), and 9–12 = ready (preparation). (See Lifestyle Management Form 3.1 in Appendix G.) This method has been incorporated into the protocol for a counseling interview/counseling session found later in this chapter and into Chapter 9, a step-by-step guide of four counseling sessions. We have found that a physical tool for this assessment is useful for beginning counselors.

Dietary adherence ruler. A similar scale can be used to assess adherence to a dietary protocol. In that case, the number 0 is equivalent to "never follow the dietary guidelines," and the number 12 corresponds to "always follow the diet." (See Lifestyle Management Form 3.1.)

Readiness-to-change scale question. Windhauser et al.[14] have described a similar method, but without a tool, by simply asking this question: "On a scale from 1 to 10,

EXERCISE 3.2—Practice Using Readiness-to-Change Assessments

Find a friend or relative willing to allow you to assess his or her motivation level regarding a needed behavior change. Use the algorithm assessment tool in Figure 3.3, the readiness-to-change ruler, Lifestyle Management Form 3.1, and the readiness-to-change questions or readiness-to-change scale question. In your journal compare and contrast the techniques, and indicate which one gave you the clearest picture of the individual's readiness to make a change.

with 1 being 'not at all' ready and 10 being 'totally ready,' where would you put yourself regarding changing the amount of fat you eat?"

Assessment of importance and confidence as well as readiness. Rollnick et al.[11] take this type of assessment a step further. In addition to checking for a general state of readiness to change, they also encourage counselors to use the same types of scales to investigate two major components of readiness, importance and confidence, in order to obtain a clearer picture of how someone feels about change.

NUTRITION COUNSELING PROTOCOLS: ANALYSIS AND FLOW OF A COUNSELING INTERVIEW/ COUNSELING SESSION

The following nutrition counseling protocols are based on the motivational algorithm for a counseling intervention presented in Figure 3.2. Table 3.1 presents an overview of the tasks and objectives covered in each phase of the algorithm. The word *interview* refers to the collection of valid and accurate data, whereas the word *counseling* implies that a counselor is assisting an individual in making life change decisions. Since nutrition counselors address both tasks in their interactions with clients, the terms have been combined.

The following narrative offers a step-by-step guide for conducting a nutrition counseling session. The guide is organized into the four phases identified in Table 3.1 and Figure 3.2. The flow of the tasks and strategies follows the motivational nutrition counseling algorithm illustrated in Figure 3.2. The guide is provided to (1) give direction to a novice nutrition counselor, (2) furnish a counseling framework which can be molded to fit individual talents and needs, and (3) supply a springboard on which to build skills. There is no perfect method for all individuals, but the guide includes what is generally considered standard in nutrition counseling.

Involving Phase

The first stage of the counseling session is the involving phase, which includes such relationship-building activities as greeting the client and establishing comfort by making small talk, opening the session by identifying the client's goals and long-term behavioral objectives, explaining the counseling process, and making the transition to the next phase.

Greeting This phase sets the stage for the development of a trusting, helping relationship. At the beginning of each session, your greeting should indicate a sense of warmth and caring. The tone of your voice and your body language should convey the message that you are happy to meet your client. This is especially important for the initial session since a first impression is a lasting impression. The manner in which you and your client will address each other should be established. In many institutional settings and among some ethnic groups the custom will be to use a formal Mr., Mrs., or

EXERCISE 3.3—Assess Importance, Confidence, and Readiness

Select five behaviors you would like to follow and list them in your journal. Use the Assessment Ruler, Lifestyle Management Form 3.1, to rank the following questions:

How do you feel right now about your readiness to make the change?
How important is it for you to make a change right now?
If you decide to change, how confident are you that you will succeed?

Put the scores in columns with the change categories as headings. For example:

	Readiness	Importance	Confidence
Use my seat belt for every car trip.	10	12	8

TABLE 3.1—Overview of the Counseling/Interview Process

PHASE	POSSIBLE TASKS[1]	OBJECTIVES
Involving	• Greeting and introduction. • Identify client's long-term behavior change objectives. • Explain rational for recommended diet. • Explain counseling process. • Set agenda.	• Establish rapport, trust, and comfort. • Communicate an ability to help. • Interact in a curious and nonjudgmental manner.
Exploration-education	• Offer educational activities. • Assess food behavior, activity patterns, and past behavior change attempts. • Explore problems, skills, and resources. • Give nonjudgmental feedback. • Elicit client response. • Assess readiness to change.	• Provide information. • Show acceptance. • Learn nature of problems and strengths. • Promote self-exploration by the client. • Clarify problems and identify strengths. • Help the client to evaluate the situation.
Resolving	• Tailor intervention to the client's motivational level.	• Help the client make decisions about behavior change. • Indicate that the client is the best judge of what will work.
Closing	• Support self-efficacy. • Review issues and strengths. • Restate goal(s). • Express appreciation. • Arrange follow-up.	• Provide support. • Provide closure.

[1]The specific tasks to be addressed are dependent on motivational level and needs of your client as well as previous interactions with your client (for example, first session or fourth session).

Miss. In less formal settings such as a health center or rehab program, calling clients by their first name may be more appropriate. If in doubt, start out using a formal address or both names, such as "Mrs. Jones" or "Sally Jones." Inquire how the client should be addressed, such as "Do you prefer to be called Mrs. Jones or Sally?" Generally you should not use the first name unless given permission.[15] Be sure to introduce anyone who is accompanying you, such as a colleague or an assistant. If there are friends or relatives with your client, be sure they are also acknowledged and warmly greeted.

Establish Comfort Hopefully before your meeting you were able to create as private and quiet an environment as possible. However, once the session has begun, paying attention to your own comfort level as well as that of your client is still important. For example, sun in the eyes can be extremely distracting, as can a loud radio in an adjacent room.

Small Talk After your greeting, it may be appropriate to engage in some small talk, depending on the setting and the amount of time you have for the session. Generally this should be limited to a question or a comment about the office or building where the meeting is taking place. This verbal exchange can aid in the development of a comfortable atmosphere, but if carried on for more than a few comments, the counseling experience may be hampered by creating a superficial atmosphere that can permeate into the rest of the session. Particularly stay away from comments unrelated to the client or counseling experience such as a current story in the news.[16]

A review of the intake records of a ninety-year old, new resident in the nursing home where I consulted indicated that this person had grown up on a farm close to where the nursing home was located. When I walked into this woman's room to introduce myself and do a diet consultation, I found the woman sitting in a corner with a scowl on her face. As I attempted to talk to her, her body language and grunts did not change until she realized what I was saying. I was telling her she had come home. I explained to her what I meant and her facial expression and body language changed completely. She then had no trouble hearing me, and we talked a little while about what it was like growing up on the farm. After that experience, I always tried to find something special about new residents from the records. It was a great way to develop rapport.

Opening—First Session A common opening after the greeting and small talk is to ask an open-ended question in a curious manner, such as "When we talked on the phone, you said the doctor told you that your blood pressure is elevated. What are you hoping to achieve in counseling?" or "What brings you here today?" or "How can I help?" This begins the process of attempting to understand your client's needs, expectations, concerns, and coping strengths. As clients clarify their needs, the direction to pursue intervention strategies becomes clearer. Some clients need time to feel truly comfortable expressing their thoughts and feelings. King[17] notes that some clients could take up to three sessions before they are relaxed enough to communicate openly. Examination of your client's topics will be more fully developed in the next phase (the exploring phase) and in subsequent sessions as well. For now emphasis should be placed on the following specific counseling approaches during this part of the encounter:

- **Relationship-building responses.** There will probably be opportunities to use several of the relationship-building statements, particularly empathizing, legitimation, and respect.
- **Reflective listening.** This skill can aid in understanding client issues.
- **Responses that indicate attentiveness and help clarify meanings.** These include paraphrasing, summarizing, clarifying, and asking open-ended questions.

The first time I ever counseled a client was in a nutrition counseling class in college. The person assigned to me volunteered to participate because his doctor told him his cholesterol was slightly elevated. The first day I saw him he made it clear that he was looking for dietary information related to this issue. I didn't actually address this concern until our fourth session. Up until that time I thought he was uncooperative and disinterested, and then I saw a new client. At that point I understood what I should have been doing from the first meeting.

Opening—Subsequent Sessions In subsequent sessions, an opening question is used to invite input, such as "Where would you like to begin?" or "How have things been going since we last talked?" This gives a client an opportunity to address any burning issues before getting into your agenda or you may decide to alter your agenda based on your client's immediate concerns.

Identifying Client's Long-Term Behavior Change Objectives (General Goals) While discussing your client's needs and expectations, a long-term goal or goals should be established. This topic needs to be covered the first session and periodically reviewed thereafter if there is long-term involvement. These can be general in nature such as to feel better or to improve the nutritional quality of food intake. If possible set at least one measurable long-term goal, such as reducing cholesterol to 190 milligrams/deciliter or keeping blood sugar levels below 160 milligrams/deciliter. Specific goals are needed for measuring outcomes. If your client comes in with a vague goal, such as improving diet quality, you can collaborate with your client after an assessment to establish more specific goals that can be evaluated. For clients who wish to lose weight, efforts should be made to avoid setting unmaintainable goal weights and to focus attention on setting goals related to healthful eating and increased activity.[18] Be sure to listen very carefully to what your client says is important to him or her. If the client received a diet prescription from a physician, provide a rationale for the diet.

Explain Program and Counseling Process—First Session Early in the session of the first meeting, your client should receive an explanation of what to expect to happen in the course of the counseling program. This would include a description of the assessment tools, a general statement regarding the issues that will be discussed, and a survey of possible intervention strategies and activities. Review frequency of meetings and telephone or e-mail contact.

Discuss confidentiality. A partnership statement is appropriate clarifying your role as a source of expertise, support, and inspiration and your intention to work collaboratively with the client to make decisions about lifestyle changes. Your comments should indicate that ultimately the client is the one that will be making decisions, for he or she will be the one to implement the change and is also the best judge of what will work. This is particularly important in cases of acute illness, such as diabetes, where dietary practices are an integral component of care and in reality it is a self-managed disease.[19]

Many programs use a counseling agreement form to verify that many of these topics have been addressed. Lifestyle Management Form 3.2 is an example of such a form (see Appendix G). Be sure to also review what your client is hoping to receive from a nutrition counselor. Here is an example of this part of the involving phase:

Counselor: *As we get into the counseling process, I'd like to share with you my hope of how we will be working together. I see myself as a source of information, support, and inspiration for you as we work together to find solutions for your issues. I hope to assist you in making an informed choice about what behaviors to change and maybe whether or not to change the behavior at all. I can help you learn about healthy options and possible strategies, but only you know best what will fit into your life and what you are willing to tackle. I would like us to work together to build on skills that you already have for dealing with your nutrition issues. How does this all sound to you? Do you see things differently?*

Discussing Weight Monitoring, If Appropriate—First Session If you have a client who would like to lose weight, discuss how to handle weight monitoring. Authorities do not agree how often dieters should weigh themselves. Weights can be taken once a day, once a week, or once a month. Generally, nutrition counselors working with uncomplicated outpatient clients should not be doing the actual weighing. The focus of your sessions should be on making changes that allow weight loss to occur, not on numbers on a scale. Since several physiological factors can affect the reading, a one- or two-pound weight loss may not be readily seen. If a counselor is weighing a client and the scale does not show a loss, this can have a negative impact on the whole counseling session. However, a long-term goal of losing a specified amount of weight would be appropriate. See Exhibit 3.1 for an alternative opinion regarding weight monitoring for clients with uncomplicated health issues.

Setting the Agenda—First Session Establish and agree with your client as to what will be covered in the counseling session. For your first session this probably means going over the flow of the session—explaining the assessment process, reviewing preliminary results, selecting a food habit to address, assessing readiness to take action and then setting a goal and plan of action, if ready, or exploring ambivalence and providing information, if not ready.

Setting the Agenda—Subsequent Sessions Before your meeting, review your notes regarding previous sessions, and prepare any educational experiences or materials that were indicated. Ask your client for any issues he or she would like to address during your meeting, explain your intentions for the session, and then come to an agreement for an agenda. At the time you do an assessment, you will establish a list of options that could be addressed. These could be reviewed in subsequent sessions to establish new priorities.

Transitioning to the Next Phase Before entering into the exploration-education phase, make a statement indicating a new direction:

Counselor: *Now that we have gone over the basics of the program, we can explore your needs in greater detail.*

EXHIBIT 3.1—Debate over Dieting

Authorities are not in agreement as to the amount of emphasis that should be placed on body weights and the concept of a "diet." Some believe the scale ought be thrown in the trash, the concept of diet abolished, and emphasis placed on body size acceptance. Some are concerned that too much importance on dieting contributes to body dissatisfaction among girls and women, which may lead to bingeing behavior, bulimia, and a host of other physiological and psychological problems.[20] Advocates of the nondiet approach maintain that healthful changes in eating and other lifestyle behaviors without "dieting" can have a positive impact on health. In the successful Kentucky Diabetes Endocrinology Center, many patients have gained good control over diabetes, blood pressure, and blood lipids without placing primary emphasis on body weight.[21]

Garfield® by Jim Davis

Exploration-Education Phase

During this phase, a nutrition counselor and client work together to understand a client's nutrition/lifestyle problem, search for strengths to help address difficulties, assess readiness to take action, and provide educational experiences. Counselors need to provide a nonjudgmental environment so clients will feel free to elaborate on pertinent issues. A counselor's verbal and nonverbal behavior should be viewed as curious rather than investigative. Responses covered in Chapter 2 that can be especially useful to advance exploration include open questions, paraphrasing, reflection, probing, and directives.

Educational Activities During your first session, assessment activities are likely to be time-consuming, resulting in too little time for involved educational activities. The main educational task should be to address health risks associated with your client's eating pattern. The educational experiences of subsequent sessions should be geared to your client's needs and desires as determined during the assessment process. See Chapter 5 for a discussion of integration of information giving in a counseling session.

Assessment—First Session Assessment is an important component of the counseling process to tailor an intervention to the needs of a client. Basic dietary and physical assessment procedures and commonly used forms in nutrition counseling are addressed in Chapter 4. While collecting information or reviewing completed forms, do not react with advice, criticism, or judgment, as this could inhibit disclosure. If at all possible, counselors should refrain from firing a series of questions to gain information. A more satisfying and valuable method would be to encourage discussions at certain points to allow clients to provide insight about their life experiences.

Often counselors attempt to have clients complete assessment or screening forms before the first session. This approach saves valuable counseling time and allows clients to focus ahead of time on issues and counseling expectations. If that is the case, when you do meet with your client, avoid rehashing exactly what is on the forms but open up for discussion the experience of filling out the questionnaires. Consider the following example:

Counselor: *Thank you for completing these forms. The information in them will be helpful as we work together to search for solutions for your food problems. I am wondering, what came to your mind as you were filling out these papers? What topics covered in these forms do you think have particular importance for your food issues? Did you feel a need to expand or clarify any of your answers? Which ones? Did it prompt you to think about what you would like to cover in our sessions together?*

Here are some appropriate topics to explore with your clients:

- Concerns about health risks associated with current eating behavior
- Concerns about changing food patterns
- Past lifestyle change successes
- Past experiences trying to change food habits by themselves, with the aid of a nutrition counselor or in an organized program

- Difficulties with making food habit changes in the past
- Strategies that worked or did not work when attempting to make a lifestyle change
- Selection of education topics to address in future sessions

If you have enough time in a counseling session (six to eight minutes), consider using "a typical day" strategy, which is similar to the diet history interview described in Chapter 4. This method encourages clients to drive the assessment discussion and to tell their "story." If the story goes on for more than eight minutes, however, the activity becomes tiring for both parties, and the counselor should intercede to speed up delivery. See Exhibit 3.2 for guidelines for using this strategy.

Some authorities suggest an alternative approach of having at least one counseling session with your client before utilizing assessment instruments to build motivation for and understanding of the assessment process. They feel that the laborious task of completing the forms begins the counseling intervention with an impersonal tone and may be an obstacle to receiving treatment at all.[9] In some facilities where the number of nutrition counseling sessions is limited, however, this procedure may not be practical.

Assessments—Subsequent Sessions Generally assessment activities are the most intense during the first session, but assessments should be made periodically to assist in setting new goals and to monitor progress. The adherence ruler discussed earlier can be used to assess how closely a client has been following a goal or a dietary protocol. Follow up the question with simple open-ended questions to gain a deeper understanding of a client's progress.

Counselor: *Please look at the picture of this ruler. What number would you pick to describe how closely you have been following your food plan?*

Tell me something about the number you chose.

How come you chose number 7?

EXHIBIT 3.2—Guidelines for Using a Typical Day Strategy

1. *Introduce the task carefully*

Sit back and relax! Ask the client a question such as: "Can you take me through a typical day in your life, so that I can understand in more detail what happens? Then you can also tell me where your eating fits in. Can you think of a recent typical day? Take me through this day from beginning to end. You got up . . ."

2. *Follow the story*

- Allow the client to paint a picture with as little interruption as possible. Listen carefully. Simple open questions are usually all you need—for example, "What happened then? How did you feel? What exactly made you feel that way?"
- Avoid imposing any of your hypotheses, ideas, or interesting questions on the story you are being told. Hold them back for a later time. This is the biggest mistake made when first using this strategy. Don't investigate problems!
- Watch the pacing. If it is a bit slow, speed things up: "Can you take us forward a bit more quickly? What happened when . . . ?" If it is a bit too fast, slow things down: "Hold on! You are going too fast. Take me back to . . . What happened . . . ?"
- If you are uncertain about details, and you are happy that you are being curious rather than investigative, then ask the client to fill them in for you.
- You know you have got it right when you are doing 10 to 15 percent of the talking, the client seems engaged in the process, and lots of interesting information about the person is emerging.

3. *Review and summarize*

A useful question at the end of your client's story is "Is there anything else at all about this picture you have painted that you would like to tell me?" Now is the time to ask probing questions to clarify any descriptions, such as "Can you tell me what kind of bread you usually use to make your lunch sandwich?" This is also a good opportunity to be honest with the client about your reaction and to provide legitimation responses wherever possible. Having listened so carefully to the client, you will now be able to explore other topics quite easily. Often this leads into an investigation for the need for general information—for example, "Is there anything about . . . that you would like to know?"

Source: Rollnick S, Mason P, Butler C, *Health Behavior Change: A Guide for Practitioners* (New York: Churchill Livingstone; 1999); © 1999, pp. 113–114. Used with permission.

Giving Nonjudgmental Feedback Assessment results should not be simply handed to a client but gone over point by point in a neutral manner. Counselors need to provide clear norms for comparison, such as a therapeutic dietary protocol or the Dietary Reference Intakes. Some assessment forms have a standard on the form. (See Lifestyle Management Form 4.3 in Appendix G.) Give only the facts, allowing the client to make the initial interpretation. If you have a great deal of feedback to provide, pause regularly to allow the client to process the information and to check for comprehension:

Counselor: *Your assessment indicates an intake of one vegetable a day. As you see, the standard recommendation based on the Food Guide Pyramid is three servings.*

Eliciting Client Thoughts about the Comparison of the Assessment to the Standard Ask simple open-ended questions to encourage a client to explore the meaning of the results through curiously asked questions and personal reflection. The ideal response would be something like "I see," "I didn't really give much thought to this before," or "I'm wondering if . . ."[11] This approach provides an opportunity for people to discover discrepancies between their condition and a standard and to make self-motivational statements as discussed in the section on motivational interviewing in Chapter 1.

Counselor: *What do you think about this information?*

Do these numbers surprise you?

I have given you a lot of information. How do you feel about what we have gone over?

Did you expect the evaluation to look different?

If your client does ask you for clarification or meaning, present the information in a nonthreatening manner by avoiding the word *you.*[11] For example, "People who have a low intake of vegetables are at a higher risk for developing several types of cancers"—not "You have an increased risk for developing several types of cancers." Nor should a counselor tell a client how he or she should feel about the feedback—for example, "You should really be concerned about these numbers. They increase your risk for developing cancer." An alarming explanation interferes with the decoding process, causing explanations to be misunderstood and increasing resistance to change.

Determining What's Next After an assessment, it is generally a good idea to summarize; include what the client is doing well, problems identified in the assessment, any self-motivational statements made by the client, and ask the client whether the summary needs any additions or corrections. Then ask your client how he or she would like to proceed. If your client asks your advice, give your impressions, provide options, and indicate that the client would know best what would work for him or her.

Assessing Readiness to Make a Change At this point your client may have clearly indicated a desire to make a behavior change, and you will need to use your judgment about making a formal assessment for readiness. If you are using a continuum scale such as the numbers on the ruler, you may decide to check for importance and confidence as well as readiness.

Counselor: *To get a better idea of how ready you are to make a food behavior change, we will use this picture of a ruler. If 0 represents not ready and 12 means ready, where would you place yourself?*

There are actually two parts to readiness, importance and confidence. I think it might be useful to look at them separately. Using the same scale, how do you feel right now about how important this change is for you? The number 0 represents not important, and 12 is very important.

The other part of readiness is confidence. If you decided to change right now, how confident do you feel about succeeding? The number 0 indicates not confident at all, and 12 represents very confident.

Resolving Phase

In the involving and exploration phases, the major objective was to assist the client in clarifying problems and identifying strengths to self and to the counselor. The direction of the remaining time of the counseling session will be determined by a client's mo-

EXERCISE 3.4—Practice Using "A Typical Day" Strategy

Review the guidelines in Exhibit 3.2 for utilizing "a typical day" strategy. Work with a colleague, taking turns role-playing counselor or client. Write your reactions to this activity in your journal. What were your thoughts and feelings while you were the counselor? The client? How would you use this activity in an actual counseling interaction? What did you learn from this activity?

tivation category. In the motivational interviewing algorithm described by Berg-Smith et al.[6] and incorporated into Figure 3.2, three preaction motivational levels are illustrated to address the needs of the majority of individuals with a dietary problem.[5] In the first and possibly the second level, the major issue related to motivation is likely to center on viewing the behavior change as important. Those who pick a higher stage or higher number on a continuum are likely to feel the behavior change is important but are struggling with confidence in ability to make a successful change.[11] Therefore, in the following analysis of the resolving phase, Level 1 counseling approaches will deal with importance, Level 2 with confidence issues, and Level 3 with selection of a goal and design of an action plan. Although the motivational levels are represented as three distinct entities, counselors need to be flexible in their approach to accommodate fluctuations in motivation level that can occur during an intervention.[9] In such cases there may need to be a cross-over in selection of counseling approaches among the three motivational levels. A summary of the counseling approaches for each level of readiness to change is presented in Table 3.2.

Level 1: Not Ready to Change (Precontemplative)

Level 1 clients have clearly indicated that they are not ready to change their behavior or are not doing well at attempting to change. Individuals have a right to decide their destiny, and a decision not to change should be respected. However, health care providers have an obligation to make clear the probable consequences of exercising their prerogative. The major goal of working with clients who fit into this category is to raise doubt about present dietary behavior; the major tasks are to raise awareness of the health/diet problems related to their dietary pattern. Often precontemplators have come to counseling because of the urgings of others, or they are sitting ducks in a hospital room as a dietitian walks in to give a consultation because of a diet order. They do not need solutions; they need to know they have a problem. The following are some strategies recommended for clients at this level of readiness to change:[6,11,22]

Raise Awareness of the Health Problem/Diet Options. Sometimes clients are not aware of the benefits of behavior change or the risks and consequences of their present dietary behavior. Others have misconceptions about the type of dietary changes that are needed. During an awareness discussion, emphasize anything positive that your client is doing that could be built on if a decision to change is made.

Counselor: *There is a lot of information in the news about dietary fat and cholesterol levels, but you may not be aware of all the other dietary factors associated with elevated cholesterol levels. I see that you enjoy eating salad and that you have soup with beans for lunch sometimes. Both of those choices could be built on to help lower your cholesterol level.*

Personalize Benefits. Clients often know that improving their diet would probably be better for them. However, they may not have given thought to how they would benefit personally or in which way they may feel better.

Counselor: *Increasing fruit and vegetable intake could be particularly beneficial to you to help lower your blood pressure and aid in your efforts to lose weight. Focusing on these foods is likely to have a positive impact on your occasional constipation problem, too.*

Ask Key Open-Ended Questions to Explore Importance of Change and Promote Change Talk. Thinking and talking about changing behavior can help elicit self-motivational statements and aid in the development of motivation to change.[11] Change talk can be elicited by utilizing key open-ended questions. Counselors should listen very carefully to the answers and concentrate on the exact meaning of what is being said. Follow up your client's answers with paraphrasing, reflective listening statements, or other open-ended questions. If you observe resistance in your follow-up, back up and use a different approach.[11] However, since the client has already indicated that there is very little desire to change, it is generally best to begin this discussion with a tentative approach by requesting permission to discuss the issue.[11]

Counselor: *Would you be willing to continue our discussion and talk about the possibility of a change in your food habits?*

Elicit Self-Motivational Statements Regarding the Importance of Changing Dietary Habits. People are often more likely to be persuaded to change, if they are presenting the arguments for change.[9]

TABLE 3.2—Resolving Phase Summary of Tailored Intervention Approach

Stage	READINESS TO CHANGE	COUNSELING APPROACH	
Precontemplation	• **Level 1**	**Goal:**	**Raise doubt about present dietary behavior.**
	Not ready	**Major task:**	**Inform and facilitate contemplation of change.**
	Ruler = 1–4	**Approach:**	• Raise awareness of the health problem/diet options. • Personalize benefits. • Ask open-ended questions to explore importance of change and to promote change talk. Elicit self-motivational statements regarding importance. Elicit identification of motivating factors. • Summarize. • Offer professional advice, if appropriate. • Express support.
Contemplation	• **Level 2**	**Goal:**	**Build confidence and increase motivation to change diet.**
	Unsure	**Major task:**	**Explore and resolve ambivalence.**
	Ruler = 5–8	**Approach:**	• Raise awareness of the benefits of changing and diet options. • Ask open-ended questions to explore confidence and promote change talk. Elicit self-motivational statements regarding confidence. Elicit identification of barriers. • Explore ambivalence by examining the pros and cons. Client identifies pros and cons of not changing. Client identifies pros and cons of changing. • Imagine the future. • Explore past successes. • Encourage support networks. • Summarize ambivalence. • Ask about next step.
Preparation / Action	• **Level 3**	**Goal:**	**Negotiate a specific plan of action.**
	Ready	**Major task:**	**Facilitate decision making.**
	Ruler = 9–12	**Approach:**	• Praise positive behaviors. • Explore change options. Elicit client's ideas for change. Look to the past. Review options that have worked for others. • Client selects an appropriate goal. • Develop action plan.

Counselor: *What do you believe will happen if you do not change the way you eat?*

What is the worst thing that could happen if you continue to eat the way you have been eating?

*When we used the ruler to evaluate how important it was to you to change your food habits, you indicated that it was somewhat important. How come you picked the number 4 instead of 1?**

*In a readiness assessment based on designated changes, the scaling questions could still be used by substituting the names of the stages for numbers. For example, "Why are you in contemplative stage rather than precontemplative?"

Elicit Identification of Motivating Factors for Changing a Diet. The objective of using the following questions is to identify what factors would make a client feel that it is important to make a dietary change.

Counselor: *What would have to be different for you to believe that it is important to change your diet?*

You indicated that changing your diet was somewhat important by choosing the number 4 on the ruler. What would cause you to view things differently and move up to the number 8?

Summaries. Summaries help reinforce what has been said, tying together various aspects of a discussion and encouraging clients to rethink their position. Give a summary of reasons not to change before giving a summary of reasons to change. Be sure to end your summary with any self-motivational statements your client may have made. Finally, ask your client whether the summary was fair and whether he or she would like to make any additions.

Counselor: *Now that the session is coming to a close, I would like to review what we covered so we can agree on where we are and where we are going. You said you came today because of pressure from your doctor and your wife. Your cholesterol readings have been high. The last one was 320. You know that people are concerned about you but you feel fine and wish people would get off your back. When we went over the types of foods that have been found to help lower cholesterol, you said you were surprised that there was more you could focus on than just fat. In fact, there were some foods that you enjoy eating that were on the review list. You thought if you did change the way you eat, that some of the people close to you wouldn't be so worried about you. Lastly, you said that there were actually several ways beans are prepared that you could eat more of. Was that a fair summary? Did I leave anything out? Where does this leave us now?*

Offer Professional Advice, If Appropriate. Well-timed and compassionate advice can aid in motivating behavior change.[9] In the ideal situation, a client asks for advice, but if that is not the case, then the counselor can ask permission to give advice. Review the guidelines in Chapter 2 for offering advice. Be sure the advice you give emphasizes that the client knows best as to what will work and the choice is up to the him or her. This would be a good time to offer educational materials, if available.

Counselor: *It is really up to you and you know best what would work for you, but it appears that there are some simple things you could do that might make a difference in your cholesterol level. For example, you enjoy bean soup and oatmeal. A good place to start could be to start having soup for lunch or dinner or oatmeal for breakfast. I have some information that you could take home to read about foods to emphasize in your meals to lower cholesterol levels. How are you feeling at this point about making a change?*

Express Support. Relationship-building skills may be ignored, and a counselor could be tempted to argue with a client, especially in the case of a serious medical condition. However, this tactic is not likely to encourage a client to move toward the action stage but may result in a stronger resistance to change. Letting the client know that you are there to offer guidance and support is likely to have a greater impact. For clients at this level of motivation, the objective is to create a doubt about their present behavior pattern, so preparing an action plan is not a likely scenario. Letting your clients know what others have done in their situation can have an impact. The fact that you and the client are not working toward making a behavior change at this time should be acknowledged, and the door should be left open for future contact.

Counselor: *I respect your decision not to change your diet. It is really up to you. I don't want to push you, but I do want to be sure you are very clear about what could happen as a result of not changing. You probably need some time to think about this. Maybe you will feel differently about this in the future. I want you to know that I will always be here to work with you to find solutions. I have met others with your problem. Most do decide to work on changing their diet. However, some do not. You are the best judge of what would work for you. Would you like for me to call you next week to talk about how you are feeling about the diet prescription? If you have any questions or need clarification about anything, do not hesitate to call me.*

Level 2: Unsure, Low Confidence During the motivation assessment, clients in this category indicated that a diet change is possible. They know the problem exists but something is needed to push the decisional balance in favor of making a change. The objective of working with people at this level is to build confidence in their ability to make a diet change, and the major task is to explore and resolve ambivalence. The following is a review of some of the approaches advocated for people who have low confidence in their ability to change.[6,11,22]

Raise Awareness of the Benefits of Changing and Diet Options. Clients at this level know they need a solution but may not have all the facts regarding the benefits of changing. They may not really know what dietary changes would have to be made. Simple facts may be all that is needed to progress to higher level of readiness to change.

Ask Key Open-Ended Questions to Explore Confidence and Promote Change Talk. The formats of these questions are similar to the ones posed for Level 1 clients; however, the objective switches from exploring importance to focusing on confidence. Again, answers to these questions should be followed up with paraphrasing, reflective listening statements, or more open-ended questions.[11]

Elicit self-motivational statements regarding confidence in ability to change dietary habits. The following illustrates key open-ended questions that promote change talk related to confidence in the ability to make a good behavior change.

Counselor: *You have indicated that you are somewhat confident that you would be able to change you diet. When we did the ruler evaluation, you picked the number 6. How come you did not pick the number 1?*

Elicit identification of barriers to changing diet. For people at this motivational level, very little thought may have been given to exactly what is keeping them from making dietary changes. By discussing their barriers, possible ways of dealing with them may be identified, and confidence in the ability to change may increase.

Counselor: *You picked the number 4 on the picture of the ruler when we were evaluating how confident you were in your ability to change your diet. What is keeping you from moving up to the number 8? How could I help you get there?*

What are your barriers to making the recommended dietary changes?

What would need to be different for you to feel you are able to make diet changes?

Explore Ambivalence by Examining the Pros and Cons. The objective of this exercise is for the client, not the counselor to identify the pros and cons related to a contemplated change. Rollnick et al.[11] suggest using the words *like* and *dislike* or *pros* and *cons* rather than *advantages* and *disadvantages* or *costs* and *benefits*. The latter words may be confusing for some individuals. You could use a balance sheet as illustrated in Table 3.3 as a visual aid and even fill in the categories. However, remember the objective of the activity is for the client to fully explain his or her thoughts and feelings. Do not let the focus of the interaction be completion of the form and thereby interfere with the flow of conversation. Start this activity by asking your client whether he or she would like to examine the pros and cons.

Counselor: *An activity that we could do that some people find useful is to explore your likes and dislikes about this issue. Would you like to do that?*

Client Identifies Pros and Cons of Present Diet. A comfortable beginning for this strategy is generally to start with what the client likes about his or her present diet. Any follow-up questions should be asked for clarification and not divert focus away from the primary subject. Listen carefully and remember key words used by your client. Before progressing to the next set of questions, summarize both sides of the position, interjecting words used by your client.[11]

Counselor: *What do you like about your present diet?*

What do you dislike about the way you are eating?

Client Identifies Pros and Cons of New or Additional Change. Likes and dislikes of the present diet often mirror those of making a change to a new diet pattern. For example, a client may like that all foods he or she enjoys can be eaten if no change is made, and one of the cons of changing is limiting some of the enjoyable foods. As a result, the conversation may have naturally flowed to the pros and cons of making a change. If not, questions can be asked to arrive at the topic and then provide a summary of the responses.

Counselor: *What do you think you would like about making the change?*

What do you think you would not like about changing your diet?

Imagine the Future. Use imagery to create a picture of a successful future assisting clients in identifying goals and hoped-for benefits. A variation of this question and one of the mainstays of solution-focused therapy is asking a client to suppose a miracle happened.[23] The final

TABLE 3.3—Balance Sheet for Someone Contemplating a Diet Change for High Blood Pressure

NO CHANGE	CHANGE
Likes (Pros) I get to eat all foods I really like. I am comfortable with my food pattern.	**Likes (Pros)** I think I will feel better. Maybe I will lose weight. Maybe I could reduce the amount of medicine I take for my blood pressure.
Dislike (Cons) I am not a good role model for my children. I do not like it that I have to take medicine for my blood pressure.	**Dislike (Cons)** I don't think I will like the foods as much. I have to get used to eating new foods. I think the new diet will be more expensive. I will have to think about what I will eat all the time.

Source: Adapted from Rollnick S, Mason P, Butler C, *Health Behavior Change: A Guide for Practitioners.* (New York: Churchill Livingstone; 1999), p. 82. Reprinted by permission of Harcourt Publishers Ltd.

question, asking whether any part of the picture is presently happening, gives clues as to resources and skills already available that can be expanded on to produce hoped-for outcomes.

Counselor: *Let us create a picture of the future. Imagine that you made all the changes necessary to lower your cholesterol level. What is the first thing you notice that is different? What else is different? How do you feel? What does your brother/wife/husband/boss/etc. see you doing? Who notices that this happened? Are any very small parts of this picture happening now?*

How would you like your diet to be in the future?

Explore Past Successes and Provide Feedback About Positive Behaviors and Abilities. By exploring successes and identifying abilities, the counselor and client can lay a foundation of existing skills that can be built on to make needed changes. One strategy for identifying successes would be to ask whether the client was ever able to accomplish the desired task. Clients should be complimented on any past or present coping abilities that have been identified. This will encourage clients to continue to make similar choices in the future.

Counselor: *What strategies have you used in the past to overcome barriers?*

Have you ever been able to go to a party and eat only one dessert when there were many available to choose among?

You are already drinking soy milk so that is a great substitution for milk to reduce your casein intake.

Encourage Support Networks. Confidence in ability to make a behavior change, or *self-efficacy,* increases when we watch and interact with others who have made the same or similar changes. Support groups can provide excellent resources for modeling. Clients should also be encouraged to share their intentions to change with others. It often brings support and assistance from associates, friends, and family.

Summarize the Ambivalence. The importance of periodic summaries has already been discussed. An effective time for providing a summary is after utilizing a variety of motivational strategies. Summarize your client's ambivalence, and ask your client how she or he would like to continue.

Counselor: *What are your options?*

How would you like to continue?

Choose a Goal, If Appropriate. If your client would like to set a goal, follow the guidelines for goal setting for Level 3 clients, and review the goal-setting process described in Chapter 4. The objective will be to specify a goal to meet the client's motivational level. This may mean buying skim milk instead of low-fat milk or taking some active steps to increase awareness, such as reading informational literature.

Level 3—Motivated, Confident, Ready Level 3 clients have indicated that they are ready to make a lifestyle change. For these clients, the nutrition counselor serves as a resource person increasing awareness of possible alternatives for solving problems. The counselor and client collaborate to select lifestyle changes to alter, clarify goals, and tailor intervention strategies to achieve goals. If possible, past successes should be utilized to find viable solutions. These strategies are outlined in Chapter 4.

Praise Positive Behaviors. To reinforce desirable behavior patterns, counselors should point them out and offer praise. Also explore what skills your client is using to accomplish the desired outcome.

Counselor: *It is so good that you use skim milk on your cereal in the morning. Has that always been the case? How did you make the switch from whole milk to skim milk?*

Closing Phase

In this phase, review with your client what has occurred during the session, including a summary of the issues, identification of strengths, and a clear restatement of goals. In addition, an expression of optimism about the future and a statement of appreciation for any obstacles overcome should be made to support self-efficacy. Plan for the next counseling encounter, which could be a phone call, e-mail, fax, or counseling session. This is also a good time to use partnership statement.

Counselor: *It was a pleasure to meet with you today. You came here because you wanted to know more about what you could do with your diet to lower your blood pressure. We reviewed your food pattern and identified several beneficial food habits, such as your use of skim milk in coffee, several servings of whole grains each day, and an adequate intake of water each day. I believe we did a good job setting a goal to eat a fruit each day for a snack at work. I am optimistic because you enjoy eating fruit and have a well-defined plan to put seven pieces of fruit in a bowl in the front of the refrigerator on Sunday, your shopping day. I am really impressed with all the research you have done about blood pressure and food and exercise. I look forward to working with you to make additional changes. Would you like me to send you an e-mail midweek for a follow-up?*

Considerations for Acute Care

Several factors need to be taken into account when counseling a patient in an acute care facility. For example, patients may have no idea that a registered dietitian was scheduled to see them. The diet consult may be the result of an internal policy and/or a guideline of an accrediting agency, or the patient's doctor may have requested the meeting. Because of recent changes in health care, patients' counseling needs present some special challenges. The physical condition of patients is likely to be more distressed than in the past and hospital stays are shorter, limiting the number of possible inpatient counseling sessions.

Here are some tips for working with patients in acute care settings:

- Each time you visit a patient, introduce yourself, verify that you have the correct patient, and explain the purpose of your visit. For example, "Are you Mary Edwards?" If the answer is yes, introduce yourself and then proceed with the greeting: "I am glad to meet you, Mrs. Edwards." It is important for the profes-

EXERCISE 3.5—Practice Using Counseling Strategies

Practice using the resolving strategies with colleagues, alternating the role of client and counselor. Each of you should choose a behavior change that you have been contemplating (such as following a walking program, increasing fiber, or flossing teeth). Select a behavior change that you feel somewhat ambivalent about implementing so that your assessment will fall into Level 1 or 2. When you are role-playing a counselor, conduct a readiness assessment and follow the suggested counseling strategies for your colleague's motivational level. Since the counselor is experimenting with a variety of new techniques, the experimenter is allowed to call time-out at any point to gather thoughts.

At the end of the designated period (usually eight to ten minutes), each person should share his or her feelings and reactions to the experience. First cover what went well, then what could have been done differently, and finally how each felt about what transpired.

Write your reactions as a client and as a counselor to this activity in your journal. Which strategies did you find useful? What did you learn from this exercise?

Sources: This activity was adapted from role-playing directions in the following sources: Rollnick S, Mason P, Butler C, *Health Behavior Change: A Guide for Practitioners* (New York: Churchill Livingstone; 1999), and Dubé C, Novack D, Goldstein M. *Faculty Syllabus & Guide Medical Interviewing* (Providence, RI: Brown University School of Medicine; 1999).

sional to explain the reason for the contact and how the patient will benefit from the consultation.

- Usually it is a good idea to ask whether this is a good time for the meeting to take place. Patients may be in too much pain or too tired to benefit from a consultation.
- Although time may be limited, do not disregard relationship-building skills and rush through the interview. A hurried atmosphere gives the impression that you do not view your discussion with the client as very important or that you do not care enough to get all the facts.[24]

INTERCULTURAL COUNSELING SKILLS

Current demographic trends in North America require that health care professionals acquire competence in counseling across cultures. Since the 1970s, the United States has been moving toward a cultural plurality, where no single ethnic group is a majority.[25] As of 1990, pluralities already exist in 186 United States counties, including those encompassing New York City and Los Angeles. The 1990 census also showed that one in four Americans has African, Asian, Hispanic, or Native American ancestry.[15] Demographers anticipate that non-Hispanic whites will become less than 50 percent of the total population by the year 2050.[25] Asian Americans are the fastest-growing minority group.[25] Review of Canadian immigrant patterns also shows substantial demographic changes. In 1980, the leading country of origin for immigrants was Great Britain, whereas by 1990 the five top countries supplying Canadian immigrants included Hong Kong, Poland, China, the Philippines, and India.[27]

All counseling occurs in a cultural context but is only referred to as **intercultural counseling** when the cultures of the counselor and the client are noticeably different. The greater the differences, the increased potential for a clash of expectations, misunderstandings of intentions and meanings, and complexity in counseling relationships.[28] Although some professionals believe existing counseling models are effective for all people and that categorizing traits of subcultural groups perpetuates stereotypes, others have attacked commonly used strategies developed by middle-class whites.[15] Some feel the interview-driven assessment reflects a sort of specificity that is both Eurocentric and androcentric.[29] Clients from certain cultural groups may have trouble relating to direct questions and believe that personal questions about background are invasive or unnecessary.[25] Clients may misinterpret the cultural context in which a question was asked, and likewise the counselor can be off base in understanding the answer. Past experience with discrimination can also predispose cultural minorities to be fearful of disclosing too much information about themselves.[29] Among some groups, family elders may need to be included in counseling sessions in order for decisions to be made. As a result, a culturally sensitive approach is essential for an effective intervention to occur.

The complexity of the cultural issues increases when we consider that individuals can identify closely with more that one cultural group. *Culture* is defined as learned patterns of thinking, feeling, and behaving that are shared by a group of people.[3] An individual can belong to several cultural groups. For example, a twenty-year-old homosexual Chinese man may have behaviors and food patterns influenced by several cultural groups. The fact is that each client must be treated as an individual and that at no time can assumptions close the door to inquiry and understanding. In addition, Anderson et al.[27] emphasize that people of a common descent are not a homogeneous group. For example, fluency in English can substantially influence life experiences for Canadian immigrants. Intercultural counseling aids can help dialogue with individuals whose cultural background is substantially different than that of the counselor. To assist in the development of cross-cultural nutrition counseling skills, Magnus[30] has developed an IQ assessment, which is presented in a modified form in Exercise 3.7.

EXERCISE 3.6—Cultural Assumptions

In my position as a nutritionist in a maternal outpatient clinic, I had many opportunities to counsel Hindu women. One day that was particularly busy, I did an assessment of a young pregnant Hindu woman. Her lab work showed low hemoglobin and hematocrit, indicative of iron deficiency anemia. I explained the finding to my client and thought I was making her feel better by stating that it is probably because she does not eat meat. My client's eyebrows rose in surprise as she said, "I am not a vegetarian."

In your journal, write a comment regarding the worldview of this nutrition counselor. If the client was a vegetarian, would the explanation have helped the client feel better? What would be an appropriate response for the counselor to make after being told that the client is not a vegetarian?

If the counselor and client do not share the same primary language, communication can be particularly challenging, but a number of educational aids are available for intercultural counseling. (See the resources at the end of this chapter and Chapters 1 and 2.) One common mistake is increasing the volume of one's voice when the client has a heavy accent. This inhibits communication and can create an even greater counseling barrier.

One challenge for working with culturally diverse individuals is the additional time required for an effective intervention. Language barriers slow communication, and the nutrition counselor may need time to become familiar with the client's culture. Also, additional time is likely to be needed before a level of trust develops between individuals of different cultures. Nutrition counselors must do their best for all involved to make the most effective use of time.[15]

Curry and Jaffe[3] have identified three tasks that need to be accomplished to achieve competence in cross-cultural counseling: (1) awareness of your cultural background, biases, and worldview; (2) knowledge about and sensitivity to other's cultural background, biases, and worldview; (3) and skill in intercultural communication. The first two components were addressed in Chapter 1.

Respondent-Driven Interview

The **respondent-driven interview**[31] is a culturally sensitive approach to gathering information. The counselor asks simple, open-ended questions to initiate conversation, prompts clients for a better understanding when necessary, but for the most part exerts little control over responses. By showing an unbiased and sincere desire to understand and accept traditional views and practices, it is hoped that clients will not fear criticism or ridicule. Exhibit 3.3 contains examples of open-ended questions to gain information to direct the flow of conversation for cross-cultural counseling encounters.[25,32]

All of the questions in Exhibit 3.3 aid in understanding an illness from a client's perspective. However, each question is not appropriate for every cross-cultural counseling encounter, so counselors must use their judgment to select suitable ones. If the counselor appreciates the powerful influence of the client's culture as well as the equally powerful culture of biomedicine, then the need for compromise and mediation becomes obvious.[32]

To allow this process to occur, Berlin and Fowkes[33] developed the **LEARN counseling model** for health care providers to elicit cultural, social, and personal information relevant to a given illness episode. This model entails the following five steps:

1. **Listen.** Again, active listening is an extremely important skill for a counselor to develop. Some factors related to listening merit emphasis for successful counseling across cultures. You should listen carefully to a client without assumptions or bias and recognize the client as the expert when it comes to information about his or her experience.[31] Not only

EXERCISE 3.7—What's Your IQ on Intercultural Nutrition Counseling?

Read the following questions and check true for those statements you believe are correct and false for those you believe are not correct.

True	False	
❑	❑	**1.** Most people who share a common language belong to the same culture.
❑	❑	**2.** Race is a good indicator of food selections.
❑	❑	**3.** When faced with a language barrier in a cross-cultural setting, children should be asked to translate.
❑	❑	**4.** One prerequisite for competent cross-cultural counseling is awareness of one's own beliefs and values.
❑	❑	**5.** The most effective way to sharpen cross-cultural nutrition counseling skills is to learn more about the eating patterns of the culturally different.
❑	❑	**6.** The most valuable skill in cross-cultural nutrition counseling is listening.
❑	❑	**7.** During a cross-cultural counseling session, the patient's nonverbal cues are a good indicator of how well the session is going.
❑	❑	**8.** It is usually harder for people to change food preferences than it is to change food practices such as food-preparation methods, portion sizes, and frequency of consumption.
❑	❑	**9.** Asking the question "How long have you been in the United States?" is a good indicator of how closely your client is following traditional food practices.

Review the answers and rationale on pages 82–83, then record your reactions to each in your journal.

EXHIBIT 3.3—Culturally Sensitive Open-Ended Questions

Questions to Understand the View and Treatment of Health Problems

1. What do you call your problem? What name do you give it?
2. What do you feel may be causing your problem?
3. Why did it start when it did?
4. What does your sickness do to your body?
5. Will you get better soon, or will it take a long time?
6. What do you fear about your sickness?
7. What problems has your sickness caused for you personally? For your family? At work?
8. What kind of treatment will work for your sickness? What results do you expect from treatment?
9. What home remedies are common for this sickness? Have you used them?
10. Are there benefits to having this illness?

Question to Understand about Traditional Healers

11. How would a healer treat your sickness? Are you using that treatment?

Questions to Understand Food Habits and Assist in Completing a Nutritional Assessment

12. Can what you eat help cure your sickness? Or make it worse?
13. Do you eat certain foods to keep healthy? To make you strong?
14. Do you avoid certain foods to prevent sickness?
15. Do you balance eating some foods with other foods?
16. Are there foods you won't eat? Why?
17. How often do you eat your ethnic foods?
18. What kinds of foods have you been eating?
19. Is there anyone else in your family who I should talk to?

Source: Modified from Kittler PG, Sucher KP, *Food and Culture in America: A Nutrition Handbook,* 2d ed. (Belmont, CA: Wadsworth; 1998), © 1998, pp. 72–73.

are you learning, but you are you are demonstrating to your clients that what they have to say is very important to you. Make sure you come to a common understanding of their issues and problems. Request clarification when necessary by saying, "I didn't quite understand that."[30] Listen carefully to how food decisions are made. Probe to find out who does the food preparation and shopping and determine whether an additional person should be included in the next counseling session. All of this information will be very important when designing intervention strategies.

❷ **Explain.** To clarify that your understanding of the issues are accurate, you should explain back to the client your perception of what has been related. For example: "You feel that diarrhea is a hot ailment, and your baby should not be given a hot food like infant formula but should drink barley water, a cool food. Did I understand you correctly?" Balancing intake of hot and cold foods is believed to help heal among several Caribbean and Asian cultures.[25] The explanation creates an opportunity to clarify any misunderstandings.

❸ **Acknowledge.** The nutrition counselor should acknowledge the similarities and differences regarding the cause and/or treatment of the problem. For example: "Both you and your doctor feel that what your baby drinks will help her feel better. You feel your baby needs a cool food like barley water, and the health care providers at this clinic feel that your baby needs a drink with minerals like Pedialyte to get better."

❹ **Recommend.** The client should be given several options that are culturally sensitive. For example, an Indian woman who is a vegetarian who wishes to lose weight might be given the following recommendations: "You could start a walking program, make rice without oil, use skim milk to make yogurt, reduce the amount of fat used to make bean dishes, or eat smaller portions of fried bread."

❺ **Negotiate.** After reviewing the options, negotiate a culturally sensitive plan of action. When the condition is life threatening or the cultural differences are enormous, Kleinman, as reported by Fadiman,[34] recommends a cultural anthropologist or a respected member of the client's community aid in the negotiation. Kleinman[32] suggests that the health practitioner decide what's critical and be willing to compromise on everything else. Look to your client to select a starting point: "Which of these options do you think would be a good place to start?" After selecting an option, discuss how that option will be implemented.

The first priority when counseling across cultures is to establish rapport in order to open lines of communication. This will happen if the counselor is following standard counseling behavior by appearing to be open, honest, respectful, nonjudgmental, and willing to learn. Once rapport is established, clients are likely to lower barriers and engage in authentic interaction. In some cultures it is inappropriate to discuss business until rapport is established.

To understand how the culturally sensitive approaches to nutrition counseling interface with the motivational nutrition counseling algorithm in Figure 3.2, a cross-cultural nutrition counseling algorithm is presented in Figure 3.4. This algorithm incorporates the response-driven interview guidelines in Exhibit 3.3 and the steps of the LEARN counseling model.

CASE STUDY 3—Nancy: Intervention at Three Levels of Motivation

Nancy is a twenty-six-year-old, overweight African American woman who was recently diagnosed with hypertension. She is five feet, four inches tall and weighs 170 pounds. She grew up in Columbia, South Carolina, and moved with her mother to northern New Jersey to live closer to her sister. Nancy has a sedentary job as the floor supervisor for an overnight mail delivery service. She lives on the second story of a two-family house with her husband, mother, and three children. Although the family is on a tight budget, they have ample money for food. Diabetes runs in the family, and there is a history of pica during Nancy's pregnancies. She is very active in church activities, which often include food. Sunday dinner is served early—2 P.M.—and is a large meal often attended by her sister and her family and other relatives. Nancy's physician prescribed a medication to lower her blood pressure and suggested that she consult with the clinic's nutritionist about her weight and diet.

The following discussion contains three possible scenarios for Nancy illustrating use of the motivational nutrition counseling algorithm at three different motivational levels. Examples of relationship-building responses are sporadically intertwined in the scenarios.

Level 1—Not Motivated, Not Ready

Nancy came into your office directly from her doctor's appointment after being diagnosed with hypertension. Nancy said her doctor told her to lose weight and go over her diet with the nutrition counselor. Nancy says she doesn't want to lose weight, she looks fine, and her husband likes her "with some meat on her." In fact, she says she doesn't want to bother with her diet, either—she has enough problems. Her husband is on disability with a bad back, her mother has been having kidney problems, and she has three small children that take up a lot of time. You reflect and justify her feelings and ask whether she wants to talk about her diet and high blood pressure since she is here. Nancy says she doesn't really know why she is in your office, she really doesn't want to talk about her diet, and besides, she has the pills to take care of her blood pressure.

You say, "I respect your decision. It is obvious that any talk about diet at this time would not be useful. You may feel differently about the problem at some time in the future. I want you to know that I will be here to assist you if you would like some help." You offer Nancy some literature about diet and blood pressure. She says she is willing to read the literature and will call you if she feels differently.

Level 2—Unsure, Low Confidence

Assessment

Nancy did not fill out assessment forms before her appointment. She says her doctor wants her to lose weight, but she doesn't believe she has a weight problem. Her husband likes her with "some meat on her." Nancy feels that taking care of three children and holding down a night job as a floor supervisor for a mail delivery service is stressful, and she doesn't need more problems in her life. Her husband is home on disability with a bad back, and the family needs the income and the benefits from her job. You say, "I hear what you are saying. You sound annoyed that you have been given another burden. You don't need any new problems. You certainly are entitled to feel this way with all the responsibilities you are shouldering. However, if you want to explore what you could do about your food

(Continued on page 80)

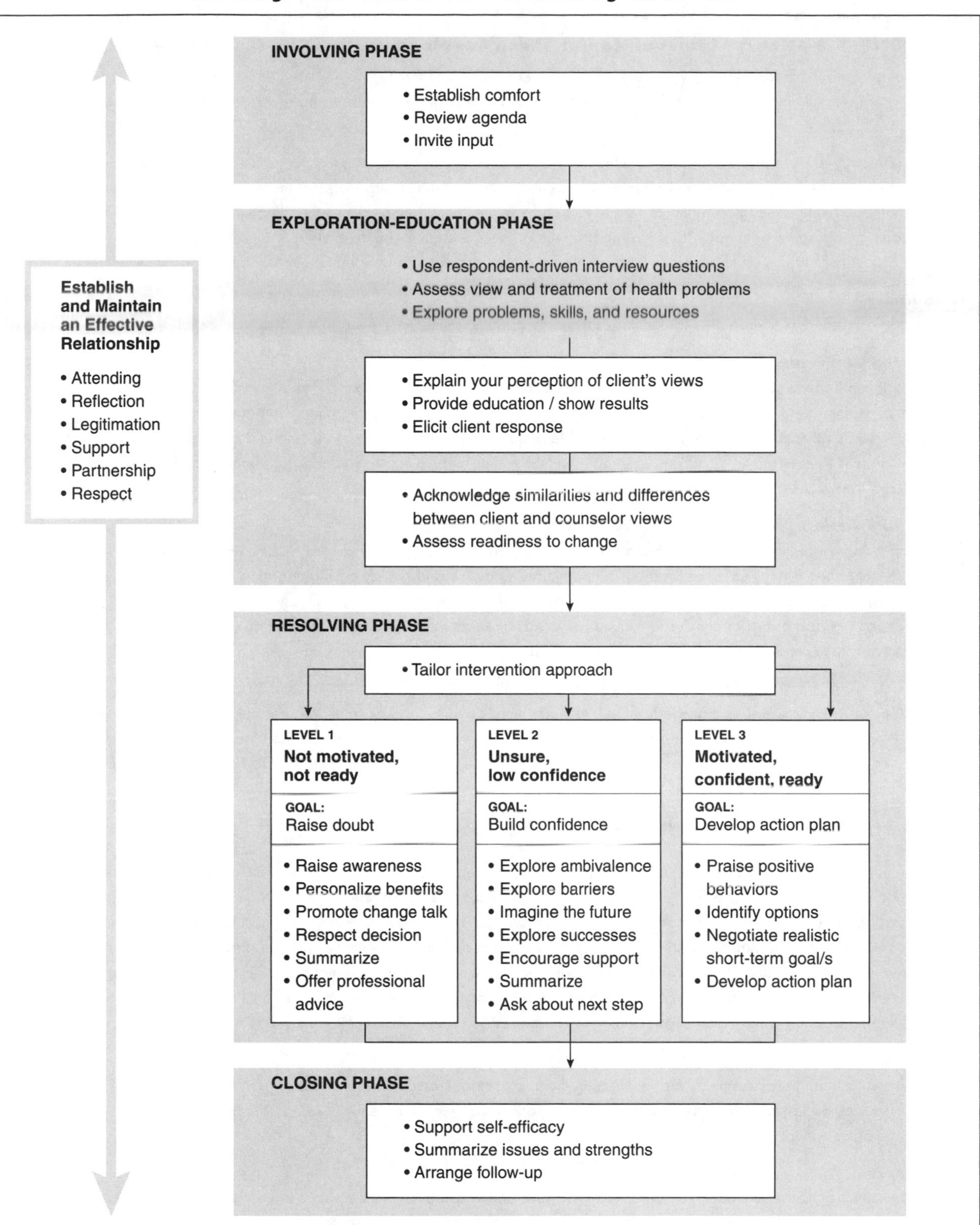

Figure 3.4 Cross-Cultural Nutritional Counseling Algorithm

Source: Adapted from Snetselaar L, *Counseling for Change.* In: Mahan LK, Escott-Stump S, eds. *Krause's Food, Nutrition, & Diet Therapy,* 10th ed. (Philadelphia: Saunders; 2000), pp. 451–462; and Berg-Smith SM, Stevens VJ, Brown KM, Van Horn L, Gernhofer N, Peters E, Greenberg R, Snetselaar L, Ahrens L, Smith K for the Dietary Intervention Study in Children (DISC) Research Group, A brief motivational intervention to improve dietary adherence in adolescents. *Health Educ Res.* 1999;14:101–112.

CASE STUDY 3—Nancy: Intervention at Three Levels of Motivation (Continued)

intake to lower your blood pressure, we could work together to set goals that would fit into your schedule. Do you want to talk about your blood pressure?" Nancy says yes. You use a typical day strategy and complete a short food frequency checklist.

Feedback

You show Nancy a list of lifestyle and food behaviors that can help lower blood pressure, compare Nancy's assessment to the list, and ask Nancy what she thinks. Nancy doesn't think it looks too good, but she is surprised that there is more that she can do than just losing weight. You explain the importance of weight loss and why her doctor emphasized the weight issue but point out that other very important diet changes can be made. "By focusing on some of them, weight loss may even occur," you add. You also point out some positive aspects of her diet, such as the collard greens for dinner and the use of skim milk in her coffee.

Readiness

When you ask Nancy her readiness to consider any of the options, she says she is about a 6 on a scale of 1 to 10.

Exploring Ambivalence

You compliment Nancy on keeping the appointment despite her ambivalence and say, "There must be a part of you that believes you should make some diet changes." You ask Nancy why she might want to make some changes. Nancy's mother also has high blood pressure, and as a result, she knows some of the problems that occur with the disease, so she is concerned. However, Nancy feels fine and has the pills to control her blood pressure. You summarize her ambivalence about making changes and ask where that leaves her. Nancy says she would like to do something.

Goals and Action Plan

You point to the list of lifestyle and food behaviors for having a positive impact on blood pressure and Nancy's diet summary and ask what appeals to her. Nancy says she would like to eat more fruit. You and Nancy go through the goal-setting process building on past experiences. Nancy's goal is to eat a banana or an orange for a snack at work. She will buy the fruit on her way home and put a sticky note on the dashboard of her car to remind her to take the fruit to work. You tell Nancy that you would like to do what you can to support her in this endeavor and ask whether it would be all right to call her. Nancy says yes.

Follow-Up

When you call Nancy, she says that she took fruit to work every day for the past week. One of her coworkers has also started bringing fruit, so Nancy feels good that she has a positive influence on someone else. You congratulate Nancy for following through on her goal and tell her that you are there to support her if she wants assistance in making more changes. You express confidence in her ability to continue making dietary changes.

Level 3—Motivated, Confident, Ready

Assessment

Between her appointment with her physician and you, Nancy completed a client assessment questionnaire and a food frequency checklist. Nancy says her doctor suggested that she see the clinic nutritionist to talk about her food intake and losing weight. You used "a typical day strategy." Improving her food selections appeals to her, but the idea of losing weight surprised her. She never thought of herself as overweight, and her husband likes her "with some meat on her." You explain why the doctor suggested weight loss and say, "I will give you some literature to read about blood pressure and weight to help you make a decision. It is one of the best things you could do to help your blood pressure. However, there are still a lot of things we can work on to help lower your blood pressure that may also result in some weight loss." You share with her the results of an analysis of the assessment forms.

Feedback

After reviewing Nancy's assessments and comparing them to a list of lifestyle and food behaviors that can help lower blood pressure, Nancy says she would like to increase her intake of fruits and vegetables. She really likes fruit but has never been in the habit of eating them. Vegetables will take some effort.

Readiness

When you ask her about her readiness to increase her fruits and vegetables, Nancy says she very much wants to make a change. On a scale of 1 to 10 she rates her confidence to succeed a 9 for fruits and a 7 for vegetables.

Exploring Ambivalence

Nancy's mother has started to experience some kidney problems from high blood pressure. If Nancy works on her diet, she thinks the whole family will benefit, especially her mother. Also, she wants to do what is best for the children. Nancy needs to be healthy. Right now her husband is home on disability with a bad back, and the family needs her to work for the income and the benefits. Nancy says she is not sure about trying to lose weight and she does have the pills.

(Continued)

CASE STUDY 3—Nancy: Intervention at Three Levels of Motivation (Continued)

Goals and Action Plan

You ask Nancy whether she has ideas about how to increase fruits and vegetables. She says her mother does all the cooking in the house and believes she would be willing to make soup with more vegetables and serve at least one vegetable with dinner. Nancy has noticed the cut-up veggie packs at the grocery and thinks they would be convenient to take to work. Nancy asks the counselor for ideas. You offer her reading material, recipes, and tips for increasing fruits and vegetables. You and Nancy go through the goal-setting process, building on past experiences. Nancy's goal is to have homemade soup with vegetables and a four-ounce glass of grape juice on waking at 3 P.M. For dinner she will have at least one serving of vegetables and will take a banana or orange and a veggie pack with her to work each day. You talk about self-monitoring and give her a food record designed to track fruit and vegetable intake. You ask whether it would be a good idea for her mother to come with Nancy for her next visit. Nancy says yes. You also ask whether you can call her in a week. Nancy says she would like that.

Follow-Up

When you call Nancy, she says that she took fruit and veggie packs to work every day for the past week. Her mother has been making soups and more vegetables for dinner. The counselor reviews her goal of eating three vegetables and two fruits each day. Nancy says she is concerned that her mother is using too much salt when cooking and asks whether her mother can come with her for their next appointment. Nancy has noticed that everyone in the house is eating more fruits and vegetables and thinks that is good. You congratulate her on her success, adding that you believe she has the ability to continue making diet changes.

EXERCISE 3.8—Case Study—Cultural Implications

After reading the preceding case study scenarios, write in your journal three family or societal strengths that should be taken into consideration during a counseling intervention with Nancy.

EXERCISE 3.9—Case Study—Counseling Strategies and Responses

- The counseling algorithm in Figure 3.2 integrates several intervention models for behavior change. Review discussions of the transtheoretical model, motivational interviewing, health behavior change method, and solution-oriented therapy discussed in Chapter 1. Give at least one example of how the methodology for each was illustrated in the case studies about Nancy.
- Review the resolving phase intervention strategies for each motivational level discussed in this chapter. Select a strategy not illustrated in the case studies that you believe may have been useful in working with Nancy at each of the three motivational levels. Explain why you believe the method would be useful. Write a statement a counselor would make when utilizing the strategy.
- Review the five relationship-building responses in Chapter 2, and underline examples of each in the three case studies about Nancy. In your journal identify the type of response that was illustrated (that is, reflection, legitimation, support, partnership, or respect), and evaluate the effectiveness of each for the particular situation that was described.

REVIEW QUESTIONS

1. Explain why the arrows for assessment, intervention, and evaluation of a client are reciprocal in the model for a nutrition counseling program in Figure 3.1.

2. Identify the two components of readiness to make a behavior change emphasized in the health behavior change method.

3. Describe three methods to assess readiness to change.

4. Describe each of the four phases of the counseling interview process.

5. How should weight monitoring generally be handled in an uncomplicated outpatient setting?

6. Explain "a typical day strategy."

7. Why should a client be allowed to do the initial evaluation after receiving feedback on an assessment?

8. Identify goals and tasks for the resolving phase for each of the three motivational levels in the motivational nutrition counseling algorithm.

9. Identify two special factors that need to be taken into consideration when counseling in an acute care setting.

10. Identify two demographic trends in North America.

11. Why do we say that all counseling occurs in a cultural context?

12. Identify the three tasks described by Curry and Jaffe that need to be accomplished to achieve competence in intercultural counseling.

13. Explain the objectives of using the respondent-driven interview questions.

14. Identify and describe the five components of the LEARN counseling model.

EXERCISE 3.7—ANSWERS

1. False—People who share a common language do not necessarily share a common culture. For example, people from Mexico, Puerto Rico, Cuba, and Spain all speak Spanish, yet their dietary patterns are not similar.

2. False—Race is not a good predictor. Race and culture are not synonymous. For example, an African American, a Nigerian immigrant, and a Jamaican are likely to have similar physical features and would be placed in the same racial category, yet they represent diverse cultures and dietary patterns.

3. False—Although children are often available to translate, they are not a good choice. They may not have adequate vocabulary, and since they are not seen as an authority figure, their credibility is compromised.

4. True—This can not be emphasized enough. Only by understanding your own beliefs and values can you respect and appreciate the beliefs and values of others. For example, a busy health care professional in the United States usually expects others to be on time for appointments. Some cultural groups have a different attitude and do not see a problem or a need for an explanation if they are twenty minutes late.

5. False—It is highly desirable to learn about foodways of various cultures, especially the dietary patterns of the groups you frequently counsel. However, there is an endless variety of cultural and religious groups that constitute American society, and it is not realistic to expect nutrition counselors to have more than a rudimentary knowledge of various food habits. When encountering an individual from an unfamiliar culture, the counselor should be willing to acknowledge the unfamiliarity and request information about his or her dietary habits. For example, the counselor could say, "I have had little experience with people of your culture, but I would like it if you could help me learn."

6. True—Only by actively listening can you learn the critical factors necessary to counsel effectively.

7. False—Nonverbal cues need to be interpreted very cautiously because meanings vary from one culture to the next. Your interpretation may be quite different from the client's intent.

8. True—This suggests that nutrition counselors should concentrate on changing food preparation methods, portion sizes, and frequency of consumption. Magnus[30] recommends the 50 percent rule when appropriate: clients can enjoy half the portion size of favorite foods or consume them half as frequently.

9. False—Although some authorities do believe this question has value,[25] others believe that little information is gained from the question because there can be considerable variation in immigrant food patterns due to acculturation, interethnic variation, social change, and social class differences. To ascertain the degree to which a client upholds ethnic foodways, Magnus[30] suggests the following question: "How often do you eat your ethnic foods?"

Source: Quiz and rationale modified from: Magnus MH. What's your IQ on cross-cultural nutrition counseling? *The Diabetes Educator.* © 1996;22:57–62. Used with permission.

ASSIGNMENT—Conducting an Interview across Cultures

Locate a volunteer who is from a culture substantially different than your own and is willing to talk to you about a health problem he or she is experiencing or has experienced or who can describe the health care practices of a particular individual from his or her culture who was ill. Depending on the accessibility of such individuals, you may need to select a person who is different from you culturally because of age or religion. The objectives of this assignment are to work on developing counseling skills, to gather information, and to learn something about the volunteer's health care beliefs and practices. The intention is not to resolve the difficulty. The volunteer may find some benefit to him or herself by clarifying his or her problem through the discussions; however, the person should not be led to believe that there will be an intervention. Therefore, only the involving phase, part of the exploration-education phase, and the closing phase of the cross-cultural counseling algorithm will be addressed in this assignment. Consider audio- or videotaping this experience for later evaluation.

PART I. Use the following interview guide/checklist to conduct the interview with your volunteer. Examples of possible counselor questions, statements, and responses are given in italics.

Preparation

- ❑ Review respondent-driven interview questions and the listen component of the LEARN model.
- ❑ Bring a completed certificate of appreciation.

Involving Phase

- ❑ Greeting
  - ❍ Verbal greeting—*I am happy to meet you.*
  - ❍ Shake hands.
  - ❍ Introduce self—*My name is Mary Smith. How should I address you?*
- ❑ Small talk, if appropriate
- ❑ Thank volunteer—*Thank you for participating in this interview.*
- ❑ Explain purpose of the interview—*This is a project I am required to do for my nutrition counseling class. The purpose of this interview is for me to work on my*

counseling skills, gather information about your health concern, and learn something about your culture, particularly how it relates to health care.

- ❑ Review the consent form (Lifestyle Management Form 3.3 in Appendix G) with your volunteer, follow the procedure for obtaining a consent in preparation of Session 1 of Chapter 9 of this text, and you and your client should sign both a client copy and a clinic copy of the form. Give the client copy to your volunteer.

Transition to Exploration Phase

- ❑ Transition statement—*Do you have any questions before we go over the interview questions?*

Exploration Phase (As you go through the interview questions, you will have to make a judgment regarding which ones are appropriate for your particular client.)

- ❑ Ask your client to describe him- or herself (age, cultural group, occupation, interests).

Cause of Illness

- ❑ Explain desire to learn about your volunteer's culture—*As we go through the questions, if there is anything you think I am missing about your health problem or treatment as it relates to your culture, please let me know.*
- ❑ *What do you call your problem? Is there any other name given to this condition in your culture?*
- ❑ *What do you feel may be causing your problem? Do you and your doctor agree about the cause?*
- ❑ *Why do you believe the problem started when it did?*
- ❑ Briefly summarize your client's perception of the cause of his or her illness to check for understanding—*From what you said, it appears that your elevated blood pressure is caused by too much blood and is the result of your fate in life and because of your family history. Your mother and father also had high blood pressure. Your family calls the problem "high blood."*
- ❑ Make a reflective statement, if appropriate.

Process of Illness

- ❑ *What does the sickness do to your body?*
- ❑ *Do you have any idea when (whether) the problem will get better?*
- ❑ *What do you fear about your sickness?*
- ❑ *What problems has your sickness caused for you personally? For your family? At work?*
- ❑ Briefly summarize your client's perception of how the illness is affecting him or her—*Let me make sure I understood you correctly. You feel that the blood pressure problem will always be with you. You are feeling good now, but both your mother and father had a stroke, and you worry that might happen to you.*
- ❑ Make a reflective statement—*Although you are feeling good, it seems to me that you are feeling somewhat fearful of the future.* (If confirmed, then use a legitimation statement.) *Considering what happened to your mother and father, it is understandable that you would feel that way.*

Treatment of Illness

- ❑ *What kind of treatment will work for your sickness? Have you been using them? What results do you expect from the treatments?*
- ❑ *Are there home remedies for this sickness? Have you used them?*
- ❑ Briefly summarize what you understand about the volunteer's treatment—*It is interesting that some of your relatives have found help from staying away from "rich" foods and trying to eat more acidic foods. But you feel more comfortable trying to do what your doctor says is best and to take your pills.*

Healers and Future

- ❑ *Are there any benefits to having this illness?*
- ❑ *Is there anyone in your family that helps make decisions about what you should be doing to treat your sickness?*
- ❑ *Are there any healers in your culture who could treat this problem? Have you used any of their treatments?*
- ❑ Make an appropriate reflective or summarization response—*Prayer is important to you. So much is being written about prayer and healing today.*

Foods and Illness

- ❑ *Can what you eat help cure your sickness or make it worse?*
- ❑ *Do you eat certain foods to keep you healthy? To make you strong?*

- ❑ *Do you avoid certain foods to prevent sickness?*
- ❑ *Do you balance eating some foods with other foods?*
- ❑ *Are there foods you won't eat? Why?*
- ❑ *How often do you eat your ethnic foods?*
- ❑ *What kinds of foods have you been eating?*
- ❑ Make an appropriate reflective or summarization response—*You are lucky you have such a supportive wife who is trying to help you eat more fruits and vegetables.*

Explore Culture and Illness—General

- ❑ Explain that you would like to learn about your volunteer's views on illness, healing, and food—*We have been focusing on your particular illness and culture. I am wondering whether people from your culture have other views about illness, healing, or food and health for other illnesses that are different than what is generally believed in this country. Can you tell me about them? Do you know anyone who has used those methods?*

Closing Phase

- ❑ Express appreciation—*Thank you very much for letting me talk to you about your blood pressure and how you are treating the problem. I learned a great deal.*
- ❑ Use a relationship-building response (respect)—*I am very impressed with all you know about your blood pressure problem and the steps you have taken to control it.*
- ❑ Express hope for the future—*I hope you have continued success with controlling your blood pressure.*
- ❑ Shake hands.
- ❑ Give a certificate of appreciation—*As a show of gratitude for your willingness to participate in this project, I have a certificate of appreciation to give to you from me and the director of the project.*

PART II. Answer the following questions in a formal typed report or in your journal. For formal reports, number and type each question and put the answers in complete sentences under the question.

1. Record the name of the person interviewed and location, time, and date of the meeting.

2. Describe the person you interviewed—age, cultural group, gender, and occupation.

3. Write a narration of the experience. There should be four titled sections in the narration: preparation, opening phase, exploration-education phase, and closing phase. Summarize what occurred in each phase.

4. Write each question in sequence, and give your volunteer's response to the question. Indicate if you did not use the question.

5. Explain the use of relationship building responses and summarizations. What do you believe was the effect of using these responses? Were you comfortable using them?

6. Explain how you believe this person's cultural orientation affects his or her perception and treatment of the illness.

7. Complete an Interview Checklist, Lifestyle Management Form 8.1—omit Resolving Phase.

8. How useful were the respondent-driven interview questions for counseling someone from a culture different than your own? Explain your answer.

9. What did you learn from this experience?

SUGGESTED READINGS, MATERIALS, AND INTERNET RESOURCES

Cross-Cultural Counseling

Pedersen PB, Draguns JG, Lonner WJ, Trimble JE, eds. *Counseling across Cultures.* 4th ed. Thousand Oaks, CA: Sage; 1996. Many issues related to counseling across cultures, with seven chapters dedicated to particular cultural groups.

Magnus M. What's your IQ on cross-cultural nutrition counseling? *The Diabetes Educator.* 1996;96:57–62.

Magnus M. *Cross-cultural Counseling: Research and Practice.* ADA's 1994 Annual Meeting, Palm Desert, CA: Convention Cassettes Unlimited; 1994.

Kittler PG, Sucher KP. *Food and Culture in America: A Nutrition Handbook.* 2d ed. Belmont, CA: West/Wadsworth; 1998.

Oregon State University Extension Web site, http://osu.orst.edu/dept/ehe/nu_diverse. Several materials related to cultural diversity, including a reading list, program planning guidelines to reach a culturally diverse audience, and links to other sites containing culturally diverse materials.

Nutrition Counseling for Medical Nutrition

Snetselaar LG. *Nutrition Counseling Skills for Medical Nutrition Therapy.* 3d ed. Gaithersburg, MD: Aspen; 1997. An abundance of counseling strategies to address requirements for specific dietary modifications.

REFERENCES

[1]Tinker LF, Heins JM, Holler HJ. Commentary and translation: 1994 nutrition recommendations for diabetes. *J Am Diet Assoc.* 1994;94:507–511.

[2]Pastors JG, Barrier P, Rich M, Gallagher S, Galligos C, Wheeler M. *Facilitating Lifestyle Change: A Resource Manual.* Chicago: American Dietetic Association; 1996.

[3]Curry KR., Jaffe A. *Nutrition Counseling & Communication Skills.* Philadelphia: Saunders; 1998.

[4]Snetselaar LG. *Nutrition Counseling Skills for Medical Nutrition Therapy.* Gaithersburg, MD: Aspen; 1997.

[5]Snetselaar L. Counseling for Change. In: Mahan LK, Escott-Stump S, eds. *Krause's Food, Nutrition, & Diet Therapy,* 10th ed. Philadelphia: Saunders; 2000:451–462.

[6]Berg-Smith SM, Stevens VJ, Brown KM, Van Horn L, Gernhofer N, Peters E, Greenberg R, Snetselaar L, Ahrens L, Smith K for the Dietary Intervention Study in Children (DISC) Research Group. A brief motivational intervention to improve dietary adherence in adolescents. *Health Educ Res.* 1999;14:101–112.

[7]Prochaska J, DiClemente C. Transtheoretical therapy: Toward a more integrative model of change. *Psychother Theory Res Practice.* 1982;61:276–288.

[8]Prochaska J, DiClemente C. Toward a comprehensive model of change. In: Miller WR, Heather N. *Treating Addictive Behaviors: Processes of Change.* New York: Plenum; 1986:3–27.

[9]Miller WR, Rollnick S. *Motivational Interviewing—Preparing People to Change Addictive Behavior.* New York: Guilford; 1991.

[10]Rollnick S, Heather N, Bell A. Negotiating behaviour change in medical settings: The development of brief motivational interviewing. *J Mental Health.* 1992;1:25–37.

[11]Rollnick S, Mason P, Butler C. *Health Behavior Change: A Guide for Practitioners.* New York: Churchill Livingstone; 1999.

[12]Miller S, Hubble M, Duncan B., eds. *Handbook of Solution-Focused Brief Therapy.* San Francisco: Jossey-Bass; 1996.

[13]Bandura A. Towards a unifying theory of behavior change. *Psychol Rev.* 1977;84:191–215.

[14]Windhauser MM, Ernst DB, Karanja NM, Crawford SW, Redican SE, Swain JF, Karimbakas JM, Champagne CM, Hoben, KP, Evans MA. Translating the dietary approaches to stop hypertension diet from research to practice: Dietary and behavior change techniques. *J Am Diet Assoc.* 1999;99(suppl):S90–S95.

[15]Holli BB, Calabrese RJ. *Communication and Education Skills for Dietetics Professionals.* 3d ed. Baltimore, MD: Williams & Wilkins; 1998.

[16]Dyer WW, Vriend J. *Counseling Techniques That Work.* Alexandria, VA: American Counseling Association; 1988.

[17]King NL. *Counseling for Health & Fitness.* Eureka, CA: Nutrition Dimension; 1999.

[18]American Dietetic Association. Position of the American Dietetic Association: Weight management. *J Am Diet Assoc.* 1997;97:71–74.

[19]Funnell MM, Anderson RM. Putting Humpty Dumpty back together again: Reintegrating the clinical and behavioral components in diabetes care and education. *Diabetes Spectrum.* 1999;12:19–23.

[20]Herrin M, Parham E, Ikeda J, White A, Branen L. Alternative viewpoint on National Institutes of Health Clinical Guidelines. *J Nutr Ed.* 1999;31:116–118.

[21]Pohl SL. Facilitating lifestyle change in people with diabetes mellitus: Perspective from a private practice. *Diabetes Spectrum.* 1999;12:28–33.

[22]Prochaska JO, Norcross JC, DiClemente CC. *Changing for Good.* New York: Avon; 1994.

[23]Hawkes D, Marsh TI, Wilgosh R. *Solution Focused Therapy a Handbook for Health Care Professionals.* Boston: Butterworth Heinemann; 1998.

[24]Whitney EN, Cataldo CB, Rolfes SR. *Understanding Normal and Clinical Nutrition.* 5th ed. Belmont, CA: West/Wadsworth; 1998.

[25]Kittler PG, Sucher KP. *Food and Culture in America: A Nutrition Handbook.* 2d ed. Belmont, CA: West/Wadsworth; 1998.

[26]DiClemente CC, Prochaska J. Toward a comprehensive, transtheoretical model of change: Stages of change and addictive behaviors. In: Miller WR, Heather N., eds. *Treating Addictive Behavior,* 2d ed. New York: Plenum; 1998.

[27]Anderson JM, Wiggins S, Rajwani R, Holbrook A, Bule C, Ng M. Living with a chronic illness: Chinese-Canadian and Euro-Canadian women with diabetes-exploring factors that influence management. *Soc Sc Med.* 1995;41:181–195.

[28]Pedersen PB, Draguns JG, Lonner WJ, Trimble JE. *Counseling Across Cultures.* 4th ed. Thousand Oaks, CA: Sage; 1996.

[29]Cormier S, Cormier B. *Interviewing Strategies for Helpers: Fundamental Skills and Cognitive Behavioral Interventions.* 4th ed. Pacific Grove, CA: Brooks/Cole; 1998.

[30]Magnus M. What's your IQ on cross-cultural nutrition counseling? *The Diabetes Educator.* 1996;96:57–62.

[31]Cassidy, CM. Walk a mile in my shoes: Culturally sensitive food-habit research. *Am J Clin Nutr.* 1994;59(suppl.):190S–197S.

[32]Kleinman A, Eisenberg L, Good B. Culture, illness, and care: Clinical lessons from anthropologic and cross-cultural research. *Ann Intern Med.* 1978;88:251–258.

[33]Berlin, EA, Fowkes, WC. A teaching framework for cross-cultural health care. Application in family practice. *West J Med.* 1983;139:934–938.

[34]Fadiman A. *The Spirit Catches You and You Fall Down.* New York: Noonday; 1997.

Chapter

4

DEVELOPING A NUTRITION CARE PLAN: PUTTING IT ALL TOGETHER

A life that hasn't a definite plan is likely to become driftwood.
—David Sarnoff

BEHAVIORAL OBJECTIVES:

- Develop goals that are specific, achievable, and measurable.
- Differentiate among anticipated results, broad goals, and specific goals.
- Design a plan of action for a goal.
- Evaluate dietary status utilizing standard assessment tools.
- Utilize common dietary assessment tools.
- Assess total energy expenditure.
- Utilize standard physical assessment methods to assess healthy weight.
- Define *overweight* and *obesity.*
- Describe functions of charting.
- Utilize SOAP and IAO documentation formats.

KEY TERMS:

- **ANDROID FAT DISTRIBUTION:** waist and upper abdominal fat accumulation; apple shape
- **BODY MASS INDEX (BMI):** preferred weight-for-height standard; determinant of health risk; predictor of mortality
- **DAILY FOOD GUIDE:** a food group plan to promote health and reduce the risk of chronic diseases; the Food Guide
- **DIETARY APPROACHES TO STOP HYPERTENSION (DASH):** an eating plan focusing on whole foods emphasizing fruits and vegetables
- **DIETARY ASSESSMENT:** evaluation of nutrient intake and food patterns
- **DIETARY REFERENCE INTAKES (DRI):** four sets of nutrient recommendations for the United States and Canada: Estimated Average Requirements, Recommended Dietary Allowances, Adequate Intakes, and Tolerable Upper Intake Levels
- **FOOD GUIDE PYRAMID:** pictorial representation of five major food groups indicating kinds and amounts of food to consume
- **GYNOID FAT DISTRIBUTION:** fat accumulation in hips and thighs; pear shape
- **HAMWI METHOD:** equation used to calculate ideal body weight
- **IAO FORMAT:** a charting by exception documentation method; issue, assessment, and outcomes
- **NUTRITIONAL ASSESSMENT:** a comprehensive analysis of an individual's dietary evaluation; medical, medication, and psychosocial history; anthropometric data; biochemical data; and physical examination
- **OBESITY:** a high amount of body fat as indicated by a weight 20 percent over the standard; a body mass index of 30 to 30.9

- **OVERWEIGHT:** a state in which body weight is 10 percent over the standard; a body mass index of 25 to 29.9
- **SOAP FORMAT:** a comprehensive documentation tool; subjective, objective, assessment, and plan
- **WAIST CIRCUMFERENCE:** a method to assess upper abdominal fat distribution
- **WAIST-TO-HIP RATIO:** ratio of waist circumference to hip circumference; assessment of abdominal fat distribution

An archer cannot hit the bull's-eye if he doesn't know where the target is.
—Anonymous

GOAL SETTING

Goal setting is a logical strategy for clients ready to make a behavior change—that is, those at Level 3 and possibly Level 2 in the motivational nutrition counseling algorithm (refer back to Figure 3.2). It helps break down complex behavior changes into small achievable steps. Successful small changes improve self-efficacy and motivate clients to continue making lifestyle alterations.

Counselors should be wary of jumping into goal setting too quickly. To be able to formulate achievable goals, the groundwork must be done. That means a counselor and client must have fully explored the nutrition issues of concern. A major objective of this whole process is for you to work in partnership with your client to develop an action plan. Your job is not to be the one setting the goals but to be sure the stated goal(s) meet the goal-setting criteria. Clients must feel a sense of ownership over the plan for goal setting to be an instrument of change. The following sections describe general guidelines for establishing goals.

Explore Change Options

While exploring viable focus areas for making a food behavior change, Berg-Smith et al.[1] emphasize the need for a counselor to remain neutral while conveying the following messages to a client:

1. A number of courses of action are available to you.
2. You are the best judge of what will work.
3. We will work together to review the options and select a course of action.

Elicit Client's Ideas for Change Your client may unequivocally voice ideas about what behavior change he or she would like to tackle. This desire may have been made quite clear during a review of the client's assessment materials. However, even if that is the case, it is better to open the topic up for discussion than to make an erroneous assumption.

Counselor: *When we went over the assessment materials, several areas were identified that could be a focus area for making a food change. You know best what would work for you. Is there one particular area that appeals to you?*

There are probably a number of options, but what do you think will work for you?

Consider Using an Options Tool For those who appear to have difficulty selecting a specific area of focus, an options tool could be useful[1] (see Figure 4.1). This tool consists of a group of several circles. While reviewing the assessment with your client, you may have identified several areas that could be addressed for behavior change. Write each topic area in one of the circles. Be sure to leave some circles blank for your client. Ask your client what else could be addressed and add any suggestions to the circles.

Counselor: *This options tool may help in the decision process. As you can see, there are a number of circles on the paper. I'd like us to work together to brainstorm ideas of what you could focus on and we will write them in the circles. As the weeks go by, we can use this tool to help us decide a new area for focusing a goal. What appeals to you the most?*

Explore Concerns Regarding a Selected Option Probe to investigate any concerns you have about an option selected by your client. Further discussion may convince you to better understand a particular choice, or the process of clarifying could alter or modify a client's choice.

Counselor: *Help me understand why you feel this is the best choice.*

This seems to be the best choice, but will this work for you?

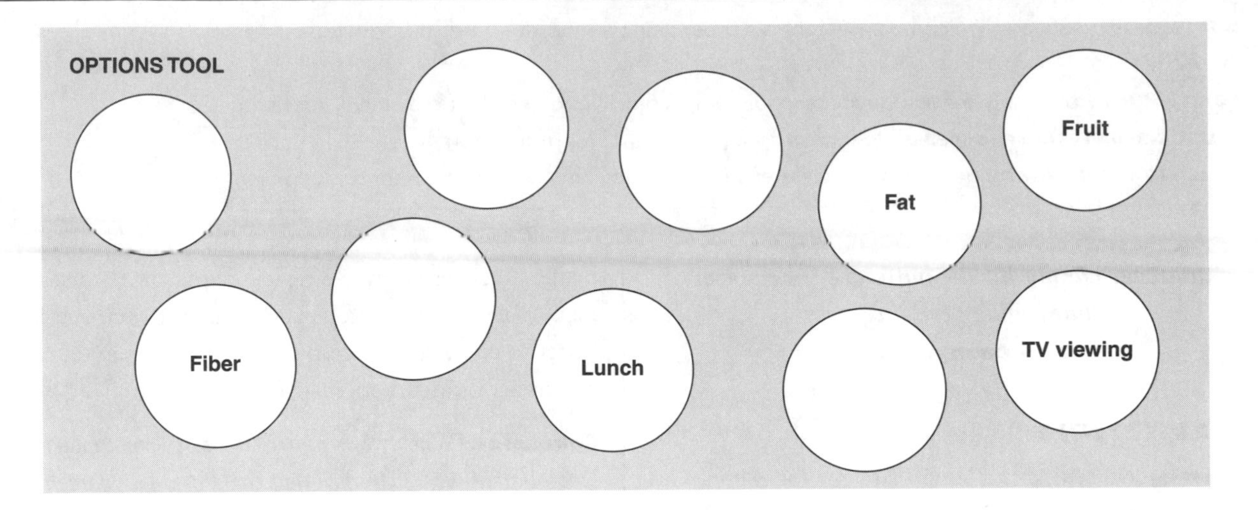

Figure 4.1 Options Tool
Source: Berg-Smith SM, Stevens VJ, Brown KM, Van Horn L, Gernhofer N, Peters E, Greenberg R, Snetselaar L, Smith K for the Dietary Intervention Study in Children (DISC) Research Group. A brief motivational intervention to improve dietary adherence in adolescents. *Health Educ Res.* ©1999;14:101–112. Adapted with permission of Oxford University Press.

Explain Goal-Setting Basics

Before beginning the process of setting a specific goal, you may wish to explain the basics of the goal-setting process to your client in order for there to be a clear understanding of the objectives. Points about goal setting to relate to your clients include that goals should

- be specific (concrete, measurable, and observable);
- answer the questions of when, where, and how often;
- be realistic (that is, small changes);
- be stated positively; and
- be under the control of the client (that is, goals should not depend upon another person).

Identify a Specific Goal from a Broadly Stated Goal

Once a broadly stated goal, such as increasing fruit intake, has been selected, the task will be to narrow the focus area down to a specific goal that is concrete, achievable, measurable, and positive. The goal should identify when, where, how often, and under what conditions the new behavior will happen. This specific goal will be one of the small steps that addresses a hoped-for behavior change and leads to the anticipated outcome (for example, reduction of high blood pressure).

Table 4.1 lists the interrelationship of these concepts. Some highly motivated individuals will be capable of implementing substantial behavior changes; however, for most people a gradual change in behavior is more likely to result in the desired outcome. Small successes lead to better self-efficacy and more ambitious goal attainment in the future. Trying to accomplish too much too soon is likely to lead to disappointment.

Use Small Talk Botman[2] suggests focusing on identification of the smallest goal that is achievable and worthwhile rather than asking how much a person can accomplish.

Counselor: *What is the smallest specific goal you believe is worth pursuing?*

Explore Past Experiences If nothing is clearly stated, then explore past experiences with the broadly stated goal. Successes of the past, no matter how small, are useful starting places for defining an achievable goal. For example, if the general goal was to increase fruits, possible questions include the following:

Counselor: *When have you eaten fruit in the past?*

When did you last eat fruit you enjoyed?

Build on Past Successes If a success was identified, try to build on that success. The objective is to identify strengths and skills the client already possesses that can be embellished to implement the new goal.

Counselor: *You said you remembered enjoying a clementine orange that was in a bowl at your mother's house. That's great because clementine oranges are in season now and can*

TABLE 4.1—Anticipated Results, Broad Goals, and Specific Goals

ANTICIPATED RESULTS	BROAD GOALS—NOT SPECIFIED ENOUGH TO BE ACHIEVABLE	SPECIFIC GOALS—CONCRETE ACHIEVABLE, MEASURABLE, POSITIVE
Decrease blood pressure.	Increase vegetables.	Eat two servings of vegetables for lunch and again at dinner.
Decrease cholesterol.	Increase fiber.	Eat oatmeal for breakfast five times this week.
Decrease cancer risk.	Increase fruit.	Take an apple or orange with me to work each day to eat during a break.
Maintain or increase bone density.	Increase calcium intake.	Prepare one kale recipe for dinner this week.

be purchased in a handy box at the grocery store. Do you believe that placing a bag or bowl of these oranges in a convenient place would help you meet your goal?

Define Goals

Sound goals are achievable, measurable, under the client's control, and positive.

Determine Achievable Goals To be achievable, goals need to be realistic and reasonable. Do not overwhelm your client by setting numerous goals at one time. Set one to three goals depending on the needs and skills of the client. Clients are the best judges of what changes are workable. One way of assessing workability of the goal is to ask your client to think about the goal in general terms. According to Botman,[2] "The more removed the discussion is from change as an immediate necessity, the more easily the client can bring to bear powers of rationality, disregard circumstances of personal history, discount pathological bias and sidestep feelings of fear and discouragement" (p. 8).

Counselor: *Would this goal make sense for someone else facing this difficulty?*

Let's imagine that we are not talking about this goal for you. You are a forty-year-old married man with two children.

Let's assume this goal is for a fifty-year-old single woman who runs a dry cleaning business. Would this goal be workable for her?

Define Measurable Goals When a client makes a specific goal, it should be clear when the goal is attained. Goal statements with vague terms such as *good, try, better, more,* or *less* cannot be measured, and clients will have trouble knowing when the goal has been reached. Also, as the need to produce outcome data to funding agencies increases, you will have a hard time reporting your results if you do not have well-defined outcomes. This requires measurable goals. You may consider working with your client to develop a goal attainment scale (GAS) with graduated levels of desired outcomes (see Chapter 8).

Set Goals over Which the Client Has Control Attainment of a goal should depend on the actions of the client, not another person. For example, "I want my husband to . . ." is not a good goal, because it focuses on someone other than the client. If a client can only accomplish the goal through the help of another person, then the goal should be abandoned and a new one sought. However, this does not mean that clients should not seek out the support of others. For instance, if the plan was for your client to walk after dinner three days next week, it would be appropriate to ask a friend or family member to participate. Your client, though, should be making plans that will occur despite another person's involvement.

Weight loss is not a behavioral goal and should never be a short-term goal, although it could be a long-term anticipated outcome. A number of physiological factors affect numbers on a weight scale over which clients have no control.

State Positive Goals What do you think and feel when you imagine a list of avoid foods on a standard diet? Chances are that you created a depressing image of deprivation and despair. A goal that takes away a pleasurable pastime and leaves an empty void is doomed to failure. Positive images create a greater likelihood of success.[3] For example, a common problem is eating high-calorie, low-nutrient dense foods while watching television. A goal of "I won't eat ice cream while watching television" is likely to leave a client feeling distressed. Better to state what will be done—for example, "Four days a week I will eat one cup of plain, low-fat yogurt with one teaspoon of preserves while watching television. Three days a week I will eat one cup of low-fat ice cream." A noneating activity, such as grooming the dog, can also be substituted if it appears workable.

A journey of a thousand miles
must begin with a single step.
—Lao-Tzu

DESIGN A PLAN OF ACTION

Once a well-defined, achievable, and measurable goal has been identified, the next step is to design a plan to implement the goal.

EXERCISE 4.1—Analyze and Rewrite Goal Statements

In your journal, record the number of the following goal statements, identify the problem(s) with the statement, and write an alternative goal to meet standard criteria.

Example
Goal: Drink more water.
Problem(s): Not measurable, not specific
Alternative goal: Take two sixteen-ounce bottles of water with me to school each day to drink throughout the day.

1. Cut back on salt.
2. Eat more fruit.
3. Try getting back to running at least twice this week.
4. Will cut down on intake of bread and cheese.
5. Increase strength training.
6. Increase calories by consuming healthy foods.
7. Lose one pound this week.

Investigate the Physical Environment

Explore anything in the physical environment that could help or may hinder achievement of the goal.

Counselor: *Your idea of placing a bowl of clementine oranges on the kitchen counter sounds great. Now let's think about how you will get the oranges there. It is usually not a good idea to rely on anyone else to achieve your goals. Is it possible for you to purchase the oranges? Where will you purchase them? When can you purchase them? Your idea of taking an orange with you as you go out the door to eat for a midmorning snack on Monday through Friday sounds like a good idea. Is there anything that could happen that would prevent that from occurring? Do you think you need any reminders, like a note in your car or bathroom?*

Examine Social Support

Explore whether there is anyone in your client's environment that can help or hinder achievement of the goal.

Counselor: *Talking about your goal with your coworker who you usually eat a snack with at the office sounds like a good idea, especially since you said she is looking to eat better, too. Maybe the two of you will be a role model for each other.*

Review the Cognitive Environment

Explore your client's cognitive environment regarding the planned goal.

Counselor: *What are you saying to yourself right now about this goal?*

Does it still seem achievable?

Are we being too ambitious to expect five days of fruit for a midmorning snack?

What will you be saying to yourself if you miss a day?

Explain Positive Coping Talk, If Necessary

If your client expresses negative judgments about self, suggest to your client to replace destructive self-talk with a positive coping talk. Explain that a problem is not a failure; it is simply part of the change process. The following is an example of an alternative self-dialogue:

Counselor: *If you find yourself berating yourself, you can substitute another dialogue, such as "I had trouble today. I learned that on days like today I am not likely to eat my* *orange because. . . . Next time I will. . . . I am on the road to a healthier lifestyle."*

Modify Goal, If Necessary

If it appears that the goal is too ambitious, be prepared to modify the goal or to completely put it aside.

Counselor: *It looks as if we got off track on this one. Let us put it on hold until we meet next week.*

Select a Tracking Technique

Explore how the client would like to keep track of the goal—journal, chart on the refrigerator, empty fruit bowl, and so forth. Tracking procedures are covered in Chapter 5.

Verbalize the Goal

When you believe the goal is clearly defined, ask your client to verbalize the goal.

Counselor: *Just to be sure we are both clear about the food goal that we are establishing for this week, could you please state the goal?*

Write the Goal Down

It has been said that goals that are not written down are just wishes. Write the goal on an index card and give it to your client. This is a good time to make a statement of support to heighten self-efficacy.

Counselor: *I feel we did a good job. I think you are ready to do this and will be able to achieve your goal this week. However, if there is any difficulty do not despair—some problems are expected.*

The preceding discussion provides a clear map for implementing the goal-setting process. For most clients who are ready to make a lifestyle change, the defined process will work very well. However, counselors need to keep an open mind, listen carefully to their clients, and remain flexible, as indicated by the accounts in the box on page 94:

EXERCISE 4.2—Write Your Own Short-Term Goal

Think about an issue, preferably a lifestyle concern, in your own life that you would like to address. In your journal, write a short-term goal that you intend to follow for the next week. Then answer the following questions:

1. Explain why your goal is specific and achievable.
2. How will you tell whether you have reached your goal?
3. Do you have direct control over the achievement of your goal?
4. Use the ruler scale, Lifestyle Management Form 3.1, to identify your motivational level (see Appendix G). Table 3.2 (see page 70) contains the ruler numbers that correspond to the readiness-to-change levels.
5. Design an action plan. Address environmental, social, and cognitive supports or hindrances.
6. How will you monitor progress of the goal?

At the end of the week describe how successful you were. Analyze why or why not. How useful was the goal-setting process for changing your behavior?

EXERCISE 4.3—Practice Goal Setting

Practice setting a goal with a colleague. Each of you should take turns taking on the role of counselor and client. When you are the client, it would be best to choose a goal to work on that you are at a high motivational level for implementation (for example, write a letter to a friend or eat a piece of fruit for breakfast). If that is not possible, then role-play. The acting counselor should work with the acting client to define a goal and work through an action plan. Be sure you address all steps of the goal-setting process. Record your impressions of this experience in your journal.

DIETARY ASSESSMENT

A **dietary assessment** is a component of a comprehensive nutritional assessment. The following list contains all of the components as identified by the American Dietetic Association:[4]

- Historical information
  - Health history
  - Drug history
  - Psychosocial history
  - Dietary assessment
- Physical examination
- Anthropometric data (body measurements, such as size, weight, and body proportions)
- Biochemical analysis (laboratory tests)

In certain situations, an additional functional capacity assessment (for example, vision and dexterity) is needed to evaluate food preparation capability. After accurately gathering all of the assessment information and carefully analyzing the results and relationships between data, the assessor makes a meaningful evaluation. A nutrition counselor may become involved in all aspects of a comprehensive nutritional assessment; however, the greatest impact will be on the dietary assessment.

Examination of information obtained from assessments can be used to accomplish the following functions:

- Furnish baseline parameters.
- Determine health and nutritional risks.
- Ascertain feasible alternatives for making dietary changes and planning interventions.
- Identify strengths and roadblocks for making lifestyle changes.
- Set priorities for making dietary changes.
- Monitor progress and success of intervention strategies.
- Document outcomes.

Several references can aid in an in-depth investigation of nutritional assessment.[5,6,7] The objective here is not to present a comprehensive discussion but to review some commonly used dietary assessment tools and procedures that can be utilized to practice nutrition counseling skills. According to Simko et al.,[8] three steps are involved in completing a dietary evaluation: food intake data collection, data analysis, and interpretation of analysis.

When goal setting with a client, I always followed the standard criteria of making sure client goals were specific, achievable, measurable, and positive and then developed an action plan. However, once I worked with a client who resisted dwelling on the specifics of implementing her goals. She asked if we could just set a rather clear goal, such as eat three vegetables a day, and not develop an action plan. Somehow this client felt a need to rebel against a specified plan when we had set one. So I followed her request, and she did well in the program. I understood the need to be flexible and to listen to clients' perception of their needs. I felt good that this client perceived that I was open to hear her concerns.

I had a twenty-year-old male client whose food intake was a nutritionist's nightmare. I believe an eight-year old boy at a birthday party would have eaten better than my client. This person ate almost exclusively high-fat, high-sugar, and high-salt convenience foods. There were no servings of vegetables (except french fries), fruits, dairy, or whole grains. We were having trouble selecting a food goal until I brought up the concept of picturing a meal plate. I was surprised that he connected to this concept because it appeared that he rarely ate what most people would call a meal. Nonetheless, his goal was to eat one meal a day in which his plate consisted of an entrée, grain, and vegetable, each taking up a third of his plate. This goal worked well.

THE BETTER HALF® **By Harris**

"I promised I'd only eat 1000 calories per day. I didn't promise anything about the NIGHTS!"

Reprinted with special permission of King Features Syndicate.

Step 1: Food Intake Data Collection

Methods and tools used to collect data to aid in understanding the kinds and amounts of food consumed and the factors influencing choice include a client assessment questionnaire, food diary/daily food record, usual diet, diet history interview, food frequency, and twenty-four-hour recall. Although these are the standard forms, a counselor working in the field of nutrition will find that a range of assessment tools are available varying in purpose, length, and complexity.[9,10] Selection of a particular instrument will depend on the objective of the interaction (the type of clientele), initial or follow-up sessions, number of planned visits, and available resources (computer, time, and so forth). Sometimes an instrument is needed to address a specific issue, such as hypertension[11] or readiness to lose weight.[12] At other times the assessment tool is employed as a general screening device and has a broader perspective, such as the "Determine Your Health" checklist developed for older Americans.[13]

While collecting data from your client, be careful not to give advice, "preach," or condemn since such remarks are likely to be interpreted as judgmental and inhibit the free flow of information. Even words of approval should be avoided since this could encourage a client to give only good answers. Some people who regularly conduct dietary assessments find it useful to tell clients that there are no wrong answers. Others tell clients not be afraid to give an answer because no one has a perfect diet, and the counselor may even give an example of a low nutrient-dense food he or she enjoys.[14]

Each type of tool has advantages and pitfalls. A summary of the strengths and limitations can be found in Table 4.2. By using more than one instrument, the probability of obtaining a clear picture of your client's nutrition strengths and problems increases. Let's take a look now at several assessment tools' features.

Client Assessment Questionnaire (Historical Data Form) Sometimes referred to as an *intake form,* client assessment questionnaires generally contain several divisions addressing information about historical data. See Lifestyle Management Form 4.1 in Appendix G for an example. The top of this form has an administrative section and is usually followed by questions related to medical history. These questions are not asked for the purpose of making a diagnosis but to ascertain any medical factors that could have a nutritional impact. The family health history portion of the form provides information about a possible tendency toward a particular health condition. This has nutritional implications for your client. For example, a family history of heart disease may warrant an emphasis on heart-healthy foods.

Drug history questions provide information about medications, herbal preparations, and nutrient supplements that may impact nutrition status. Care should be taken to check for any interactions of food and drugs that can alter the effectiveness of a drug and/or a client's nutritional status.

A section on socioeconomic history furnishes valuable information about your clients' support systems, family settings, or significant others—any of which can play a role in their ability to make successful diet changes. This information can especially be helpful during the goal-setting process when exploring with your clients whether there are particular individuals in their lives that may interfere or help them achieve their goals.

The food pattern section, often referred to as *dietary history,* contains questions about food preferences and food selection variables that influence food intake. This knowledge will be particularly helpful for prioritizing goals and designing interventions.

A final section requesting clients to identify nutrition issues they would like to explore helps in planning the educational component of future sessions.

These forms can be tailored to meet counseling needs for specific clientele. For example, if serving mainly low-income individuals or students living in a dormitory, you would have a greater need for questions about accessibility to refrigeration and cooking facilities. A form to be used with eating disorder clients could contain specific questions about laxative use or purging, and a form for weight control clients may have a request to detail weight history.

Portion Size Many of the following methods require clients to estimate or measure their portion sizes. Some short videos are available that can be used as an instruction tool for clients who will be keeping food records. (See the resources at the end of the chapter.) Several aids have been employed to help respondents recall portion sizes for retrospective data collection. These include two- and three- dimensional food models; various

TABLE 4.2—Summary of Methods, Strengths, and Limitations of Selected Diet Assessment Tools/Procedures

METHOD	STRENGTHS	LIMITATIONS
Client assessment questionnaire/historical data form: a preliminary nutritional assessment form usually divided into sections for administrative data, medical history, medication data, psychosocial history, and food patterns	• Provides clues to strengths and potential barriers	• May seem invasive, may not be culturally sensitive
Food diary/daily food record: a written record of an individual's food and beverages consumed over a period of time, usually three to seven days	• Does not depend on memory • Provides accurate intake data • Provides information about food habits	• Requires literacy • Requires a motivated client • Recording process may influence food intake • Requires ability to measure/judge portion sizes • Time-consuming
Twenty-four-hour recall: a dietary assessment method in which an individual is requested to recall all food and beverages consumed in a twenty-four-hour period	• Quick • Easy to administer • No burden for respondent • Does not influence usual diet • Literacy not required	• Relies on memory • May not represent usual diet • Requires ability to judge portion sizes • Underreporting/overreporting occurs
Food frequency: a method of analyzing a diet based on how often foods are consumed (i.e., servings per day/week/month/year)	• Furnishes overall picture of diet • Not affected by season	• Requires ability to judge portion sizes • No meal pattern data
Usual diet: clients are led through a series of questions to describe the typical foods consumed in a day	• May be more of a typical representation than a twenty-four-hour recall	• Not useful if diet pattern varies considerably
Diet history interview: a conversational assessment method in which clients are asked to review their normal day's eating pattern	• Provides clarification of issues	• Relies on memory • Requires interview training

shapes of cardboard or plastic household cups, bowls, plates, glasses, and spoons; life-size photographs; graduated measuring spoons and cups for liquid and dry ingredients; and a ruler. Containers with two to three cups of dried beans, rice, or dry cereal can also be helpful for estimating portion sizes, as can premeasured plastic or net bags of beans in sizes equal to one cup, one-half cup, and one-quarter cup.[14] Here are some other portion size equivalents, from the International Food Information Council:

Commonly Used Estimates of Portion Sizes

One-half cup fruit, vegetable, cooked cereal, pasta, or rice = a small fist
Three ounces cooked meat, poultry, or fish = a deck of cards
One tortilla = a small (seven-inch) plate
Half bagel = the width of a large coffee lid
One muffin = a large egg
One teaspoon of margarine or butter = a thumb tip
Two tablespoons of peanut butter = a golf ball
A small baked potato = a computer mouse
One pancake or waffle = a four-inch CD
One medium apple or orange = a baseball
Four small cookies (such as vanilla wafers) = four casino chips
1.5 ounces of cheese = six dice

Food Diary/Daily Food Record To employ this method, a client records food and liquid intake along with preparation method as it occurs for a specified period, generally three to seven days.[15] Sometimes additional information is recorded such as time, place, activities, social setting, degree of hunger, and emotional state. A limitation of this tool for assessment is the impact recording can have on food intake. The hassle of writing a food item in a journal could discourage consumption of some foods, and the activity of recording encourages a person to take time to evaluate the particular choice. As a result, food diaries can be used as an intervention technique to alter food habits as discussed under journaling in Chapter 5. Review of food records is especially useful for both the counselor and client to gain insight into the client's eating lifestyle. Identification of positive behaviors may help identify skills that merit expansion to help solve problem areas.

Clients need to be given directions for completing food record forms and guidelines for measuring, weighing, and estimating portion sizes. Since accuracy is thought to decline if weighing all food items is requested (clients are less likely to eat some foods due to the burden of weighing), household measures are generally considered acceptable.[6] See Lifestyle Management Form 4.2 in Appendix G for an example of a food diary recording form.

Usual Intake Form The usual intake form gives a counselor an idea of a client's typical daily pattern of food intake. This form is simple and generally not time-consuming to complete. However, its usefulness will be limited for clients whose intake varies widely from day to day. In such cases, answering general questions would be difficult, and another assessment tool should be used.

The assessor begins by inquiring into the client's first food or drink of the day. This line of questioning continues until a daily pattern has emerged. The counselor must refrain from asking leading questions that may influence answers. Probing questions should be asked to ascertain the nutrient characteristics of the items consumed. For example, if a sandwich is generally consumed for lunch, investigate type of bread, filling, and condiments. See Lifestyle Management Form 4.3 in Appendix G for an example of a usual intake form.

Diet History Interview A diet history interview is similar to asking clients about their usual diets; however, the emphasis is on minimizing questions and allowing clients to tell their "stories." Clients are invited to give an account of a normal day's eating pattern, with the counselor utilizing attending skills and interrupting a little as possible. In this respect, the technique is similar to "a typical day" strategy covered in Chapter 3. After the narrative, the counselor selectively chooses follow-up questions to obtain only new and relevant information.

The conversational emphasis of this approach interfaces very will with motivational counseling protocol (Chapter 3) as well as the culturally sensitive respondent-driven interview (Chapter 3). During the process of conducting a diet history interview, a counselor could use the Twenty-Four-Hour Recall/Usual Intake Form (Lifestyle Management Form 4.3) to record a client's diet pattern. However, the act of completing the form should not be allowed to interfere with your clients' ability to relate their stories. The conversational nature of this approach will be disrupted if clients are asked to repeat something, and attending skills will be less than

EXERCISE 4.4—Estimating Portion Sizes

Estimate the amount of liquid, cereal, and beans in various-sized cups, glasses, bowls, and plates set up by your instructor. In addition, estimate the serving sizes of each food item in both a standard and "large portion" TV dinners. Measure the quantities and compare your findings to the recommended Food Guide Pyramid servings. Record your reaction to this activity in your journal.

adequate if a counselor has eyes on a piece of paper for most of the interview. Quickly jotting down notes during the story would probably work well. Alternatively, the usual diet form could be filled out after the client has related his or her story while probing questions are used for clarification. See Exhibit 4.1 for a protocol of the method.

Food Frequency Checklist The food frequency checklist is an assessment tool containing lists of food grouped according to similarity in nutrient quality and quantity. They are designed to be either read to clients by an interviewer or distributed in printed form for self-administration. The form contains a set of response options to be checked off that indicate how often certain foods are consumed (for example, by the day, week, month, and so forth). Some questionnaires are designed to consider one or two specific nutrients, such as calcium or fat and cholesterol,[16,17] whereas others are comprehensive in scope.[18,19] Food frequency questionnaires vary in the amount of detail requested regarding serving size and preparation methods. If they are too short, the knowledge gained is limited. If they are too long, the process of completing the form is tedious and accuracy can decline.

Overall, this method is easy for most people to use. This questionnaire helps counselors evaluate diet in terms of how often certain foods and food groups are eaten. Food groups not eaten very often or omitted are indications of dietary imbalances. Close attention should also be given to frequently consumed foods as to their nutritional desirability. (See Lifestyle Management Form 4.4 in Appendix G for an example.)

EXHIBIT 4.1—Protocol for Diet History Interview

Diet History

The purpose of the diet history is to obtain an account of a person's usual food intake. Structurally it takes the form of a description of meals consumed throughout the day accompanied by a food frequency cross-check. One way of looking at the first component of the diet history is as a story with a beginning (usually breakfast) and end (usually supper or evening snack). Use of the narrative approach means that participants are given the opportunity to finish their story before they are asked any questions. In this way, the flow of participants' information giving is not interrupted (but what they say is acknowledged and supported by the interviewer). Additional comments, not necessarily on food per se, made during this description may provide the interviewer insights for questions or discussion later. When introducing the diet history, the interviewer refers to the notion of usual, meaning within the past couple of months, and of a time sequence for the description, such as the duration of the day. Participants are asked to provide a general pattern and then point out variations.

Interview Protocol

- Explain the purpose of the interview. Advise the participant that you are seeking a description of usual eating patterns and suggest that she or he start with the beginning of the day.
- If the participant begins with the first meal of the day and uses time references or meal sequences of the day to progress with the description, do not interrupt the story; merely indicate that you are listening (nod, write, say "hmm" or "yes").
- If the participant stops at intervals along the way waiting for you to respond, provide narrative support to continue—for example, "Was that all for breakfast?" or "Do you have anything after that?"
- If the participant volunteers explanations for why or how she or he consumes certain foods, acknowledge the explanations in a supportive, nonjudgmental way, but keep the account on track.
- When the participant has reached the end of the day, look at what you have noted and identify areas for which you need more detail. This will depend on the purpose for taking the history. Ask specific, strategic questions.
- If the participant responds to a topic with "It depends," be sure to encourage all possible variations on that topic (usually a meal description).
- If the participant says "probably" in defining amounts of foods, use visual aids to assist in the estimation process.
- Summarize the overall pattern of the diet and ask whether there is a great deal of variation in this pattern. Note the variation.
- Proceed with a food frequency checklist and questions on food preparation.
- Ask the participant if there is anything else she (or he) would like to add to what she has told you and if she or he thinks you have a true reflection of her or his usual eating pattern.

Source: Tapsell LC, Brenninger V, Barnard J, Applying conversation analysis to foster accurate reporting in the diet history interview. © 2000, The American Dietetic Association. Reprinted by permission from the *Journal of the American Dietetic Association,* 100:818–824.

EXERCISE 4.5—Practice Gathering Information for a Dietary Assessment

Complete a client assessment questionnaire and a food frequency questionnaire (Lifestyle Management Forms 4.1 and 4.4 in Appendix G) based on your own diet history and food habits. Exchange forms with a colleague and take turns acting as a counselor. Gather information using the following interview guide. The collected data will be evaluated in Exercise 4.7.

- ❑ Ask your client whether she or he has any nutritional concerns.
- ❑ Review the completed Client Assessment Questionnaire, Lifestyle Management Form 4.1—*I am wondering, what came to your mind as you were filling out this form? What topics covered in this form do you think have particular importance for your food issues?* Look over the form and ask for clarification where appropriate. Your client may have already covered relevant issues in response to your previous open questions.
- ❑ Conduct a diet history interview. Follow the protocol in Exhibit 4.1. While your client is telling you his or her story, fill in the 24-Hour Recall/Usual Diet Form, Lifestyle Management Form 4.3.
- ❑ Summarize.
- ❑ Review the completed Food Frequency Questionnaire, Lifestyle Management Form 4.4. Clarify portion sizes using food models, if needed—*Thank you for completing the Food Frequency Questionnaire. I am wondering, what came to your mind as you were filling out this form? Did you feel a need to clarify or expand on anything while you were completing this form?*

In your journal, write your impressions of this experience as a client and as a counselor. The data you collected in the exercise will be analyzed in Exercise 4.7

Twenty-Four-Hour Recall In this method, the interviewer asks the client to recall all foods, beverages, and nutritional supplements consumed, including amounts and preparation methods, over a twenty-four-hour period. Counselors can define the period of time from midnight to midnight of the previous day or the past twenty-four hours.[6] The starting point can be the most recent or the most distant of the twenty-four-hour period. The Twenty-Four-Hour Recall Form is similar to a Usual Diet Form, and these two tools have been combined in Lifestyle Management Form 4.3.

An advantage of this method is that it is easy to administer and requires little effort on the part of the client. However, one day may not be representative of a person's usual intake. This difficulty can be overcome if the twenty-four-hour recall is administered on several nonconsecutive days, including both weekdays and weekend days.[20] Although this form can be self-administered, the accuracy increases if counselors assist their clients in recalling their food consumption and portion sizes.[10] The following are some components of an effective twenty-four-hour recall:

- Do not ask leading questions, such as those assuming a meal was eaten. Refrain from prompts, such as "What did you eat for breakfast?" Better to ask, "What liquids or foods were first consumed after awakening today?"
- Ask probing questions. For example, "You said you had a lot of butter on your toast. How much is a lot? What kind of bread was used to make the toast? Was anything else put on the toast besides butter?"
- Ask sequential questions about the day's activities, travels, and encounters with others to help clients recall foods consumed. Inquire if any foods were consumed during meal preparation or clean-up or during the middle of the night.
- Use portion size estimation tools to improve the accuracy of the answers.
- Research has shown that certain food items are frequently missed in twenty-four-hour recalls—crackers, breads, rolls, tortillas; hot or cold cereals; cheese added as a topping on vegetables or on a sandwich; chips, candy, nuts, seeds; fruit eaten with meals or as a snack; coffee, tea, soft drinks, juices; and beer, wine cocktails, brandy, any other drinks made with liquor. Sugerman et al.[10] suggest going over this list before completing the recall to be sure none of the items were missed.
- Another aid to increasing retrieval of memory is the *multiple-pass procedure.*[21] At first clients are asked to

EXERCISE 4.6—Conduct a Twenty-Four-Hour Recall

Using Lifestyle Management Form 4.3, take turns with a colleague administering a twenty-four-hour recall. Use visual aids to help estimate portion sizes. In your journal, write your impressions of the experience.

recall foods eaten in sequence during a twenty-four-hour period, but details such as serving size and condiments are not requested. The whole procedure is repeated again in another "pass" of the twenty-four-hour period, but this time probing questions regarding type, amounts, additions/toppings, and preparation method are included. In a final pass, the assessor reads through the record and asks whether there any additions or corrections.

Culturally Appropriate Assessment Instruments

There is a critical need for the development of culturally specific techniques and tools to conduct nutritional assessments.[22] Depending on communication difficulties and cultural feelings about invasiveness, a counselor may find a qualitative rather than a quantitative approach to yield greater success.[23] To establish trust during the first session, consider using "a typical day technique" reviewed in Chapter 3 or the diet history interview data collection method covered in this chapter. These approaches eliminate the need to differentiate meals or categorize food items. A request for additional information such as frequency and portion size could be delayed until the next meeting.[22]

Step 2: Data Analysis

After dietary information is collected, it needs to be analyzed for food groups and/or components of food, such as energy, nutrients, or phytochemicals.

- **Food group evaluations** can generally be done quickly, making immediate feedback possible. Some forms have the standards on the collection form or as an attachment to the assessment form allowing for a speedy evaluation. See Lifestyle Management Forms 4.3, 4.4, and 4.5.
- **Food component analysis** is rather time-consuming, so generally feedback cannot be given the same day data are collected. Nutrients can be analyzed from food composition tables, the U.S. Department of Agriculture (USDA) Nutrient Database (http://www.nal.usda.gov/fnic/foodcomp/), or with the aid of a nutritional analysis software program.

Step 3: Interpretation of Analysis

Interpretation of analysis of dietary information is done by comparing the data analysis to a standard. Computer programs automatically execute both steps 2 and 3—that is, analyze and interpret, generally for food groups and nutrients. The following describes the most commonly used standards:

- The **Daily Food Guide (Food Guide Pyramid)** found in Appendix A (http://warp.nal.usda.gov:80/fnic/dga/). This is a pictorial representation of five major food groups indicating kinds and amounts of food to consume and can be used to assess diets. The Twenty-Four-Hour Recall/Usual Diet Form, Lifestyle Management Form 4.3, has the Daily Food Guide servings as part of the form to enable a rapid evaluation. This is particularly valuable when immediate feedback is desired.
- Recommended Dietary Intakes (RDI), commonly used standards when assessment of specific nutrients is desired, can be found on the inside cover. They are divided into four categories: Estimated Average Requirements (EARs), Recommended Dietary Allowances (RDAs), Adequate Intakes (AIs), and Tolerable Upper Intake Levels (ULs). For a description of these divisions, see Table 4.3 and the USDA Web site (http://www.nal.usda.gov/fnic/dga/).
- For clients who desire a more ambitious dietary regimen, comparisons could be made to the DASH **(Dietary Approaches to Stop Hypertension)** eating plan, particularly if you are working with an individual who has elevated blood pressure (http://dash.bwh.harvard.edu/dashdiet.html). The DASH Food Plan is a heart-healthy regimen rich in fruits, vegetables, fiber, and low-fat dairy foods and low in saturated and total fat. See Appendix B.
- Canada's Food Guide to Healthy Eating for People Four Years and Older (www.hc-sc.gc.ca/hppb/nutrition/pube/foodguid/foodguide) can also be used to assess diet for general good health.

ENERGY DETERMINATIONS

Nutrition counselors may need to estimate *total energy expenditure* (TEE) of their clients for a variety of therapeutic reasons, including planning a weight loss program. There are three components making up TEE: *resting energy expenditure* (REE), the *thermic effect of food* (TEF), and *energy expended in physical activity* (EEPA).[24] Generally only resting energy and physical activity energy are calculated in counseling interventions. The thermic effect

TABLE 4.3—Recommended Dietary Intake Terms

TERM	DEFINITION
Adequate Intake	Amount of a nutrient that maintains a function; used when recommended dietary allowance (RDA) cannot be determined
Recommended Dietary Allowances (RDAs)	The amount of a nutrient covering the needs of nearly all healthy individuals
Estimated Average Requirements (EARs)	Amount of a nutrient estimated to meet the requirement of half the healthy people in a given age and sex group
Tolerable Upper Intake Level (UL)	Maximum level of a nutrient that appears safe

of food is often omitted because the inherent error factor of the total equation is greater than the amount that would be added due to the thermic effect value.[18] The following sections outline the steps for calculating TEE.

Step 1: Calculate REE.* Several standard formulas can be used to estimate REE, all providing about the same values. The Harris-Benedict equations are commonly used in the United States; however, some authorities report that they can overestimate REE, particularly for obese individuals.[25] The World Health Organization (WHO) equations provide an alternative method. In addition, a simplified formula has been developed that is not considered as accurate as the more involved equations. The formula is useful when a quick estimate of REE is needed.

Note that 1 centimeter = 0.3937 inches, and 1 inch = 2.54 centimeters; 1 kilogram = 2.2 pounds, and 1 pound = 0.4536 kilograms.

Harris-Benedict Equations

Women: REE (kilocalories) = 655 + 9.56 weight (kilograms) + 1.85 height (centimeters) − 4.68 (age)

Men: REE (kilocalories) = 66.5 + 13.75 weight (kilograms) + 5.0 height (centimeters) − 6.78 (age)

World Health Organization Equations

Women (18–30 years): REE = (14.7 × weight [kilograms]) + 496

Women (30–60 years): REE = (8.7 × weight [kilograms]) + 829

Men (18–30 years): REE = (15.3 × weight [kilograms]) + 679

Men (30–60 years): REE = (11.6 × weight [kilograms]) + 879

EXERCISE 4.7—Data Analysis and Interpretation

To complete this activity, work with the same colleague you paired with for Exercise 4.5. Complete the Food Group Feedback Form (Lifestyle Management Form 4.5).

- ❑ Review the feedback form, point by point, in a nonjudgmental manner with your client. Compare the standards to your volunteer's food intake. *As you can see your usual vegetable intake is two servings a day, and the Food Guide Pyramid suggests three to five servings a day, and the DASH Food Plan protocol is four to five a day.* Continue in this vein until you have gone over all the findings.
- ❑ Ask your client his or her impression of the evaluation. *What do you think about this information?*
- ❑ If your client expresses interest in making a change, use the Assessment Ruler (Lifestyle Management Form 3.1).
- ❑ Summarize.

Write your reactions to this exercise in your journal. What did you learn from this experience that you would like to incorporate or change when working with future clients?

*Shape Up America guidelines suggest using the Mifflin–St. Jeor formulas to estimate twenty-four-hour REE (http://shapeup.org):
Women: REE = 10 × weight (kilograms) + 6.25 × height (centimeters) − 5 age (years) − 161
Men: REE = 10 × weight (kilograms) + 6.25 × height (centimeters) − 5 age (years) + 5

Quick Estimate Equations[27]

Women: REE = kilograms × 23 = kilocalories/day

Men: REE = kilograms × 24 = kilocalories/day

Step 2: Multiply REE by a factor for physical activity level (PAL). Generally a PAL of 1.5 for women and 1.6 for men is selected for those engaged in light activity.[26] Table 4.4 provides factors for physical activity level based on doubly labeled water studies, a highly accurate measure of energy expenditure using deuterium and oxygen isotopes.[24]

Step 3: Multiply REE times PAL to obtain the estimated TEE (kilocalories/day) to maintain weight.

REE × PAL = TEE

Step 4: If weight loss is desired, subtract 500 kilocalories/day to obtain adjusted caloric intake required to achieve weight loss of approximately one pound per week.

EXERCISE 4.8—Calculate Your Total Energy Expenditure (TEE)

Calculate your TEE using the three different REE equations and the simplified method described in this chapter.

REE		PAL		TEE
Mifflin–St. Jeor =		×		=
WHO =		×		=
Harris-Benedict =		×		=
Quick estimate =		×		=

Compare and contrast the various methods in your journal.

PHYSICAL ASSESSMENT AND HEALTHY WEIGHT STANDARDS

Since weight issues are related to many of the major health problems in North America, nutrition counselors often need to address **overweight** or **obesity** concerns. The prevalence of these

TABLE 4.4—Physical Activity Level Factors

LIFESTYLE AND LEVEL OF ACTIVITY	FACTOR FOR PAL
Chair-bound or bed-bound	1.2
Seated work with no option of moving around and little or no strenuous leisure activity	1.4–1.5
Seated work with discretion and requirement to move around but little or no strenuous leisure activity	1.6–1.7
Standing work (e.g., housework, shop assistant)	1.8–1.9
Significant amounts of sport or strenuous leisure activity (30–60 minutes four to five times per week)	+0.3 (increment)
Strenuous work or highly active leisure	2.0–2.4

Source: Johnson RK, Energy. In L. Kathleen Mahan & Sylvia Escott-Stump (Eds.), *Krause's Food, Nutrition, and Diet Therapy* (pp. 19–30). Philadelphia: Saunders, 2000; adapted from Shetty PS, et al. Energy requirements of adult: An update on basal metabolic rate (BMRs) and physical activity levels (PALs). *Eur J Clin Nutr.* 50(suppl 1):S11, 1996.

conditions is increasing in the United States.[28] The third National Health and Examination Survey (NHANES III), conducted from 1988 to 1994, found 54.9 percent of United States adults aged twenty years and older to be overweight or obese. The following sections describe commonly used methods and standards for assessing weight.

Weight-for-Height Tables

Even though authorities disagree about the usefulness of available tables, they have been widely used to determine desirable weight. Historically the Metropolitan Life Insurance tables (Appendix D) have been employed to determine appropriate weight for height based on gender and frame size. See Appendix C for this table and a description of a method to determine frame size.

The USDA/DHHS (Department of Health and Human Services) Dietary Guidelines table (Table 4.5) provides weight ranges for adults based on height but not gender, frame size, or age. Higher weights in ranges are for individuals who have more muscle and bone, such as athletes and men. For average individuals, a midpoint can be used to determine a desirable weight. Overweight is calculated as 10 percent over the standard, obesity is 20 percent over, and underweight is 10 percent under the standard.[28]

Hamwi Method

The **Hamwi method** is frequently used to calculate ideal weight for height. See Exhibit 4.2 on page 104 for this efficient method.

Body Mass Index

Body mass index (BMI) is the preferred weight-for-height standard and is used as a determinant of health risk and a predictor of mortality. It can be determined from existing tables, equations, or Web sites. A BMI chart can be found in Appendix E.

The standard calculation is based on metric units, but BMI can be estimated from another equation using pounds and inches:

BMI = weight (kilograms) ÷ height (meters) squared

(1 pound = 0.4536 kilogram)

(1 inch = 2.54 centimeters = 0.0254 meter)

BMI =(weight [pounds] ÷ height [inches]^2) × 703

TABLE 4.5—Dietary Guidelines: Healthy Weight Ranges for Men and Women

HEIGHT*	WEIGHT (IN POUNDS)†	
	MIDPOINT	RANGE
4′10″	105	91–119
4′11″	109	94–124
5′0″	112	97–128
5′1″	116	101–113
5′2″	120	104–137
5′3″	124	107–141
5′4″	128	111–146
5′5″	132	114–150
5′6″	136	118–155
5′7″	140	121–160
5′8″	144	125–164
5′9″	149	129–169
5′10″	153	132–174
5′11″	157	136–179
6′0″	162	140–184
6′1″	166	144–189
6′2″	171	148–195
6′3″	176	152–200
6′4″	180	156–205
6′5″	185	160–211

*Without shoes.
†Without clothes.
Source: *Report of the Dietary Guidelines Advisory Committee on the Dietary Guidelines for Americans.* Washington, DC: Government Printing Office, 1995.

EXHIBIT 4.2—Hamwi Method for Calculating Ideal Body Weight

Females: One hundred pounds for the first five feet of height; add five pounds for every inch over five feet, or subtract five pounds for every inch less than five feet.

Males: One hundred six pounds for the first five feet of height; add six pounds for every inch over five feet, or subtract six pounds for every in less than five feet.

In each case, add 10 percent for a large frame size or subtract 10 percent for a small frame size.

Example: Calculation of ideal body weight for a small-boned, five feet, six inch woman:

100 pounds for the first 5 feet = 100 pounds
5 pounds for every inch over 5 feet = 6 × 5 = 30 pounds
100 pounds + 30 pounds = 130 pounds
10 percent of 130 pounds = 13 pounds
Ideal body weight range is 117–143 pounds.
Ideal body weight for small-boned woman is 130 − 13 = 117 pounds.

See the following Web sites for electronic calculation of BMI:

- National Heart, Lung, and Blood Institute Obesity Education Initiative: http://www.nhlbi.nih.gov/about/oei/index.htm
- Shape Up America: www.shapeup.org
- Mayo Clinic Health Oasis: http://mayohealth.org/mayo/9707/htm/weight.htm

In general, a desirable BMI is 19 up to 25, with a midpoint of 22. The risk for developing associated morbidities or diseases such as hypertension, high blood cholesterol, type 2 diabetes, and coronary heart disease begins to climb above the desirable range (25).[28] Some authorities extend the optimal BMI based on mortality risk to 27.[27] Mortality risk increases for both underweight and overweight (Figure 4.2). In the United States, the average BMI of adults is 26.3.[27] NIH Expert Panel[28] weight classifications based on BMI areas follows:

Weight Classification	BMI
Underweight	Below 18.5
Normal	18.5–24.9
Overweight	25.0–29.9
Obesity	30.0 and above

BMI cannot be used as a standard to estimate body fat risk for pregnant and lactating women, highly muscular people, growing children, or adults over age sixty-five. It can be used to conveniently calculate an individual's goal weight. See Exhibit 4.3.

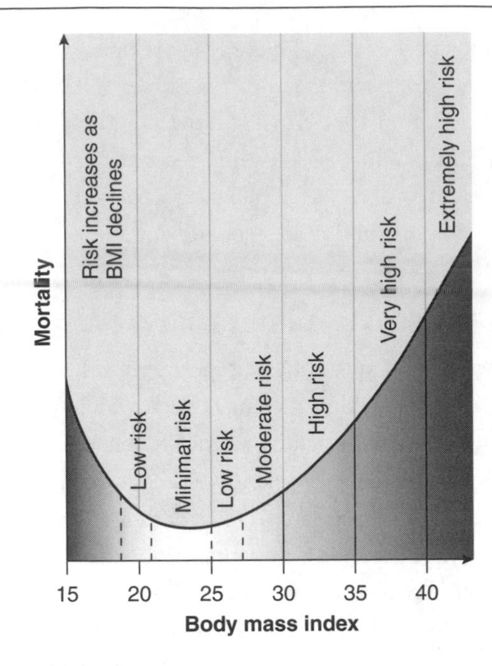

Figure 4.2 Body Mass Index and Mortality

Waist Circumference

Waist circumference is an indicator of upper abdominal fat (stomach area) accumulation and is used to assess central or upper-body obesity. People with this type of obesity are sometimes referred to as "apples" or as having **android** (man-like) **fat distribution,** and they are at an increased risk for heart disease, stroke, diabetes, hypertension, and some types of cancer. As a general rule, risk of central obesity-related health problems increases for women with a waist circumference greater than thirty-five inches and for men greater than forty inches.[28] Smoking, high alcohol intake, and menopause tend to increase abdominal fat, and exercise tends to decrease it.[27] People who accumulate fat in their hips and thighs are not as susceptible to the obesity-related diseases, but their fat is more resistant to breaking down from calorie deprivation and exercise. These people are often referred to as having a pear shape or **gynoid** (woman-like) **fat distribution.**

EXERCISE 4.9—Calculate Ideal Body Weight Using the Hamwi Method

In your journal, show calculations for the ideal body weight of a large-boned male who is five feet, seven inches tall.

EXHIBIT 4.3—BMI Chart Used to Determine Desirable Weight

Find your client's height and run your finger along the corresponding horizontal line until you come to the weight that matches the desired BMI, such as 24. That weight will be the desired weight, which could be used as a goal weight. For example, look at the BMI chart in Appendix E. Suppose a man was five feet, six inches tall and wanted to know a healthy weight. You begin at the horizontal line corresponding to that height and run your finger along the line until your reach the weight number that vertically corresponds to 24. That weight is 148 pounds. If the person's weight is 158 pounds, that would mean he would need to lose 10 pounds to achieve a BMI of 24.

Waist-to-Hip Ratio

Waist-to-hip ratio (WHR) is another method used to assess abdominal fat distribution by measuring both waist and hip circumference and calculating the ratio. This procedure is not thought to provide any additional information over waist circumference alone.[28] A WHR of 1.0 or greater in men and 0.8 in women is indicative of android obesity and an increased risk of obesity-related disease. For example, a woman with a thirty-six-inch waist and forty-eight-inch hips would have a ratio of 0.75. This measurement would not indicate an increased risk for obesity-related diseases.

According to the National Institutes of Health (NIH) *Clinical Guidelines,*[28] BMI and waist circumference provide practitioners with the most accessible and accurate measurements of degree of overweight and obesity. The guidelines also suggest body weight measurements alone be used to follow weight loss and to determine efficacy of weight loss interventions.

DOCUMENTATION/CHARTING

After completing a counseling session, the next step is to reflect, evaluate, document, and plan. The amount of time available for this step will depend on the setting of the counseling session. Lifestyle Management Form 4.7, Client Concerns and Strengths Log (see Appendix G), can aid in that endeavor by guiding reflection on the concerns and strengths expressed by your client and identified by you. This activity is particularly useful to a novice counselor who may become bogged down on concerns over which the client has very little control. This reflection could be shared with your client during the next session, or the activity may help you to form the framework for future counseling sessions. The specific criterion in the documentation depends on institutional standards or the setting of the intervention. Charting can provide the following benefits:[5,8]

- Evidence of care
- Demonstration of accountability in meeting legal, regulatory and professional standards
- A basis for evaluation and planning to ensure quality care
- Documentation for legal protection of clients,

EXERCISE 4.10—Assess Your Healthy Weight

Complete the chart for yourself and transfer the material to your journal.

MEASUREMENTS	STANDARD
Actual weight =	Insurance table (Appendix D) = Dietary Guidelines (Table 4.5) = Hamwi ideal body weight (Exhibit 4.2) = BMI desirable weight (Exhibit 4.3) =
Body mass index =	Desirable = 19–25
Waist circumference =	High risk: males, >102 centimeters (40 inches); females, > 88 centimeters (35 inches)
Waist-to-hip ratio =	Increased risk = males, ≥1.0; females, ≥ 0.8

Some authorities believe that North American health officials often assess health through a thinness lens. What is your impression of that statement in light of the various methods you just used to assess your healthy weight?

practitioners and facility
- A tool for communication among health care team members
- Justification for third-party reimbursement

Since charting in medical records is considered a legal document, care must be taken to provide clear, well-written notes. The development of automated computer systems continues to grow and is likely to become commonplace. The following are some general guidelines for documentation of counseling sessions:[5,8]

- Notes should be concise. Goals and plans should not be embodied in a lengthy narrative. Physicians are more likely to respond to dietitian recommendations when goals and plans are easily identified.[25]
- Entries should be clear and legibly written in blue or back pen. The number of facilities using electronic charting is expected to increase and offers numerous advantages, such as a reduction of duplication and repetition and an increase in care management tools, including alerts or reminders.[26]
- Documentation should be accurate for ongoing referencing.
- Entries should be appropriate and pertinent. Personal opinions and criticisms should be avoided.[8]
- Notes should be in chronological order, leaving no blank spaces.
- Entries should be made as soon as possible after the encounter.
- Date and sign all entries with full name and credentials.

There are a variety of documentation styles.[5] Each institution defines a charting format for its facility. Many of the formats are similar in that they present objective data, provide an assessment, and end with a plan of action and expected outcomes. Since the Joint Commission of Accreditation of Healthcare Organizations (JCAHO) standards have been emphasizing outcomes, charting in medical facilities is likely to embrace a format that clearly defines outcomes.[27] The **SOAP** (subjective, objective, assessment, and plan) and **IAO** (issues, actions, and outcomes) **formats** are commonly used and are appropriate to utilize in a lifestyle management program.

SOAP Format

Because of the comprehensive nature of this format, new practitioners find practice with this method particularly useful for developing charting skills. The case study addendum in this chapter contains three samples of SOAP notes for each of the three levels of motivation case studies in Chapter 3. Each segment of this format contains the following components:

S (Subjective):
- Information relayed to you from the client or the family.
- Citations do not need to be in complete sentences. Since the notes are under *S*, the assumption is that the information came from the client and there is no need to begin a sentence with "Client says."
- Entries may include information about physical activity, weight patterns, appetite changes, socioeconomic conditions, and so forth, that may be relevant to the client's nutrition issues.

O (Objective):
- Information generally comes from charts and laboratory reports and includes factual, scientific information that can be proven.
- Citations do not need to be in complete sentences.
- Examples of possible information include; age, sex, diagnosis, nutritionally pertinent medications, anthropometrics, lab data, height, weight, ideal body weight, changes in weight, and diet order.

A (Assessment):
- This is your interpretation of the client's status based on subjective and objective information.
- Information should be written in complete sentences as a paragraph.
- The following can be the format for this entry:

— Begin with a statement summarizing the client's nutritional status and/or concerns.
— Reflect on subjective and objective data and their impact on concerns including possible problems and/or difficulties with self-management and the effect of medications on nutritional status.
— Provide possible approaches and interventions.
— Assess degree of readiness, comprehension of information provided, and previous goal achievement.

P (Plan):
- Notations are generally short, concise statements written in complete sentences that can include the following:

— Long-term goals and specific, measurable short-term goals
— Need for additional diagnostic data—assessments, lab work, consultations

— Therapeutic plans—changes in nutrition care plan, diet prescription, supplement recommendations
— Educational plans to address dietary issues
— Description of education intervention during the counseling session

IAO Format

This concise method of charting clearly identifies a care plan and decreases time needed for documentation. It is an example of charting by exception in which only significant findings or exception to the norms are documented.[28] The issues identified must be those that can be appropriately addressed by the nutrition counselor. As an aid to identifying suitable issues, see the Client Concerns and Strengths Log, Exhibit 4.4 (Lifestyle Management Form 4.7 in Appendix G). Outcomes statements are clearly delineated in accord with JCAHO nutrition care standards.[32] The focus of this documentation is on the client rather than the process or "tasks" of care.[29] For an example of the IAO format, see the case study in this chapter.

Each segment of the IAO format contains the following components:

I (Issue): Log all nutrition related client concerns that can be realistically addressed.

A (Assessment): Identify specific interventions that address the related issue.

O (Outcomes): This category identifies expected outcomes of each action. They are the same as specific goals. They must be tangible and measurable so that anyone reading the chart can recognize the objectives of the intervention. They should also state time limits for achievement

Follow-up: Sometimes a follow-up column is included to indicate the progress of the anticipated outcomes, and they are labeled as ongoing, achieved, or unrealistic. This is a valuable addition for assessment because it indicates the progress made on achieving goals.

EXHIBIT 4.4—Client Concerns and Strengths Log

1. List all concerns expressed by your client or identified by you.

too little time
children responsibilities—NC
too little exercise
big family dinners every Sunday
eating while watching television
no planning for meals
nonsupportive husband—NC
low intake of fiber, fruits and vegetables
little dairy, does not like, low intake of calcium
frequent consumption of fast food
constipation
little knowledge about the role of nutrition in treating hypertension

2. Write "NC" next to of all concerns over which you or your client have no control.
3. Categorize the remaining concerns that you and your client can address to set realistic goals.

NUTRITIONAL*	BEHAVIORAL	EXERCISE
Low intake of fiber, fruits and vegetables	No planning for meals	Too little exercise
Frequent consumption of fast food	Eating while watching television	
Little dairy, low intake of calcium	Too little time	
	Big family dinners every Sunday	
	Frequent trips to fast-food establishments	

*Address food pattern, frequency, and variety concerns, if appropriate.

(Continued)

EXHIBIT 4.4—Client Concerns and Strengths Log (Continued)

4. List strengths and skills that could be utilized to set goals that are applicable to the above concerns (for example, organizational skills, knowledge of calories and food groups, cooking skills, regular activity).

- *has an exercise bike; enjoyed it at one time*
- *a walking club in client's church*
- *tasted soy milk once; liked it*
- *enjoys oatmeal, whole-grain crackers*
- *enjoys apples, dried apricots, cherries, carrot and celery sticks*
- *good organizational skill*
- *cooking/good preparation knowledge*
- *good support system—Mom*
- *has taken children to high school track to play while client ran; children enjoyed*
- *her mother has made vegetable platters for the family dinners; would probably do so more frequently if the request was made*
- *shredded carrots in canned spaghetti sauce once; children didn't seem to mind*

5. Categorize the strengths and skills in the following chart:

NUTRITIONAL	BEHAVIORAL	EXERCISE
Soy milk	Mother makes veg. platters	Exercise bike
Oatmeal, whole-grain crackers		High school track
Apples, dried apricots, cherries		Walking club available
Shredded carrots in spaghetti sauce		

6. What strengths and skills can be used to address the concerns? List them in the following chart.

STRENGTHS AND SKILLS	CONCERNS	POSSIBLE INTERVENTION STRATEGIES
Likes oatmeal, whole-grain crackers	Too little fiber	Make oatmeal with cut-up apples
Likes apples, dried apricots, cherries	Low fruit intake	Make oatmeal with cut-up apples
Likes carrots and celery sticks	Low vegetable intake	Take veggie packs to work for a snack Ask Mom to put vegetables in soup
Interest in cooking Shredded carrots in spaghetti sauce Supportive mother	Limited variety High sodium intake	Heart-healthy cooking classes with Mom Cooking demonstrations Include mother in next counseling session
Good organizational skills	No planning for meals	Plan week's menus on one hour break

EXERCISE 4.11—Evaluate Documentation Methods

Review the two methods, SOAP and IAO, presented for documenting interactions with Nancy in the case study on pages 109–110. Compare and contrast the two methods. Which method do you believe would be more useful? Explain.

CASE STUDY 4—Nancy: Documentation at Three Levels of Motivation

The following contains a follow-up of the three scenarios of Nancy at different motivational levels in Chapter 3's case study. It is very important to write your notes about a counseling intervention as soon as possible after the session. Included here are examples of SOAP notes for each of the scenarios. Next is a four-week example of IAO charting for Nancy at motivational Level 3 (see p. 110).

Level 1—Not Motivated, Not Ready

S Doctor wants me to lose weight, don't want to; husband likes me with "meat on my bones"; too many personal problems to worry about dieting.

O 5′4″, 170#, 35# overweight

A Client referred by physician for diet interventions for weight loss to reduce hypertension. She expresses no interest in learning about diet options, as she feels overwhelmed by personal problems. Her feelings were acknowledged, and intervention was limited to provision of literature.

P Rx: Provision of business card for future referral
Provision of literature on diet and hypertension
Goal: Increased awareness of the role of diet and hypertension
Referral in next 3 months for further counseling

Level 2—Unsure, Low Confidence

S Doctor wants me to lose weight, don't want to; husband likes me with "meat on my bones"; too many personal problems to worry about dieting.

O 5′4″, 170#, 35# overweight, hypertension

A Referred by physician for diet interventions for weight loss secondary to hypertension. Her ambivalence to commit to a counseling program due to stress and personal responsibilities was acknowledged. She exhibits several lifestyle concerns that negatively impact her health and is somewhat ready to address these issues (motivation level of 6 on a scale of 1–10). Discussion of the role of diet and hypertension as well as the possibility of realistically incorporating interventions into her lifestyle was well received. Focusing on positive actions, involvement in realistic goal setting, and stressing the relationship of her health to meeting her family responsibilities are expected to increase her degree of readiness.

P Rx: Client education
1. Diet and hypertension
2. DASH Food Plan
3. Seeking a support system—work/home

Will follow up with telephone call in 1 week

Goals: 1. Eat a banana or orange daily at work for a snack

Level 3—Motivated, Confident, Ready

S Doctor wants me to lose weight; husband likes me with "meat on my bones"; never really thought of losing weight to help blood pressure; willing to try.

O 5′4″, 170#, 35# overweight, hypertension

A Referred by physician for diet interventions for weight loss secondary to hypertension. Assessment forms indicate that she has various lifestyle concerns that are influencing her health. She is motivated by her need to remain healthy to meet her family responsibilities as well as the desire to prevent long-term complications as are being experienced by her mother. Empowering her to set realistic and achievable goals for realistic lifestyle changes should further motivate her.

P Rx: Client education
1. Diet and hypertension educational materials
2. DASH Food Plan and role of balanced diet
3. Seeking a support system—work/home
4. Food journaling for increased awareness of intake

Will follow up with telephone call in 1 week

Goals:
1. Fruit and vegetable intake of 3 servings/day for 1 week
   Eat a banana or orange daily at work for a snack
   Use pre-packaged cut-up veggies
   One serving of vegetables at dinner
   Homemade soup with vegetables and a 4-ounce glass of grape juice at 3 P.M.
2. Increase awareness of food intake and diet variety

Client Progress Report **Name:** ____________________

DATE	ISSUE	ACTION	OUTCOME (ANTICIPATED)	FOLLOW-UP* 1/15	1/29	2/9
1/5/00	Hypertension	Provide education materials	Pt. understands diet	O	O	O
	35 lbs. overweight Sedentary lifestyle Constipation	Banana or orange for snack Carry veggie packs to work 5 × week	Fiber intake of 15–20 g 2 servings of veg./day	C A	A A	
	Lack of knowledge about nutrition and hypertension Low intake of fruits and veg.	1 veg. with dinner, 5 × week 6 oz. grape juice for breakfast	2 servings of fruit/day Free of constipation 5 lb. weight loss	A O O	A A O	 A O
1/15/00	Low intake of whole grains 33 lbs. overweight Lack of awareness of food intake Abundance of fast food	Track food intake, journal 2 × week Walk for 10 minutes 3 × week 1 serving oatmeal 4 × week with 2 tbsp. All Bran cereal	Pt. increases knowledge of eating patterns 30 minutes moderate exercise/week		O A	O
1/29/00	32 lbs. overweight Low intake of calcium	Track food intake, journal 3 × week Walk for 15 minutes 3 × week Vegetable soup 4 × week Drink calcium-fortified soy milk, 8 oz. /day	3 servings veg./day 45 minutes moderate exercise/week 500 mg calcium/day			A A A
2/9/00	31 lbs. overweight High intake of sodium Fast-food lunches	8 oz. low-fat yogurt for lunch 3 × week Handful of salt free nuts for lunch 5 × week	1,000 mg calcium/day			

*O = ongoing; A = achieved; U = unrealistic.

REVIEW QUESTIONS

1. While engaging in the goal-setting process, what are three messages a counselor should convey to a client?

2. What is an options tool, and how should it be used?

3. Identify three strategies for identifying a specific goal from a broadly stated goal (focus area).

4. Give an example of an anticipated result, broad goal, and a specific goal.

5. Identify three environmental factors that should be taken into consideration when designing a plan of action to implement a goal.

6. Identify the four basic components of a comprehensive nutritional assessment.

7. Identify eight possible functions of a dietary assessment.

8. Name the three steps involved in completing a dietary evaluation.

9. Describe client assessment questionnaires, food diaries, twenty-four-hour recalls, diet history interviews, food frequencies, and usual diet analysis. Explain the advantages and disadvantages of each.

10. Identify three standards that could be used to assess dietary intake for relatively healthy individuals.

11. Name the three components that make up total energy expenditure.

12. Why is thermic effect of food often omitted when calculating total energy expenditure?

13. Identify two weight-for-height tables commonly used for assessment.

14. Explain the Hamwi method.

15. What is the range of desirable BMI?

16. Name two methods commonly used to assess upper abdominal fat accumulation.

17. Explain SOAP documentation method.

18. Explain IAO documentation method.

ASSIGNMENT—Nutritional Assessment

In this assignment you will complete a nutritional assessment, give feedback regarding dietary evaluation, discuss broad general goals if your volunteer wishes, and document the intervention using the SOAP and IOA methods. Since the objective is for you to gain experience in performing common nutritional assessment procedures, you will be completing more tasks than would normally be done in one intervention. There should be no intention on your part to resolve difficulties. Volunteer clients may find some benefit to themselves by clarifying their problems through discussions and feedback; however, the participants should not be led to believe that there will be an intervention. If the person wishes to explore additional nutrition counseling, then a referral to an appropriate health care professional can be made, or the volunteer can be directed to the American Dietetic Association Web site search service (http://www.eatright.org/) for help in finding a nutrition counselor.

Only the involving phase, exploration-education phase, and the closing phase of the nutrition counseling motivational algorithm will be addressed in this assignment. Ask a colleague, friend, or relative who is willing to have an assessment to work with you. Complete the following assessment forms and activities:

PART I. Use the following interview guide/checklist to conduct the interview/assessment with your volunteer. Examples of possible counselor questions, statements, and responses are given in italics. Since the focus of this assignment is nutritional assessment, most relationship building responses have been omitted from the checklist. You are encouraged to use these responses when appropriate. See Chapter 3's assignment for suggested responses.

Preparation

❑ Review the following procedures/guidelines:

- ❍ Obtaining a consent in preparation for session 1 of Chapter 9 of this text.
- ❍ "A typical day strategy" (Chapter 3)
- ❍ Protocol for a diet history interview (Chapter 4)
- ❍ Anthropometric measurement protocols (Appendix C)

- ❍ Hamwi method for calculating ideal body weight (Exhibit 4.2)
- ❍ Daily Food Guide—Food Guide Pyramid (Appendix A) and DASH Food Plan (Appendix B)

❑ Give copies of the Client Assessment Questionnaire and Food Frequency Questionnaire (Lifestyle Management Forms 4.1 and 4.4) to your volunteer to complete before your meeting.

❑ Bring copies of the following forms:

- ❍ Lifestyle Management 3.1, Assessment Ruler
- ❍ Lifestyle Management Forms 4.1, in case your client does not bring
- ❍ Lifestyle Management Form 4.3, Twenty-Four-Hour Recall/Usual Diet Form
- ❍ Lifestyle Management Forms 4.4, in case your client does not bring
- ❍ Lifestyle Management Form 4.5, Food Group Feedback Form
- ❍ Lifestyle Management Form 4.6, Anthropometric Feedback Form
- ❍ Lifestyle Management Forms 3.3, Student Nutrition Counseling Assignment Agreement Form, duplicate copies
- ❍ Completed Certificate of Appreciation
- ❍ Dietary Guidelines Healthy Weight Ranges for Men and Women (Table 4.5)
- ❍ Metropolitan Life Insurance Weight for Height Tables (Appendix D)

❑ Bring a tape measure and a calculator.

❑ Minimize distractions.

❑ Review the motivational nutrition counseling algorithm (Figure 3.2).

❑ Bring visuals to estimate portion size.

Involving Phase

❑ Greeting

❑ Thank volunteer—*Thank you for participating in this interview.*

❑ Set agenda. Explain purpose of the interview—*This is a project I am required to do for my nutrition counseling class. The purpose of this interview is for me to work on my counseling skills, complete a nutritional assessment, give feedback to you regarding your diet, and explore your interest in making dietary changes. I will be taking some physical measurements, and I can give you feedback regarding the standards for your age. These physical standards are a guide for desirable weight; however, other factors such as susceptibility to kidney stones or osteoporosis should be considered before embarking on a change in weight.*

❑ Review the consent form with your volunteer, follow the procedure for obtaining a consent in preparation for session 1 of Chapter 9 of this text, and you and your volunteer should sign both a client copy and a clinic copy of the form. Give the client copy to your volunteer. The clinic copy should be handed in with this report.

Transition to Exploration Phase

❑ Transition statement—*Do you have any questions before we begin the assessment procedure?*

Exploration-Education Phase

❑ Ask your volunteer to describe him- or herself (age, cultural group, occupation, interests)

❑ Ask your client whether he or she has any nutritional concerns.

❑ Review the completed Client Assessment Questionnaire, Lifestyle Management Form 4.1—*Thank you for completing the Client Assessment Questionnaire. I am wondering, what came to your mind as you were filling out this form? What topics covered in this form do you think have particular importance for your food issues?*

Look over the form and ask for clarification where appropriate.

❑ Health history: Inquire whether the client had any nutrition concerns related to health history responses—*I see you stated that you have a family history of heart disease and high cholesterol. Has this influenced your food selections in anyway?*

- ❑ Drug history: If your volunteer is taking a medication, ask the purpose for taking it and if s/he is aware of any nutritional implications of the drug.
- ❑ Socioeconomic history: Comment on highlights of the responses of this section—*I see you frequently eat at fast-food restaurants. Is this just a habit, or is it something you really enjoy doing?*
- ❑ Diet history: Ask for clarification of any significant reporting—*You wrote that you don't eat fruits. Is there a particular reason?*
- ❑ Utilize a typical day strategy—*Can you take me through a typical day in your life so I can understand more fully what happens and tell me where eating fits into the picture? Take me through this day from the beginning to the end.* While your client is telling you his or her story, fill in the Twenty-Four-Hour Recall/Usual Diet Form, Lifestyle Management Form 4.3.
  - ❍ Summarize
  - ❍ Clarify portion sizes and preparation methods in order to complete Lifestyle Management Form 4.3.
  - ❍ Complete food group evaluation servings on Lifestyle Management Form 4.3. Extrapolate client's serving sizes to food group serving sizes.
- ❑ Review the completed Food Frequency Questionnaire, Lifestyle Management Form 4.4. Clarify portion sizes using food models, if needed—*Thank you for completing the Food Frequency Questionnaire. I am wondering, what came to your mind as you were filling out this form? Did you feel a need to clarify or expand on anything while you were completing this form?*
- ❑ Measure or ask your volunteer's height and weight (Appendix D).
- ❑ Measure your volunteer's wrist and waist circumference (Appendix C).
- ❑ Calculate ideal body weight using the Hamwi method (Exhibit 4.2).
- ❑ Complete the Food Group Feedback Form, Lifestyle Management Form 4.5, and Anthropometric Feedback Form, Lifestyle Management Form 4.6.

Provide Feedback

- ❑ Review both feedback forms, point by point, in a nonjudgmental manner with your client. Compare the standards to your volunteer's food intake or to his or her anthropometric findings—*As you can see, your usual vegetable intake is two servings a day, and the Food Guide Pyramid suggests three to five servings a day, and the DASH Food Plan protocol is four to five a day. Your body mass index is 26, and the desirable numbers range from 19 to 25.* Continue in this vein until you have gone over all the findings.
- ❑ Clarify when needed. Your client is likely to ask about the DASH Food Plan, BMI, and others. Be sure you are familiar with the standards so that you can provide educated answers. Again, your answers should not indicate judgment. Avoid the word you—*Body mass index numbers are based on height and weight. Authorities have found that people who have a body mass index between 19 and 25 have a lower risk of developing high blood pressure, high blood cholesterol, diabetes, and coronary heart disease.*
- ❑ Ask your client his or her impression of the evaluation—*What do you think about this information?* Give your opinion if requested.
- ❑ If your client expresses interest in making a change, use the ruler to evaluate willingness to make a change.
- ❑ Summarize.

Closing Phase

- ❑ Express appreciation—*Thank you very much for volunteering for this project.*
- ❑ Give certificate of appreciation—*As a show of gratitude for your willingness to participate in this project, I have a certificate of appreciation to give to you from me and the director of the project.*

PART II. Answer the following questions in a formal typed report or in your journal. Number and type each question and put the answers in complete sentences under the question.

1. Record the name of the person interviewed and location, time, and date of the meeting.

2. Describe the person you interviewed—age, cultural group, gender, occupation.

3. Write a narration of the experience. There should be four titled sections to the narration—preparation, opening phase, exploration-education phase, and closing phase. Summarize what occurred in each phase.

4. Complete a Client Concerns and Strengths Log, Lifestyle Management Form 4.7.

5. Chart your experience twice using the SOAP and IAO format. Use the Client Progress Report, Lifestyle Management Form 4.8, for the IAO charting.

6. Complete an Interview Checklist, Lifestyle Management Form 8.1. Do not fill out portion C of the checklist.

7. What did you learn from this experience?

8. Attach completed copies of Lifestyle Management Forms 4.1, 4.3, 4.4, 4.5, 4.6, 4.8, 3.3, and 8.1.

SUGGESTED READINGS, MATERIALS, AND INTERNET RESOURCES

Nutritional and/or Dietary Assessment

Grant A, DeHoog S. *Nutrition Assessment Support and Management.* 5th ed. Seattle, WA: Grant & DeHoog; 1999.

How to Keep a Food Diary. National Health Video, Inc. Behavior Modification Series. Order by calling (213) 472-2275.

Lee DR, Nieman D. *Nutritional Assessment.* 2d ed. St. Louis, MO: Mosby; 1996.

Simko MD, Cowell C. *Nutrition Assessment: A Comprehensive Guide for Planning Intervention.* 2d ed. Gaithersburg, MD: Aspen; 1995.

Thomas PR (Ed.). *Weighing the Options—Criteria for Evaluating Weight-Management Programs.* Washington, DC: National Academy Press; 1995. A useful assessment instrument to point out potential problems with motivation and attitudes toward weight loss diets and exercise, and a psychological assessment to determine the need for a psychological referral.

The 24-Hour Food Recall. 1998. Oklahoma State University, 315 HES Building, Stillwater, OK 74078-6163; (405) 744-6283. A $35 video and instruction manual to promote twenty-four-hour recall skill development.

REFERENCES

[1] Berg-Smith SM, Stevens VJ, Brown KM, Van Horn L, Gernhofer N, Peters E, Greenberg R, Snetselaar L, Ahrens L, Smith K. Dietary Intervention Study in Children (DISC) Research Group. A brief motivational intervention to improve dietary adherence in adolescents. *Health Educ Res.* 1999;14:101–112.

[2]Botman JA. Seven steps toward behavioral compliance. *Behav Health Treat.* 1997;2:8.

[3]Laquatra I, Danish SJ. Practitioner counseling skill in weight management. In: Dalton S, ed. *Overweight and Weight Management: The Health Professional's Guide to Understanding and Practice.* Gaithersburg, MD: Aspen; 1997:348–371.

[4]Council on Practice, Quality Management Committee. ADA's definitions for nutrition screening and nutrition assessment. *J Am Diet Assoc.* 1994;94:838–839.

[5]Grant A, DeHoog S. *Nutrition Assessment Support and Management.* 5th ed. Seattle, WA: Grant & DeHoog; 1999.

[6]Lee DR, Nieman D. *Nutritional Assessment.* 2d ed. St. Louis, MO: Mosby; 1996.

[7]Thompson FE, Byer T. Dietary assessment resource manual. *J Nutr.* 1994;124 (suppl):S2296–S2298.

[8]Simko MD, Cowell C, Gilbride JA. *Nutrition Assessment A Comprehensive Guide for Planning Intervention.* 2d ed. Gaithersburg, MD: Aspen; 1995.

[9]Roman-Shriver CR. The nutritional assessment. In: Israel D, Moores S. eds. *Beyond Nutrition Counseling: Achieving Positive Outcomes Through Nutrition Therapy.* Chicago: American Dietetic Association; 1996:15–30.

[10]Sugerman SB, Eissenstat B, Srinith U. Dietary assessment for cardiovascular disease risk determination and treatment. In: Kris-Etherton P, Burns JH, eds. *Cardiovascular Nutrition.* Chicago: American Dietetic Association; 1998:39–71.

[11]Windhauser MM, Ernst DB, Karanja NM, Crawford SW, Redican SE, Swain JF, Karimbakas JM, Champagne CM, Hoben, KP, Evans MA. Translating the dietary approaches to stop hypertension diet from research to practice: Dietary and behavior change techniques. *J Am Diet Assoc.* 1999;99(suppl):S90–S95.

[12]Thomas PR, ed. *Weighing the Options Criteria for Evaluating Weight-Management Programs.* Washington, DC: National Academy Press; 1995.

[13]White KV, Dwyer JT, Posner BM, Ham RJ, Lipschitz DA, Wellman NS. Nutrition screening initiative: Development and implementation of the public awareness checklist and screening tools. *J Am Diet Assoc.* 1992;92:163–167.

[14]Joyce EH, Williams GS. *The 24-Hour Food Recall: An Essential Tool in Nutrition Education.* Stillwater: Oklahoma State University; 1998.

[15]Hammond, KA. Dietary and clinical assessment. In: Mahan L K, Escott-Stump S, eds. *Krause's Food, Nutrition, and Diet Therapy.* 10th ed. Philadelphia: Saunders; 2000.

[16]Ammerman AS, Haines PS, DeVellis RF, Strogatz DS, Keyserling TC, Simpson RJ, Siscovick DS. A brief dietary assessment to guide cholesterol reduction in low-income individuals: Design and validation. *J Am Diet Assoc.* 1991;91:1385–1390.

[17]Cummings SR, Block G, McHenry K, Baron RB. Evaluation of two food frequency methods of measuring dietary calcium intake. *Am J Epidemiol.* 1987;126:796–802.

[18]Block G, Hartman AM, Dresser CM, Carroll MD, Gannon J, Gardner L. A data-based approach to diet questionnaire design and testing. *Am J Epidemiol.* 1986;124:453–469.

[19]Willet WC, Reynolds RD, Cottrell-Hoehner S, Sampson L, Browne ML. Validation of a semi-quantitative food frequency questionnaire: Comparison with a 1-year diet record. *J Am Diet Assoc.* 1987;87:434–447.

[20]Whitney ER, Cataldo CB, Rolfes SR. *Understanding Normal and Clinical Nutrition.* 5th ed. Belmont, CA: West/Wadsworth; 1998.

[21]Jonnalagadda SS, Mitchell DC, Smiciklas-Wright H, Meaker KB, Van Heel N, Karmally W, Ershow AG, Kris-Etherton PM. Accuracy of energy intake data estimated by a multiple-pass, 24-hour dietary recall technique. *J Am Diet Assoc.* 2000; 100:303–308,311.

[22]Kittler PG, Sucher KP. *Food and Culture in America: A Nutrition Handbook.* 2d ed. Belmont, CA: West/Wadsworth; 1998.

[23]Cassidy CM. Walk a mile in my shoes: Culturally sensitive food-habit research. *Am J Clin Nutr.* 1994;59(suppl):S190–S197.

[24]Johnson RK. Energy. In: Mahan LK, Escott-Stump S, eds. *Krause's Food, Nutrition, and Diet Therapy.* 10th ed. Philadelphia: Saunders; 2000:19–30.

[25]Insel PM, Turner RE, Ross D. *Nutrition.* Boston: Jones & Bartlett; 2001.

[26]Shape Up America and American Obesity Association. *Guidance for Treatment of Adult Obesity.* Bethesda, MD: Shape Up America; 1997–1998.

[27]Whitney ER, Rolfes SR. *Understanding Nutrition.* 8th ed. Belmont, CA: West/Wadsworth; 1999.

[28]National Institutes of Health (NIH) Obesity Health Initiative. *Clinical Guidelines on the Identification, Evaluation, and Treatment of Overweight and Obesity in Adults,* NIH Publication No. 98-4083. Washington, DC: U.S. Department of Health and Human Services; 1998.

[29]Klein CJ, Bosworth JB, Wiles CE. Physicians prefer goal-oriented note format more than three to one over other outcome-focused documentation. *J Am Diet Assoc.* 1997;97:1306–1310.

[30]Brylinsky C. The nutritional care process. In: Mahan LK, Escott-Stump S, eds. *Krause's Food, Nutrition, and Diet Therapy.* 10th ed. Philadelphia: Saunders; 2000:431–451.

[31]Curry KR, Jaffe A. *Nutrition Counseling and Communication Skills.* Philadelphia: Saunders; 1998.

[32]Charles EJ. Charting by exception: A solution to the challenge of the 1996 JCAHO's nutrition care standards. *J Am Diet Assoc.* 1997;97(suppl 2):S131–S138.

Chapter

5

PROMOTING CHANGE TO FACILITATE SELF-MANAGEMENT

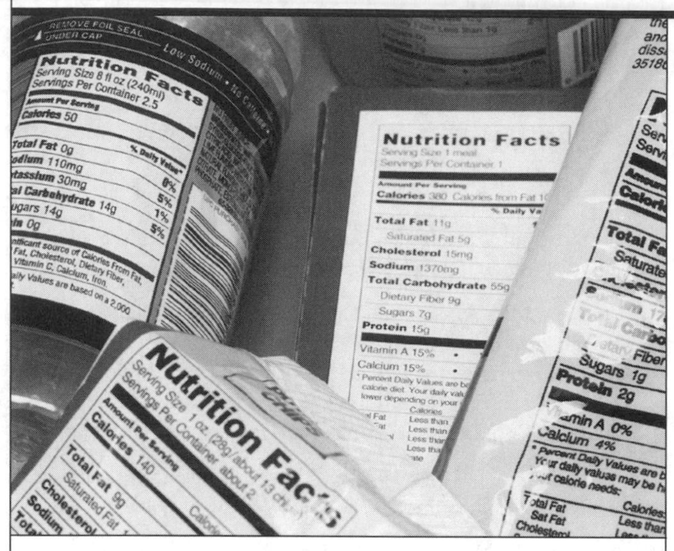

Habit is habit, and not to be flung
out of the window by any man,
but coaxed downstairs a step at a time.
—Mark Twain

BEHAVIORAL OBJECTIVES:

- Utilize common food management tools.
- Identify various methods used to track food behavior.
- Identify five major assumptions regarding adult learners.
- Explain the integrated approach to adult education.
- Identify effective ways to enhance education in a nutrition counseling session.
- Discuss factors affecting dietary adherence.
- Explain the ABCs of eating behavior.
- Utilize common behavior change strategies.

KEY TERMS:

- **ABCS OF BEHAVIOR:** antecedent, behavior, and consequence; used to describe behavior chains
- **BARRIERS:** obstacles that hinder accomplishment of a goal
- **BEHAVIOR CHAINS:** a sequence of events that explains recurrence of behavior
- **CONTRACT:** a formal agreement to implement a goal
- **COUNTERING:** substituting healthy responses for problem behaviors
- **CUE MANAGEMENT (STIMULUS CONTROL):** addresses antecedents of a behavior chain; technique involves using cues to increase or decrease a particular behavior
- **DAILY FOOD GUIDE:** sorts food into five major food groups
- **DASH FOOD PLAN:** food group developed to lower high blood pressure
- **EXCHANGE LISTS:** food management tools that organize foods by their proportions of carbohydrate, protein, and fat
- **JOURNALING:** a tracking method used to analyze and modify behavior
- **MODELING:** behaviors learned by observing and imitating others
- **REINFORCEMENT** or **REWARDS:** tangible or intangible incentives to encourage a behavior change

STRATEGIES TO PROMOTE CHANGE

Nutrition counselors need to become familiar with a variety of strategies for promoting behavior change to meet their clients' needs. Since counselees differ in their predisposition and ability to use certain strategies, it is best to have several alternatives available. Also, some clients may need additional options for tackling difficult situations, particularly as circumstances change. For example, in the beginning of a dietary behavior change, it may be possible to avoid exposure to foods or occasions where temptations exist. That condition is not likely to continue indefinitely and clients should be ready with several coping mechanisms. To explain the need to learn new strategies, the following script has been developed by Tsoh et al.:[1]

Counselor: *Having more than one or two strategies is very important for you. What we are trying to do is expand your toolbox, so that if one tool does not work, you will have another. In many situations, you may require a combination of various tools. (p. 25)*

Food Management Tools

A variety of tools are available to aid in the management of eating behavior (see Table 5.1). Both professional and commercial organizations have developed techniques to help design food plans and track eating behavior. Clients differ in their desirability for structure and their ability to work with specific tools. To determine whether a diet plan more involved than goal setting is advisable, you must explore the issue with your client. To encourage clients to think about this topic, the

TABLE 5.1—Comparison of Food Management Tools

METHOD	ADVANTAGES	DISADVANTAGES
Detailed menus/meal plans	• Clearly defined • Useful for someone who expresses a need for a lot of structure • Useful for someone who has complex dietary requirements who is not ready or not capable of following a food group plan	• Does not allow for spontaneous events • Food items needed for the plan may not be available • May be difficult to design so as to complement a client's lifestyle
Exchange lists	• Offers choices • Provides structure • Allows for variety • Meal pattern is individualized	• May be too complex for some individuals
Daily Food Guide/Food Guide Pyramid	• Easy to understand • Flexible • Pyramid depicts proportions of food groups	• Does not indicate servings of whole grains
DASH Food Plan	• Easy to understand • Flexible • Rich in various nutrients believed to benefit good health	• Some of the foods may not be part of a client's usual intake (e.g., nuts, beans, and seeds)
Goal setting	• Easy to understand • Flexible • Designed to take into consideration a client's lifestyle	• Approach may move too slowly when aggressive intervention is needed

Client Assessment Questionnaire (Lifestyle Management Form 4.1 in Appendix G), has questions asking clients regarding the amount of structure they desire.

One of my clients was an elderly female who had kidney disease and had a number of factors in her diet that had to be monitored. It was my responsibility to design a discharge diet plan for her. She was a very nice person, but she told me she would never be able to understand the diet since she had never been good with numbers. We scheduled to have her son meet with us for another time and all together we designed a detailed food plan.

Just tell me exactly what to eat. I want a detailed plan. Sometimes clients are overwhelmed with the diagnosis of a new disease or have been frustrated with previous attempts to follow a therapeutic diet plan on their own, and they are looking for a great deal of structure. Help in designing up to twelve weeks of meal plans complete with recipes and a shopping list can be found on several Web sites. Many of these sites can develop meal plans for common dietary restrictions, such as low sodium, low fat, and high fiber. (See the resources at the end of the chapter.) Care should be taken to design a plan based as much as possible on your client's food preferences and lifestyle patterns. In fact, the best way to use Web sites with predesigned meal plans is to have your client sit at the computer with you to work out a week's plan while discussing upcoming life events for the next seven days. If you have a capable client, this activity can be repeated at weekly intervals until the client can take over complete responsibility.

One of my clients was a man who had been told by his doctor to lose weight. He said he had been on diets before and did not want a diet where he had to choose something from columns or groups. He said, " Just tell me exactly what to eat." I told him my concerns about boredom or the need to make adjustments due to events or food availability. He was adamant and his wife, who was also at our meeting, assured me that she would be sure to always have needed food accessible. During this meeting, we were only able to design three days' worth of meals. At our next appointment, which was in two weeks, he was happy to report a loss of 6 pounds. but he said that he really wanted more variety. We developed three more days of detailed eating plans that provided some flexibility for fruits and vegetables. As he became familiar with the new eating pattern and serving sizes, his program evolved into a more general, flexible plan based on food groups. Actually this ended up being just the type of food group plan he said he did not want. This client seemed to need a highly structured plan to begin a program.

I want a lot of structure but freedom to select foods. The Exchange Lists for Meal Planning in Appendix F would probably work for someone who wants a lot of structure but freedom to make some choices. The American Diabetes Association and the American Dietetic Association have developed two versions of this system: one is oriented to diabetic meal planning, and the other can be applied for more general food management needs. The exchange system booklet geared to nondiabetic meal planning stresses healthful eating, regular activity, goal setting, and lifestyle change. **Exchange lists** are organized into three main groups:

- Carbohydrate group—includes starch, fruit, milk, other carbohydrates, and vegetable exchanges
- Meat and meat substitutes group—includes very lean, lean, medium-fat, and high-fat exchanges
- Fat group—includes monounsaturated, polyunsaturated, and saturated fat exchanges

Modifications of the exchange system are often available in clinical settings to address a variety of medical needs. For example, exchanges can be altered to take into account intake of potassium, fluid, or protein. See the resources at the end of the chapter for ordering exchange system materials.

I want some structure and freedom to select foods. Either the Daily Food Guide (Food Guide Pyramid) found in Appendix A (see also http://www.nal.usda.gov/fnic/dga) or the DASH Food Plan found in Appendix B (http://dash.bwh.harvard.edu/ dashdiet.html) can be used to design a flexible but somewhat structured program. The **Daily Food Guide** organizes food into groups based primarily on their protein, vitamin, and mineral content. The **DASH Food Plan** is similar to the Food Guide Pyramid, but recommendations provide for an additional food group consisting of legumes, seeds, and nuts, and there is a lower limit of servings for the meat, poultry, and fish group. The DASH Food Plan is an example of a way to accomplish a higher fiber intake through increased selections of fruit and vegetable groups. It was developed for individuals with high blood pressure; however, it provides a healthy food plan for anyone.

A number of educational resources are available for either protocol and can be obtained from their respective Web sites, or see the resources section at the end of the chapter for ordering materials. Lifestyle Management Form 4.3 contains an assessment section for the Daily Food Guide, and Forms 4.4 and 4.5 are useful for both the DASH Food Plan and the Daily Food Guide dietary evaluation.

One of my clients who used journaling successfully stated, "Some people say keeping a diary is too much work, too stressful, but I found it much more stressful before journaling when I was eating well and exercising and gaining weight. Now that we worked out a diary method that works for me, it does not take much of my time, and I am able to eat well and have foods that I enjoy."

I don't want to follow a diet. I just want to eat better. For clients who do not want any type of structured eating plan, lifestyle changes can be made solely through goal setting. This approach complements nondieting programs for weight management, which have a significant voice among nutrition professionals.[2] Advocates of this approach maintain that diets to lose weight are not reliable and could lead to health problems and eating disorders.[3] The goals of these programs are to focus on healthier eating, enjoyable physical activity, and a positive self-image.[4] However, the nondiet approach can be applied to a variety of dietary objectives that do not focus on weight issues, such as reducing sodium intake or developing a vegetarian lifestyle.

Tracking

No matter what food management system has been selected, your clients should be encouraged to keep track of their progress. Without a self-monitoring method, it becomes difficult to evaluate attainment of food behavior goals.

Tracking methods vary in complexity. The method selected should depend on your client's ability to work with structure and details. Some people enjoy recording everything they eat and reviewing the material at the end of the day; others become quickly frustrated with the procedure. Discuss with your client several options.

Journaling Journaling is a tracking method that has consistently been shown to be effective in altering behavior in general and food habits in particular.[5,6] Recording behavior becomes a self-management tool by increasing awareness and providing a "time-out" for making a decision. The most effective time to record food intake is immediately before or after eating. The short time spent on reflection may well result in taking an action in keeping with one's lifestyle objectives.

A number of factors related to food intake can be recorded in a journal depending on the client's motivation and journaling objectives. At the very least a food diary would include a list of foods consumed and may also contain portion size, calories, and/or fat grams. Some behavior management programs have encouraged participants to record time, place, the presence of others, mood, thoughts, concerns, degree of hunger, and activities at the time of food consumption.[7] As can be expected, client resistance is likely to mount with increasing requirements for journal entries. However, the need to record the psychosocial variables probably declines for most individuals after a few weeks.

Analysis of the records can help counselors and clients develop new goals. Streit et al.[8] found the following procedures successful in guiding clients who chose journaling to help manage their food intake:

- **Provide training.** Clients gain a better understanding of the process when hands-on activities are included as well as complete instructions of how to keep a journal. Although clients should be encouraged to measure what they eat whenever possible for at least a few weeks, the training should include a review of estimation of portion size since measuring will not always be possible. An education method found to be successful in one study included a review of participants' food diary entries of a sample meal by dietitians.[8]
- **Use estimates.** Sometimes clients can become very frustrated with writing down exact amounts of food consumed and then calculating caloric content and/or percentage of calories from fat, especially when consuming mixed dishes that the client did not prepare. Therefore, to reduce anxiety, clients should understand that approximations are acceptable.
- **Set meaningful and achievable goals.** Clients are more likely to be successful in record keeping if the first goal is modest, such as two days a week, and then gradually increase journaling activity in subsequent weeks. Streit et al.[8] found successful dietary

changes with as few as two days of records per week. However, more frequent journaling is associated with even greater food behavior change.[9]

The anxiety of record keeping is likely to be reduced if there is a degree of flexibility enabling clients to feel they are "off the hook" for complicated meals or stressful days.

- **Provide a variety of record-keeping options.** Some clients prefer to use a pocket calendar with enough space to record a days worth of food intake. Others like to use 8½ by 11-inch forms. At times clients will devise their own record-keeping system.
- **Provide nonjudgmental feedback.** Clients should always be praised for their journaling work, and counselors should always review their clients' journals with no hint of criticism. Client thoughts regarding the journaling activity should be sought. In a study reported by Streit et al.,[8] the most effective technique for motivating participants to keep food records was to have dietitians review the journals and return them with brief comments. Clients were also encouraged to write questions in their diaries.

I am so pleased when clients keep food records because I know it is one of the best ways to change and maintain food behavior. I always review the food records with my clients, and I often put stars in the journal on entries that indicate that their goals were kept.

Journaling Alternatives There is more than one approach to journaling. Counselors and clients should think creatively to find ways to make it work. Here are some ideas:

- **Checking off.** For clients following food group plans or the exchange system, checking off boxes or crossing out slashes throughout the day can be used. Names of exchanges or food groups could be written on a form, or the names could be abbreviated and written each day in a pocket calendar. This methodology may appeal to some who resist writing; however, there would be a reduction of information collected for evaluation.
- **Messaging.** Hand-held, voice-activated recorders or leaving messages on a confidential voice mail system has been reported to work successfully.[10]
- **Using art.** Drawing pictures, scribbling, or choosing colors has been used to assist clients in getting in touch with their feelings or moods while consuming food.[10]
- **Empty bowl.** Put desired food objectives in a visible spot such as a bowl in the kitchen at the beginning of the day or week. Goals are assessed according to the amount in the bowl at the end of the day or week.
- **Electronic note pads.** Tracking in electronic note pads can be used for clients comfortable with the technology.

Clients should be encouraged to develop systems. Consider the first-person account at the bottom of the page about a client who did just that.

COMPONENTS OF EFFECTIVE INTERVENTIONS

You cannot teach a man anything;
you can only help him to find it within himself.
—Galileo

Education

The primary step to changing dietary behavior and maintaining dietary objectives is education.[11] Clients must understand why dietary change is important and be informed of pertinent nutrition information

One of my clients who had made many successful changes in her food behavior developed a creative system that allowed her to eat all the foods she enjoyed. This knowledgeable client found that after menopause and chemotherapy for breast cancer, her weight steadily climbed despite exercising regularly and eating well. She bought index cards of various colors. Each color represented a particular food group that was pertinent to her food goals—dairy, fish, legumes, nuts, grains, vegetables, fruits, and other. On the appropriate color card, she wrote the name and quantity of the food item that she planned to eat. On the top of the card, the amount of calories, calcium, fat, fiber, and sodium in the particular food was recorded. Home-prepared food combinations were grouped on one card. See Exhibit 5.1 for an example of the card she prepared for her oatmeal breakfast. She reported:

I made cards for frequently consumed individual foods and meals. Often in the morning, I select the cards that I will need for the day. If there is no card for a particular food I plan to eat, I make a new card. As I eat a food, I move the corresponding card into an envelope. At midday or near the end of the day, I review what I have eaten and plan the rest of the day accordingly. This process only takes me a couple of minutes. At the end of the day I record my calorie intake and exercise in a small calendar and calculate my calcium intake. Then I take a calcium supplement if needed. Lately I also have been writing in short notation the names of the foods I ate during the day. Sometimes I skip a day of food journaling if it is just too difficult to figure out calories. In the past I had tried keeping a piece of paper or a small booklet with me to write down the foods I was eating and to keep track of calories. Using the cards has worked much better for me.

EXHIBIT 5.1—Food Group and Nutrient Tracking Card

Grain Group (nuts, dairy, fruit)	***136 / mg sodium***	***9 / g fiber***	***17 / g protein***	***13 / g fat***	***300 / mg calcium***	***506 calories energy***
¾ c dry oatmeal	—	6	7.5	4	—	225
1 c 1% milk	125	—	8	2.5	300	110
½ t honey	—	—	—	—	—	20
½ apple	—	2	—	—	—	40
2 t brown sugar	—	—	—	—	—	30
1 T ground flax seed	10	—	—	2	—	30
1 T chopped mixed nuts	1	1	1.5	4.5	6	51

to be capable of making informed decisions to change their behaviors. In a study designed to measure the key determinants of satisfaction with diet counseling, patients identified knowledge along with facilitative skills as the most important components of their counseling experience.[12] Patient education was an integral part of the successful Diabetes Control and Complication Trial.[13] In an assessment of the avoidability of long-term complications of diabetes, Nicolucci et al.[14] found a fourfold increased risk of major complications in patients who never received diabetes education.

Adults as Learners Fundamental to the process of educating adults is understanding the learning needs of mature individuals. Characterizing adult learning process as *andragogy* (as compared to pedagogy), Knowles et al.[15] have identified five major assumptions regarding adult learner needs:

- Adults think of themselves as *self-directed* learners, versus children, who are dependent learners. An adult expects to be treated as an independent, responsible person capable of making informed decisions. During counseling, the self-concept of adults is threatened if they are treated as dependent children, incapable of making important decisions. Adults are not likely to respond well to a reprimand or being "talked down to." To promote a collaborative approach, adults should be involved in formulating learning objectives.
- Adult learners bring a vast amount of *experience* into the learning environment. Nutrition counselors can enhance learning by recognizing past experiences as significant resources to be expanded upon. Clients who have followed a low-fat diet for many years may feel annoyed if their past experiences and knowledge are not taken into account. This emphasizes the need for an assessment. Also, incorporating life experiences is believed to enhance the motivation to learn.[16]
- Readiness to learn is dependent on the acceptance of the *necessity to learn*. In contrast to children, who are expected to learn what adults chose for them, adults are not likely to put in the time and effort unless they are convinced that a need exists.

BEETLE BAILEY

Reprinted with special permission of King Features Syndicate.

- Adult learners enter into a learning experience with a *problem-centered, present-orientation focus.* Children learn about subjects that do not necessarily have a clear application in the present; however, adults in a counseling environment are not likely to respond to learning activities unless they are geared to solving a problem or to performing a task.
- Adult learners tend to be *intrinsically motivated.* Potent motivators for adults include internal pressures such as wanting a better quality of life or increased self-esteem, as compared to children, who are usually extrinsically motivated by teachers, parents, and others. Nutrition counselors need to make clear to clients how the quality of their lives is likely to improve with making lifestyle changes. To help clients to get in touch with their personal motivators and counseling goals, sometimes counselors have asked them to write a personal mission statement for counseling.

Effective Education Strategies The education component of nutrition counseling is no longer viewed as simply disseminating information to increase knowledge. The objective is to focus on addressing the educational needs of the client and increasing knowledge in a manner that facilitates behavior change. In recognition of this concept, recent versions of the manual for organizations seeking program recognition from the American Diabetes Association now substitute the term *self-management education* for *education.*[17]

Educational targets have been linked to specific educational interventions in Table 5.2. Funnell and Anderson[18] advocate an integrated approach to education so clients can make informed decisions about their behaviors. This means addressing psychosocial concerns and initiating behavior change strategies before pouring educational content into an "empty bucket." This methodology interfaces very well with the counseling objectives covered in Chapter 3 and includes the following:

- Review role of client as self-manager and role of counselor as a source of expertise, support, and inspiration.
- Elicit client concerns and questions.
  - Discuss clients' experiences and understanding of their condition.
  - Identify what the client wants from the counselor.
  - Ascertain educational topics client would like addressed.
  - Explore behaviors the client wishes to alter.
- Present information to address concerns and questions.
- Discuss strategies to address the behavioral aspects of the concerns.

I hear and I forget,
I see and I remember,
I do and I understand.
—Confucius

TABLE 5.2—Linkage of Educational Targets and Interventions

TARGET	INTERVENTION
Knowledge, beliefs	*Didactic education:* increasing awareness of risks and benefits; helping clients know how to make appropriate self-care decisions
Skills	*Demonstration/feedback:* showing how to execute skills; observing performance, correcting errors
Intentions	*Goal setting:* establishing specific and appropriate goals that are ambitious but realistic; behavioral contracting to increase commitment
Barriers	*Problem solving:* helping clients find ways to overcome barriers to implementing intentions
Self-efficacy, burnout	*Support/counseling:* helping clients maintain positive emotional well-being

Source: Adapted from Peyrot M, Behavior change in diabetes education. *Diabetes Educator.* 1999;25(suppl 6):62–73. Used with permission.

To enhance the educational impact and to support different learning styles, varied approaches and active learning experiences can be incorporated into counseling sessions. See Exhibit 5.2 for a list of interactive approaches that have been used successfully in nutrition counseling sessions. Holli and Calabrese's[19] analysis of progressive strengtheners for remembering gives support for an interactive educational approach. They found that people remember what they:

Read	10 percent
Hear	20 percent
See	30 percent
Both see and hear	50 percent
Say	70 percent
Both say and do	90 percent

EXHIBIT 5.2—Interactive Educational Experiences

Here are some hands-on experiences that counselors can share with clients:

- Grocery store tour
- Cooperative cooking
- Cafeteria meal
- Fitness trail walk
- Trip to a gym
- Practice selecting items from a menu—circle high-fat foods on a menu
- Interpret food labels—compare the labels of two similar products
- Jointly modify recipes (have client bring recipes)
- Create menus
- Measure and weigh portion sizes
- Analyze blood glucose records of previous clients
- Role playing
- Simulations

Getting the Message Across

- **Avoid technical jargon.** Calibrate your use of technical terms to the background of your client. Generally technical jargon, such as *hypertension* for high blood pressure, should be avoided. Use low-literacy materials when appropriate.
- **Simplify directions.** Concise, straightforward instructions with information about actual choices (such as items or brands) are more likely to be followed than complex regimens.[20]
- **Incorporate self-help materials.** To support the educational process, nutrition counseling programs have successfully incorporated self-help materials, such as workbook activities, for highly motivated clients who report few external stressors.[21] See the resources at the end of the chapter for some self-help references.
- **Repeat important points several times.** Explain important points in several ways and vary learning experiences.
- **Limit the number of learning objectives per session.** Too many learning objectives produce information overload, dilute important messages, and cause confusion.
- **Organize material in a logical manner.** Generally, the first third of an information giving session is remembered best.[22] Use organizing terminology, such as "We will go over three ways to reduce cholesterol. First, . . ."
- **Check for understanding.** When giving factual data, be sure that the client understands what you are saying, especially before starting to cover a new topic. Watch for verbal and nonverbal cues or ask a question, such as "Do you understand?" or "Do you have any questions?"
- **Incorporate significant others.**
- **Utilize visuals.** Discussions of important concepts can be supplemented with anatomical models, videotapes, displays, pictures, diagrams, charts, audiotapes, and so forth.
- **Provide meaningful support materials.** To help memory and encourage the processing of information after leaving a counseling session, supportive reading material can be beneficial. This is particularly important when the client is feeling stressed. Studies of patient education have shown that clients typically forget half the information presented to them within five minutes. However, clients should not be confounded with an abundance of fact sheets, brochures, recipes, and coupons. Feeling overwhelmed may lead to an inability to take action. Handouts should be geared to a particular educational objective and tailored to your client's needs. The lifestyle behavior change program at Health Partners, Inc., in Minneapolis leaves large blank spaces on their handouts for dietitians to write personalized messages for their clients.[23]
- **Disperse information over a period of time.** A planned educational experience can be designated for part of a counseling session (for instance, portion size activities); however, education can be introduced or

reinforced throughout a session when the need or opportunity arises. King[10] refers to these responses as *sound bites*. Used appropriately, they can affirm a client's behavior. For example, a client who enjoys eating chocolate could be told that chocolate contains caffeine and a chemical that enhances the feeling of well-being. However, a counselor should be sure of the educational value of the sound bite since overuse of this method could interfere with the progress of counseling.

- **Use stories, examples, personal accounts, and comparisons.** These are aids to enhance an educational experience or to understand complex material. Exhibit 5.3 provides an example of using a comparison. The learning value of educational aids will be improved when they mesh with a client's cultural orientation.

Positive or Negative Approach

The issue as to whether a health risk message regarding dietary behavior should be cast in a positive or a negative light has not been resolved. For example, the focus on increasing fruit and vegetable intake could be an emphasis on decreasing the risk of cancer or on looking and feeling better. Brownell and Cohen[11] point out that a moderate amount of fear appears useful for motivating a client to make a change. There would be no reason to change if there is too little fear; however, too much fear provokes denial and encourages attention to be directed elsewhere. Snetselaar[22] notes that clients should not be protected from negative information and that they have a right to all relevant facts. As a counselor, you will have to evaluate the situation as too how much emphasis should be placed on the negative aspects of your client's condition.

Once your client has started to implement a lifestyle change, there should be a strong emphasis on the positive. Berry and Krummel[20] advocate a positive approach with clients by emphasizing foods to include rather than those that should be avoided. Also, they advocate positive reinforcement by acknowledging, praising, and encouraging clients when they make desirable changes, no matter how small.

EXHIBIT 5.3—Example of Using a Comparison

Nancy: I don't even think anything is wrong. I think they might have made a mistake. I feel good.

Counselor: Actually, with high blood pressure you may not feel a thing. But that doesn't mean there isn't anything going on that can't eventually make you feel bad. When you were growing up and in school, did you ever get a callus on your finger from writing all the time?

Nancy: Yeah. As a matter of fact, I have a callus there now. I write orders all night long, and I have this ugly hard spot from my pen.

Counselor: That hard spot is from the pressure of your pen pressing against your finger while you are writing. The pressure causes scarring or hard tissue to form. A similar type of thing happens to your blood vessels when you have high blood pressure for years. They harden—it is called arteriosclerosis.

EXERCISE 5.1—Enhancing Learning

Write in your journal what enhances your learning experiences. Identify two things you do to aid your memory.

SUPPORTING SELF-MANAGEMENT

Following diet orders has been a challenge since the beginning of time. Traditionally adherence or compliance has been defined as following advice, recommendations, diet orders, or a prescribed regimen. A better definition, reflecting today's cooperative approach toward nutrition counseling, would be to describe the degree to which an individual's dietary behaviors coincide with the dietary objectives as set by clients in collaboration with their health practitioners. This definition takes into consideration a more positive and accurate change process.

Although the words *adherence* and *compliance* are often used interchangeably, the term *dietary adherence* is generally preferable. For many, the word *compliance* conjures up an image of an authoritarian counselor dictating dietary orders and expecting obedience. More recently the term *concordance* has been suggested.[24]

Journal articles about compliance often cite that 50 percent of patients adhere to long-term drug regimens, and even a smaller percentage follow advice to change their lifestyle.[25] Such statistics have prompted investigation to find new ways of interacting with clients in the health arena. Historically, clients have simply been given information and advice and then labeled as noncompliant if they did not follow the advice. It was assumed that such difficult clients had their own good reasons or were not motivated due to a character flaw. This old standard is no longer acceptable, and authorities have been searching and analyzing motivational theories and approaches to find ways to improve health behavior practices.

Individualization of Therapy

Nutrition counseling should be tailored to meet client needs, goals, and living arrangements to enhance adherence.[12,21] Clients want to eat foods that taste good and develop a dietary pattern that can realistically fit into their lifestyle. King and Gibney[26] found that dietary advice to lower fat intake was more successful when existing eating frequency patterns were taken into consideration.

Dietitians can generally find ways to include favorite foods into a diet pattern.[27] Positive emphasis can be placed on which foods to add or substitute, rather than which foods should be avoided.[28] Evaluation of dietary satisfaction in the Modification of Diet in Renal Disease Study found that study participants who enjoyed eating their diets with the level of protein they were allowed were more likely to adhere to the regimen.[29]

Although food habits are highly resistant to change, it is encouraging that once a new habit has been fully adopted, our taste preferences can change.[30] For example, a person who formerly preferred the taste of whole milk may well prefer the taste of skim milk after consuming it over a period of time. Also, frequently people who follow a sodium-restricted diet come to prefer low-salt foods to higher-salt foods.

Generally, gradual stepwise modifications are recommended for changing dietary patterns that will endure.[31] Slower changes allow for the nutrition counselor to help tailor adjustments to the taste preferences of the client. However, some highly motivated clients may be capable of making substantial changes. Barnard et al.[32] reviewed thirty published research trials designed to reduce risk of heart disease and found that studies setting relatively strict limits of fat intake achieved a greater degree of dietary change than those with modest dietary goals.

Length and Frequency of Counseling Sessions

Length of visits and number of sessions needed to produce favorable outcomes vary with complexity of the problems. One study of non-insulin-dependent diabetes mellitus (NIDDM) patients found that any contact with a dietitian produced better medical outcomes than no contact.[33] Nutrition practice guidelines for medical nutrition therapy for NIDDM patients include at least three sessions after diagnosis and an ongoing visit once every six to twelve months.[34] Four to five sessions have been reported as necessary for cardiac patients to achieve desired results.[20] King[10] notes that in her private practice she may see clients over a period of one to five years to help them fully achieve their goals.

Perception of Quality of Care

Perception of quality of care is highly related to adherence.[35] A warm, caring environment created by the counselor, staff, and physical setting creates a setting conducive to counseling.[19] The same counselor should be seen at each visit. The exact physical surroundings may vary since nutrition counseling sessions are conducted in a variety of settings, including clinics, private offices, fitness centers, hospital rooms, and work site locations. Attempt to arrange a meeting place that is attractive, comfortable, quiet, well ventilated, adequately lighted, and private.[20] Be sure sturdy seating is available for large individuals. Do not allow big obstacles, such as a desk, to be a barrier between a counselor and a client. If a desk is a necessity, then have the client sit alongside it, not behind it. If a table is needed to view materials, a round table is preferable because it avoids the head-of-the-table position.[19]

The essential consideration is to provide an environment as free of distractions as possible.[36] If the meeting place is not ideal, search for innovative ways to rearrange the environment. In a clinic, creative placement of a bookcase or a plant could help define space and give the illusion of privacy. In a hospital room, this may mean pulling a privacy curtain, asking the client to turn off the television, and pulling a chair near the patient's bed. Every effort should be made not to stand while a client reclines in a bed. If the patient is ambulatory, a conference or counseling room may be available for meetings.[37]

Arrangements should be made so phone calls and staff members do not disturb sessions. If a phone call or a colleague does interrupt a session, make every effort to discontinue the intrusion and then make an apology. Finally, clients should not be made to wait for long periods of time.

The health care team can contribute to the perception of care through an interdisciplinary approach. The team should present a single, unified treatment plan and meet on a regular basis to maintain good communication with each other.[38] Sometimes clients will question several health care members about their treatment plan because they want assurance that what they are being asked to do is really necessary.[20]

Nonadherence Counselor Issues

Working with a client who is successfully making healthful lifestyle changes is a joy; however, sometimes clients will not change, leaving the counselor feeling frustrated or even angry. It is irrational thinking on the part of a counselor to assume personal responsibility for a client's inaction. Counselors are there to assist their clients, but clients have the responsibility to make changes.[38,39] King[10] emphasizes, "We are not here to fix our clients but to inform, guide, encourage, understand and support them" (p. 4).

Changing dietary behavior is a complicated process requiring numerous lifestyle adjustments that often interfere with pleasurable activities compromising a client's motivational level. A counselor may view the required changes as extremely important, but the client may not have the same priorities. Counselors need to be willing to accept the motivational level of their clients and work with them accordingly.

Another factor to consider is that sometimes the benefits of an interaction with a nutrition counselor will not be immediately observable. Sigman-Grant[40] emphasizes that change involves a series of processes requiring time as an important dimension. The interaction with a nutrition counselor may be one of several important events that eventually lead to a significant change.

BEHAVIOR CHANGE STRATEGIES

Although educational components of a counseling session are essential, they are usually not sufficient to maintain a substantial food behavior change. A number of behavioral change strategies can be incorporated into a counseling program. The selection of a particular strategy to be used with a client will depend on his or her motivational stage; resources; lifestyle, educational, and emotional needs; as well as the expertise of the counselor. Behavior change strategies can be addressed in terms of the **ABCs of behavior,** which are included in social cognitive theory:[19,41]

A: Antecedent (stimulus, cue, trigger)
B: Behavior (response, eating)
C: Consequence (punishment, reward)

Let's take a closer look at each of these components:

- **Antecedents.** Encountering antecedents to eating occurs normally throughout the day. For example, time of day or the feeling of hunger in our stomach stimulates our desire to eat so we eat. Usually in nutrition counseling, we are particularly interested in cues that trigger unconscious eating or consumption of large quantities of certain types of food. Behavior change strategies addressing antecedents often concentrate on physical availability of food (cookie jar), social (parties), emotional (stress), or psychological (motivation; destructive thought patterns). Behavior change strategies dealing with antecedents can focus on avoiding the cue (remove cookies from the house) or altering the cue (cover a piece of cake with pepper).
- **Behavior.** Strategies dealing with the behavioral response to an antecedent may address the actual act of eating (speed), physical (eat in one place), emotional (do not clean your plate), awareness (pay attention to eating; no TV), or attractiveness (sparkling water in a wine glass with a slice of lemon). Behavior change strategies may focus on providing a substitute for eating (take a walk).
- **Consequences.** Consequences can be positive reinforcers or punishment; such as a reward or losing a privilege.

The sequence of events from antecedent to consequence is referred to as a **behavior chain;** see Exhibit 5.4 for some examples. Behavioral strategies can address all aspects of a behavior chain or can zero in on one aspect. Caban et al.[21] have determined that tailoring counseling approaches to client needs improves outcomes. Specifically, they found that clients who are highly motivated and reported few external stressors were likely to be receptive to nutritional and physical activity interventions. On the other hand, those who reported more external stressors in the initial assessment benefited from relaxation training, stress management, and problem-solving approaches in the beginning of treatment.

EXHIBIT 5.4—Behavior Chain Examples

Negative Consequence

Juanita comes home from work very hungry. She walks into the kitchen to fix dinner. There is a bag of her favorite potato chips on the counter (antecedent). Juanita tells herself she will eat just a few and opens the bag, leaving it on the counter as she prepares dinner. She ends up eating the whole bag of chips (behavior). When dinner is ready, Juanita picks at her food because she is not very hungry. She is frustrated because she knows a bag of potato chips is not very nutritious and she is not getting enough vegetables and fiber in her diet (consequence).

Positive Consequence

Juanita comes home from work very hungry. There is an orange and walking clothes and shoes prominently displayed on a chair in the kitchen (antecedent). Juanita eats the orange, changes her clothes, and goes out for a half-hour walk (behavior). When she returns, Juanita is in good spirits, and she prepares dinner. She eats a healthy meal with whole grains and vegetables; she is content (consequence).

Asking clients to reflect on their behavior can give clues as to what behaviors are in the greatest need of change. One excellent method for evaluation, mentioned earlier, is a complete diary of food behavior. Intense journaling activity can be very demanding, but clients may be willing to do an extensive recording for a few days or one week. Some clients may enjoy the intense analytical activity and want to continue this form of journaling for an extended period. See Lifestyle Management Form 5.1 (in Appendix G) for an example of a food diary form for intensive journaling.

Cue Management (Stimulus Control)

Cue management (stimulus control), deals with the antecedent component of a behavior chain by prearranging cues to increase a desired response or to suppress a detrimental one. An effective method of introducing the concept of cue management to a client is by using an analogy (see Exhibit 5.5).

The first task for using this method will be to identify cues. Sometimes triggers that stimulate a particular food response are obvious (such as a stash of candy bars in an office desk), but other times the cues are inconspicuous. When this is the case, a few days or weeks of intensive journaling by using a form such as Lifestyle Management Form 5.1 could help identify stimuli. In the case of decreasing the occurrence of a detrimental food stimuli, the objective in cue management is to alter the triggering stimuli or reduce exposure to it.[41] Table 5.3 lists some strategies for controlling common physical environment and social cues. A discussion of regulating stress and cognitive antecedents can be found in Chapter 6.

Once a strategy is developed to counteract a cue, it is generally a good idea to talk through or visualize with your client how the strategy will be implemented. The scheme should include reminders to perform the new activity, such as the following:

- Sticky notes left in strategic places*—computer, dashboard, refrigerator
- Notes on a calendar
- Cartoons or jingles posted in a book or briefcase used everyday
- Entries on a daily to-do list

Remember when using this technique to look for cues that produce beneficial behaviors as well as cues that encourage detrimental behaviors. If the focus can be on increasing the occurrence of positive behavior producing cues, the transition to a healthier lifestyle is likely to go more smoothly with seemingly less effort. As desirable behaviors make a greater impact on one's lifestyle, the time available to engage in the undesirable

EXHIBIT 5.5—Example of Using an Analogy to Introduce Cue Management

Counselor: Imagine that one day you are driving your [use the name of client's dream automobile] and pass through a big pothole. Luckily, you have a sturdy car and it was not damaged. What would you do if you were to go to the same destination again?

Client: Avoid the pothole. Drive slowly. Take an alternative route.

Counselor: Very good! You simply drive differently. We are trying to apply the same idea here, which is to learn to identify and anticipate potential "potholes" in your road to meeting your food goal so that you can be prepared to "drive" differently.

Source: *Behavioral Medicine* 1997;23:15–28. Adapted with permission of the Helen Dwight Reid Educational Foundation. Published by Heldref Publications, 1319 Eighteenth St., NW, Washington, DC 20036-1802. Copyright © 1997.

*Note that the location of the notes should be periodically changed, or else they become "part of the scenery" and lose their influencing effect.

TABLE 5.3—Strategies to Control Environmental Cues

STIMULI	SOLUTION
Physical Environment	
Location: Some people eat in many places of their home, and each have become cues for eating—bed, TV room, kitchen sink, or refrigerator.	• Designate one place for eating all meals and snacks. Besides reducing location cues, the need to travel to a particular spot provides extra time to think about whether the food actually needs to be eaten.
Activities: Engaging in activities such as reading or driving a car while eating results in producing additional cues and encourages unconscious eating.	• Only eat when eating. With fewer distractions, focus can be placed on the pleasures of eating and on the degree of fullness.
Shopping: Degree of hunger can influence food purchases and make consumers more receptive to store cues, which encourage impulse buying.	• Do not shop when hungry. • Use a shopping list. • Bring only enough money to purchase foods on the shopping list.
Serving: Attractiveness of food can be a cue to eat and encourage food consumption. This is particularly important if foods that need to be consumed are not well liked.	• Serve foods that are being encouraged in an attractive manner; for example put a slice of lemon in a water glass or use parsley for a pleasing effect. • Serve foods in desired portion size and put away the rest. For example, count out the desired number of nuts into an attractive bowl and replace the storage bag in the refrigerator. Take the bowl to the designated place to eat. • Learn to say "No thank you" to offers of food.
Reminders: Physical cues can be used as reminders to perform a behavioral action.	• Use sticky notes, cards, or other reminders to eat certain foods.
Social Environment	
Social events: Many people identify social gatherings as a particularly difficult time. There is often an array of tempting foods, and if alcohol is consumed, defenses decline.	• Plan and rehearse how to deal with temptations at an upcoming event. • Do not go to social events hungry. Have a snack ahead of time, such as an apple or orange. • Drink water and plan to have one alcoholic drink, if any. • Volunteer to bring a vegetable platter or other acceptable food for your needs. • Do not stand near food. Move food to a distant location if it is placed near you.
Social support: Family and friends can help or hinder with progress in meeting dietary objectives.	• See Chapter 6 for a discussion of ways to enhance social support for lifestyle changes.

(Continued)

TABLE 5.3—Strategies to Control Environmental Cues (Continued)

STIMULI	SOLUTION
Eating Behavior	
Food in mouth: Eating quickly may be a habit that developed.	• Put utensils down between mouthfuls. • Chew fully before swallowing. • Take small bites. • Take a break during the meal. • Swallow each bite before adding any more food.
Unconscious cues: A person may not be aware of the triggers that encourage the desire to eat certain foods.	• Move to a new place at the table or a new room.
Food on plate: There may be a need to clean the plate.	• Set a goal to leave food on the plate.
Storage of food: Usually food stored in sight, such as counters and tables, provides cues to eat.	• Store foods that should be avoided out of sight in inconvenient places. • Foods that should be encouraged should be highly visible. • If possible, do not bring foods into the home that should be discouraged. • Do not store undesirable foods in an office desk.

is automatically reduced. Sometimes counselors place so much emphasis on solving problems and getting rid of undesirable food behaviors that a search for desirable cues is neglected.

Countering

Countering, a technique of exchanging healthy responses for problem behaviors, addresses the B (behavior) portion of the behavior chain.[42] If an individual simply stops a pattern, there is likely to be a void unless a new behavior is substituted. For example, a client may consume two cups of ice cream each night after dinner in front of the television. The transition to changing that behavior will go more smoothly if an alternative behavior is planned, such as riding an exercise bike, stretching, or mending. Ordinarily countering is a usable behavior intervention strategy when the objective is to find an activity to substitute for eating altogether or when an acceptable food alternative is available. General categories of substitutions include the following:

- Foods that are acceptable or a healthier alternative, such as baked chicken for fried.
- Active diversions, such as knitting, writing, playing an instrument, working on a puzzle, or cleaning.
- Physical activities, such as walking, stretching, or weight lifting.
- Relaxation activities, such as deep breathing, yoga, prayer, or progressive muscle relaxation. (Relaxation is covered in Chapter 6 under stress management.)

EXERCISE 5.2—Behavior Chains

Record in your journal one of your frequent behavior chains. Identify the cue (trigger), specify the behavior, and describe the consequence. Is this a behavior that you would like to continue? If so, is there a way to encourage the occurrence of the cue? If not, is there a way to reduce the incidence of the trigger? Explain.

Reinforcement: Rewards

For some people, tracking and cue management interventions are adequate to develop and sustain a desired behavior; however, others need an added incentive to regulate and strengthen behavior.[43] **Reinforcement**

EXERCISE 5.3—Identifying Cues and Exploring Countering

List five common cues to unconscious eating or eating undesirable foods. Next to that list, write a countering behavior that would be acceptable to you if the cue was applicable to you. For example:

Cue	Behavior	Alternative
Movies	*Large popcorn with butter*	*Small popcorn without butter*

behavior change strategies provide incentives by addressing the end of the behavior chain—consequences. **Rewards** provide positive consequences, generally thought to be more effective than negative consequences, such as punishment.[42] Generally, there is an inherent reward of feeling good following the accomplishment of achieving a goal, resisting temptation, or substituting alternative behaviors for undesirable behaviors. In fact, clients should be encouraged to get in touch with those feelings and compliment themselves with positive self-talk, such as "I am doing great!" or "I can make this work." Although self-compliments are beneficial, more tangible rewards can provide additional motivation during challenging times—for instance, in the initial attempts of making a lifestyle change.

Rewards can take many forms, and a client may need some time to think over potential options before selecting a workable reward. Usually rewards are luxury items (purchase of clothing) or pleasurable experiences (relaxing bath, doing nothing, reading a book). A technique for accumulating small rewards could be considered, such as placing pennies in a jar each time a particular behavior is accomplished. When the desired amount is collected, the money could then be used to buy something special. Rewards can be self-administered or given to a client by a significant other.

Exhibit 5.6 provides guided questions to help you find meaningful reinforcers with your client. If you believe your client would benefit from rewards, you should discuss the behavior change strategy with him or her and evaluate the response. If there is hesitation on the client's part, it may be best to allow time to think about the concept until your next meeting.

The following are some major factors to consider when establishing rewards:

- Rewards should be individualized.
- Rewards should be well defined—what and how much.
- Rewards should be timed to come after the behavior, not before.
- Rewards should be planned to come as soon as possible after the behavior.

Contracting

Rewards are often used in conjunction with **contracting.** A contract documents an agreement between a counselor and a client to implement a particular goal. The contract can cover short- or long-term goals. An example of a behavioral contract can be found in Exhibit 5.7 and Lifestyle Management Form 5.2 (see Appendix G).

Contracts should be used when clients want structure and accountability,[10] and they should be recorded

Garfield ® by Jim Davis

EXHIBIT 5.6—Questions for Identifying Reinforcers

1. What activities do you enjoy?
2. What are your favorite times of the day?
3. Are there healthy foods that you particularly like to eat?
4. Do you have any hobbies?
5. What do you find relaxing?
6. When do you have fun?
7. Is there something you would hate to lose?
8. What are things you enjoy doing each day that you would hate to give up?
9. Is there a present you would like someone to give to you?
10. Is there something special you would like someone to do with you?

Source: Adapted from Holli BB, Calabrese RJ, *Communication and Education Skills for Dietetics Professionals*. 3d ed. Baltimore, MD: Williams & Wilkins; © 1998, p. 78. Used with permission.

because written contracts are more powerful than oral ones.[42] The following is a list of factors to consider when using contracting:

- Clients should define their intended behavior change for the contract. The counselor should never impose a goal.
- Behavioral goals should be clearly defined. The same conditions apply as those described in Chapter 4 for goal setting. The goal statement should answer what will happen, how often the behavior will occur, and when it will take place.[41]
- Time limits for reaching the goals should be delineated.
- Reinforcers should be stated. They can be rewards or punishments. Generally rewards are considered to have a greater impact and should be immediately available after the intended behavior is performed. If the contract covers a long-term goal, a reward that requires greater effort, such as a trip or shopping, can be considered. Otherwise, rewards that do not put an extra burden on a client should be sought. For example, a contract for an individual who enjoys coffee in the morning can be as simple as, "I do not drink coffee for breakfast until I have eaten a piece of fruit—no fruit, no coffee."
- Signing and dating the contract can add reinforcement to a client's commitment.[7] Usually the contract is signed by only the counselor and the client; however, there are times when consideration should be given to having the document signed by a support person, such as a spouse or friend.[22]

I generally encourage my clients to identify positive reinforcements for their behavioral contracts. However, one time I had a client insist that her contract stipulate that she would give her ex-husband a hundred dollars if she did not accomplish her task. Needless to say, she followed through on her goals.

Encouragement

Encouragement is generally well received; however, a client's past experience trying to accomplish a desired behavior change can influence the impact of encouraging remarks. Also, the effect will vary with the credibility,

EXHIBIT 5.7—Sample Behavioral Contract

Counseling Agreement

Name ______________________________ **Date** ______________

My plan is to do the following:

____

____

This activity will be accomplished by ______________________________

____

My reward will be ______________________________

______________________ ______________________
Client signature Date

______________________ ______________________
Counselor signature Date

EXERCISE 5.4—Make a Contract

The purpose of this activity is to give you direct experience with contracting. Think about a behavior that you would like to change or a task you would like to accomplish within the next two weeks. Use Lifestyle Management Form 5.2 to complete the contract. Have a colleague discuss the contract with you and sign it. Since the goal you will be setting is not part of a total counseling program and you may not be ready to take action on a major behavior change you are contemplating, the goal you establish should be modest. Some examples of modest goals could be brushing teeth after breakfast, giving the dog a bath, or writing a letter to a friend. The goal should be something that you believe you need a little push to accomplish. At the end of two weeks, reflect on the experience and document any problems or successes you encountered in your journal.

trustworthiness, and prestige of the person giving the words of encouragement.[19] Some counselors have provided encouragement creatively using e-mails, cards, notes, hugs, voice mail, personalized signs, notes, or recorded tapes. One counselor has clients recording their own relaxation tapes with their own words of encouragement to listen to before going to bed.

I had a religious client who consumed a lot of low-nutrient-dense foods in her bedroom. I made a small sign for her nightstand reminding her of her goals and telling her I would be thinking about her and praying for her. When my client went on vacation, she was worried about how she would handle her food goals. I gave her an audiotape I recorded for her trip with some personal words of encouragement, a few supportive sayings I took from a book on motivation, and a prayer. This client did well, and I believe the encouragement I provided was helpful.

Goal Setting

Achievement of goals provides a pathway to actually performing the new behavior. Breaking down desirable behavior patterns into small achievable steps provides a series of successes and an improvement in self-efficacy. As the saying goes, "Nothing breeds success like success." Each success raises mastery expectations. We covered the process of goal setting in Chapter 4.

Modeling

Some of our behaviors are learned by observing and imitating others. By observing others accomplishing a goal similar to their own, clients' beliefs in their ability to imitate the behavior increases. This process is referred to as **modeling.** Videotapes, lay counselors, written testimonials, success stories, counseling buddies, and role playing are possible models that can be used in a counseling environment to increase self-efficacy. Outside counseling sessions, models who have prestige, status, or expertise are more likely to influence behavior than those who do not have those characteristics.[7] For example, when Oprah Winfrey used a diet drink to lose weight, sales of the product soared. However, most often people with whom we identify closely have the greatest influence on our behavior. We are more likely to imitate an individual who is similar to our age, gender, culture, and other qualities.

One of my clients relayed a story of going to a pancake house for breakfast with friends. She had not been to this type of restaurant in years and assumed that she would order waffles or pancakes and sausages for her meal. However, her friend ordered a broccoli omelet made with an egg substitute. This observation influenced my client, and she decided that she would imitate her friend and order the same thing.

Modeling can have a greater impact if a client can practice the observed behavior under supervision and then receive immediate feedback.[41] For example, clients could select acceptable items from a restaurant menu after watching a video with people making desirable choices.

Barriers Counseling

Barriers are obstacles or roadblocks to achieving a desired lifestyle change. The first task is to identify barriers and then develop strategies to minimize or eliminate their impact. (See Table 5.4.)

Perceived barriers come in many forms. These could include taste preferences, access to proper foods, ease of preparation, complexity of the diet, social support, social eating, eating out, expense, job or family pressures, and time constraints. An investigation to identify barriers of middle-aged patients who experienced myocardial infarction found social and work situations, the price of food, and situations in which large amounts of food are

EXERCISE 5.5—Exploring Encouragement

Describe in your journal a time when words of encouragement were particularly useful to you to accomplish a task. Explain anything you can bring from the experience that could help you in your nutrition counseling endeavors.

TABLE 5.4—Overcoming Barriers to Change

CLASSIFICATION	BARRIER	STRATEGY
Lack of knowledge	The client does not know what foods can lower blood pressure.	Provide pamphlets, videos, Internet sites, and grocery store tour.
Lack of skill	Limited cooking ability	Shared cooking; demonstrations
Lack of risk taking	Afraid of hurting mother's feelings. The client's mother prepares a large dinner every Sunday. There are seldom foods available that meet dietary objectives.	Explore ambivalence to taking action to make a request for the mother to make acceptable alternatives. Use imagery or role playing to practice making a request.

EXERCISE 5.6—Identifying Models

Divide into groups and discuss the potential impact of a counselor who is not at "ideal body weight." Do you believe a counselor who does not appear to be what is generally considered fit can be an effective nutrition counselor? Record your thoughts in your journal.

available to be the most frequently reported challenging conditions.[44] Holli and Calabrese[19] state that characteristics of the dietary regiment are the most important factors in adherence; of these, complexity is the most significant. Sometimes demographic variables have been found to have an impact on barriers, too. For example, low income, low education, and being male have been found to increase barriers to consumption of fruits and vegetables.[45]

In their discussion of goal achievement, Laquatra and Danish[46] identify four major obstacles: lack of knowledge, a lack of skill, the inability or fear of appropriate risk taking, and lack of adequate social support. Some of the most frequently named barriers can be addressed by an increase

CASE STUDY 5—Mary: Busy Overweight College Student and Mother

Mary is a thirty-four-year-old university student presently carrying twelve credits. She is on campus four days a week from 9:00 A.M. to 2:00 P.M. She is very committed to her schoolwork and currently has a 3.4 cumulative average.

Mary is married and has three children ages six, eight and a half, and ten. She wants to be a good role model for her children, setting an example as to the importance of education. Her husband is a pharmaceutical sales representative, and due to the nature of his job he is away from home frequently. This means that the bulk of child care responsibility is hers.

Mary has come to the Lifestyle Management Program seeking assistance for weight loss and because she really does not feel as good as she thinks she should. She is 5 feet, 6 inches tall and weighs 155 pounds. Mary completed a client assessment questionnaire and a food frequency form in advance, and you review them when she arrives. The exploration phase of your interview allows Mary to elaborate on her lifestyle that has created and perpetuates her diet concerns.

Mary is short on time in the morning because she must make her children's lunches, get them dressed, drive them to school, and make a 9:00 class. She does not eat breakfast until 10:30, when she grabs something in the Student Center. Lunch consists of a vending machine snack that she grabs on her way to the parking lot, and she eats it on the forty-minute drive home. She drives directly to the elementary school to pick up her children, since they do not get bused.

Because she has such small children, Mary feels the need to keep a supply of snack foods on hand. Upon returning home, it is time to oversee her children's homework and give them a snack. By the time Mary's starts to make dinner, she is starving and finds herself picking at whatever is around simply to keep her sanity while she is helping with homework and separating some occasional boxing matches. Because her husband is away so frequently, she must clean up the dishes and get the children showered and into bed herself. By the time everyone is settled down and bedtime stories are finished, Mary is exhausted. It is about 9:00 before she is freed up to do her own homework and study.

Typically, she first puts in a load of wash and folds the clothes from the dryer, puts on her robe, prepares herself a snack, and settles down to do schoolwork. It is at this time that Mary really feels the loneliness and starts to feel sorry for herself. She expresses dismay that in a few hours she will have to get up and do it all over again. After about an hour of snacking and studying, she puts in some more laundry and heads to bed.

EXERCISE 5.7—Exploring Barriers

Interview someone who is attempting to make a dietary change. Record in your journal the difficulties the person is encountering. How do your findings compare to your readings? What have you learned about barriers?

in knowledge. Nutrition counselors can often aid clients in finding ways to make needed dietary changes palatable, assessable, and convenient. However, counseling time should not be devoted to giving unneeded information if your client already has an adequate knowledge base. Lack of skill can be in areas of mental or physical skills including assertiveness, decision making, positive self-talk, time management, label reading, or estimation of portion sizes. Clients who have an inability or fear of appropriate risk taking are afraid of the negative consequences of taking action. For example, they could be afraid of hurting a friend's feelings by refusing an offer of food or afraid of dealing with the consequences of disruption for friends and family. Requesting help from family and friends or joining a support group can address social support concerns. See Chapter 6 for elaboration of this topic.

REVIEW QUESTIONS

1. Four food management tools were reviewed in this chapter. Describe them, indicating the advantages and disadvantages of each.

2. What are the benefits of tracking food behavior goals?

3. Identify five ways a nutrition counselor can support a client's decision to keep a food journal.

4. List the five Knowles assumptions regarding adult learning, and explain how they can be addressed in a nutrition counseling situation.

5. Explain the integrated approach to learning advocated by Funnell and Anderson.

6. Explain the ABCs of eating behavior and behavior chains. What are their importance in nutrition counseling?

7. Identify four factors to consider when guiding clients in the use of rewards.

8. Explain the following behavior change strategies: cue management, countering, rewards, modeling, barriers counseling, contracting, and encouragement.

EXERCISE 5.8—Intervention Strategies for Mary

Complete the following questions after reading the preceding case study.

1. Complete a Client Concerns and Strengths Log, Lifestyle Management Form 4.7, for Mary.
2. What are the most crucial factors (barriers) influencing Mary's potential adherences to any plan?
3. What type of food management tool guide (menus/meal plans, exchange system, Food Guide Pyramid, goal setting) do you think would benefit Mary? Why?
4. What system of tracking do you believe she would select?
5. An assessment would identify educational issues Mary would like to address. What educational materials and interactive experiences would you like to incorporate in counseling sessions with Mary?
6. Review the five major assumptions of adult learners. How would these assumptions influence your interactions with Mary?
7. Describe how positive or negative your educational messages would be with Mary.
8. Write one food behavior chain for Mary and at least one possible method of breaking the chain.

Antecedent	Behavior	Consequences

9. Review the following behavior change strategies, indicate for each whether you would consider using the strategy with Mary, and explain why or why not.

Countering
Contracting
Reinforcement (rewards)
Encouragement
Goal setting
Modeling
Barriers counseling

ASSIGNMENT—Food Management Tool Utilization

The objective of this assignment is to investigate, develop, and utilize the four food management tool options and tracking methods covered in this chapter.

1. Develop a food plan for three days for each of the four tools. There will be twelve days of plans altogether. Use the following directions:

a. Use one of the Web sites in the suggested Internet resources to develop three days of menus/meal plans for a specialized diet (for example, low sodium, high fiber, low fat, vegetarian). Choose a calorie level appropriate for yourself and address your food preferences and lifestyle needs. Record all of these factors in your final report. Also hand in copies of all three days of menus/meal plans.

b. Decide on an appropriate calorie level and a percentage of fat, carbohydrate, and protein for yourself and use Exchange Lists for Meal Planning to develop a food plan. See the Appendix F for Exchange List resources. In your final report, hand in a copy of the food pattern you developed along with a copy of a completed Template for Calculations to Develop an Eating Plan based on Exchange Lists for Weight Management.

c. Use the DASH Food Plan educational materials in Appendix B and the Web site to develop a dietary pattern for yourself based on servings of food groups. Hand in a copy of the diet plan with your final report.

d. Review the goal-setting process described in Chapter 4, and determine an appropriate food goal for yourself. Write that goal in your final report.

2. Follow each of the plans for three days and use the designated tracking system identified for each.

a. For the detailed menus/meal plan program use Lifestyle Management Form 4.2 to record your food intake.

b. For the exchange system plan, track consumption of exchanges by crossing out slashes for each exchange. See the following example:

Exchange Food Plan	Cross out slashes as exchange is consumed
Starch								Starch X X X X			
Fruit				Fruit X X	
Milk (Reduced-fat)			Milk (Reduced-fat) X	
Other carbohydrates				Other carbohydrates X		
Vegetable						Vegetable X X			
Meat					Meat X X		
Fat						Fat X X			

c. Use the DASH Food Plan tracking form to track your DASH Food Plan found in Appendix B.

d. Develop your own tracking method for your food goal plan (marks in a calendar, tear off stubs of a sticky note, empty fruit bowl, and so forth), and hand in a copy or describe the tracking method in your final report.

3. Write an evaluation of your experiences by answering the following questions:

a. Describe the advantages and disadvantages of each eating plan guide.

b. Describe the advantages and disadvantages of each tacking method.

c. What barriers did you encounter in trying to achieve your dietary objectives?

d. What did you learn from this experience?

SUGGESTED READINGS, MATERIALS, AND INTERNET RESOURCES

Meal Plans and Recipes

American Medical Association Web site (www.ama-assn.org). Search for General Health for a collection of healthy recipes.

Shape Up America Web site (www.shapeup.org/sua). A personalized meal plan can be developed in the Cyberkitchen.

Cyberdiet (http://www.cyberdiet.com/). A partner in the mediconsult.com Network. This Web site offers a variety of health-enhancing assessments and guides, including information about label reading, eating out (sample menus of twenty-six fast-food and thirteen ethnic restaurants), and examples of recipe modifications. A personal menu plan can be constructed complete with recipes and a shopping list.

Meals for You Web site (http://www.mealsforyou.com/). Packed with recipes and meal plans. Calorie controlled, vegetarian, low carbohydrate, high fiber, and low sodium are just a sample of the meal plans that can be designed. Also, if a client desires to increase intake of a particular food, such as tofu or cabbage, recipes can be selected accordingly. An aisle-by-aisle shopping list can also be printed out, even for the client's specific grocery store.

Food Guides

Exchange Lists for Meal Planning. This booklet can be ordered from the American Dietetic Association from the Web site, www.eatright.org; write to 216 W. Jackson Boulevard, Chicago, IL 60606-6995; or call customer service at (800) 877-1600, ext. 5000, for a catalog.

Facts about the DASH Food Plan. This booklet (NIH Publication No. 98-4082) contains a description of the diet, sample food plans, and guidelines for making the DASH Food Plan part of an individual's lifestyle. Order from the National Institutes of Health, National Heart, Lung, and Blood Institute, NHLBI Information Center, PO Box 30105, Bethesda, MD, 20824-0105; (301) 592-8573.

Food Guide Pyramid. This guide can guide selection of nutrients needed (without too many calories) and reduce the fat, cholesterol, sugar, sodium, or alcohol. It costs $1 and can be ordered from the Federal Consumer Information Center, Department www, Pueblo, CO, 81009; call (888) 878-3256.

Self-Help Resources

Duyff RL. *Monthly Nutrition Companion: 31 Days to a Healthier Lifestyle.* New York: Wiley; 1997. This book can be ordered from the American Dietetic Association, www.eatright.org/catalog or (800) 877-1600, ext. 5000.

Kostas GG. *The Balancing Act: Nutrition and Weight Guide.* This book can be ordered from The Balancing Act Nutrition and Weight Guide, P.O. Box 671281, Dallas, TX 75367-8281; (214) 239-7223.

Wylie-Tosett J, Swencionis C, Caban A, Friedler A, Schaeffer N. *The Complete Weight Loss Workbook: Proven Techniques for Controlling Weight Related Health Problems.* Alexandria, VA: American Diabetes Association; 1997. Order from the Web at http://www.diabetes.org.

REFERENCES

[1]Tsoh JY, McClure JB, Skaar KL, Weter DW, Cinciripini PM, Prokhorov AV, Friedman K, Gritz E. Components of effective intervention. *Behav Med.* 1997;23:15–28.

[2]Herrin M, Parham E, Ikeda J, White A, Branen L. Alternative viewpoint on National Institutes of Health clinical guidelines. *J Nutr Educ.* 1999;31:116–118.

[3]Higgins L, Gray W. What do anti-dieting programs achieve? A review of research. *Aust J Nutr Diet.* 1999;56:128–136.

[4]Omichindki L, Harrison KR. *Nondiet Weight Management: A Lifestyle Approach to Health & Fitness.* Eureka, CA: Nutrition Dimension; 1998.

[5]Foreyt JP, Goodrick GK. Attributes of successful approaches to weight loss and control. *Appl Prev Psychol.* 1994;3:209–215.

[6]Shape Up America and American Obesity Association. *Guidance for Treatment of Adult Obesity.* Bethesda, MD: Shape Up America; 1997–1998.

[7]Williams S. Behavior modification. In: Holli BB, Calabrese RJ, *Communication and Education Skills for Dietetics Professionals.* 3d ed. Baltimore, MD: Williams & Wilkins; 1998.

[8]Streit KJ, Stevens NH, Stevens VJ, Rossner J. Food records: A predictor and modifier of weight change in a long-term weight loss program. *J Am Diet Assoc.* 1991;91.213–216.

[9]Stevens VJ, Rossner J, Hyg MS, Greenlick M, Stevens N, Frankel HM, Craddick S. Freedom from fat: A contemporary multi-component weight loss program for the general population of obese adults. *J Am Diet Assoc.* 1989;89:1254–1258.

[10]King NL. *Counseling for Health & Fitness.* Eureka, CA: Nutrition Dimension; 1999.

[11]Brownell KD, Cohen R. Adherence to dietary regimens: Components of effective intervention. *Behav Med.* 1995;20:155–165.

[12]Trudeau E, Dube L. Moderators and determinants of satisfaction with diet counseling for patients consuming a therapeutic diet. *J Am Diet Assoc.* 1995;95:34–39.

[13]Pohl SL. Facilitating lifestyle change in people with diabetes mellitus: Perspective from a private practice. *Diabetes Spectrum.* 1999;12:28–33.

[14]Nicolucci A, Cavalier D, Scorpiglione N, Carinci F, Capani F, Tognoni G, Benedetti MM. A comprehensive assessment of the avoidability of long-term complications of diabetes: A case-control study. *Diabetes Care.* 1996;19:927–933.

[15]Knowles MS, Holton EF, Swanson RA, Holton E. *The Adult Learner: The Definitive Classic in Adult Education and Human Resource Development.* 5th ed. Houston, TX: Gulf; 1998.

[16]Walker EA. Characteristics of the adult learner. *Diabetes Educator.* 1999;25:16–24.

[17]American Diabetes Association. *Meeting the Standards A Manual for Completing the American Diabetes Association Application for Recognition of Diabetes Self-Management Education Programs.* Alexandria, VA; Author; 1999.

[18]Funnell MM, Anderson RM. Putting Humpty Dumpty back together again: Reintegrating the clinical and behavioral components in diabetes care and education. *Diabetes Spectrum.* 1999;12:19–23.

[19]Holli BB, Calabrese RJ. *Communication and Education Skills for Dietetics Professionals.* 3d ed. Baltimore, MD: Williams & Wilkins; 1998.

[20]Berry M, Krummel D. Promoting dietary adherence. In: Kris-Etherton P, Burns JH, eds. *Cardiovascular Nutrition: Strategies and Tools for Disease Management and Prevention.* Chicago: American Dietetic Association; 1998:203–215.

[21]Caban A, Johnson P, Marseille D, Wylie-Rosett J. Tailoring a lifestyle change approach and resources to the patient. *Diabetes Spectrum.* 1999;12:33–38.

[22]Snetselaar LG. *Nutrition Counseling Skills for Medical Nutrition Therapy.* Gaithersburg, MD: Aspen; 1997.

[23]Gehling E. *Model Program: Promoting Lifestyle Behavior Changes—A Counseling Approach.* University of Minnesota Educational Videos from the 1999 National Maternal Nutrition Intensive Course. Minneapolis: University of Minnesota School of Public Health; 1999.

[24]Mullen, Patricia D. Compliance becomes concordance: Making a change in terminology produce a change in behaviour. *Br Med J.* 1997;314:69–70.

[25]Butler C, Rollnick S, Stott N. The practitioner, the patient and resistance to change: Recent ideas on compliance. *Can Med Assoc J.* 1996;154:1357–1362.

[26]King S, Gibney M. Dietary advice to reduce fat intake is more successful when it does not restrict habitual eating patterns. *J Am Diet Assoc.* 1999;99:685–689.

[27]Milas NC, Nowalk M, Akpele L, Castaldo L, Coyne T, Doroshenko L, Kigawa L, Korzec-Ramirez D, Scherch L, Snetselaar L. Factors associated with adherence to the dietary protein intervention in the Modification of Diet in Renal Disease Study. *J Am Diet Assoc;* 1995;95,1295–1300.

[28]Windhauser MM., Evans MA, McCullough ML, Swain JF, Lin PH, Hobe KP, Plaisted CS, Karanja NM, Vollmer WM. Dietary adherence in the dietary approaches to stop hypertension trial. *J Am Diet Assoc.* 1999;99:S76–S83.

[29]Coyne T, Olson M, Bradham K, Garcon, M, Gregory P, Scherch L. Dietary satisfaction correlated with adherence in the Modification of Diet in Renal Disease Study. *J Am Diet Assoc.* 1995;95:1301–1307.

[30]Mattes, Richard D. (1993). Fat preference and adherence to a reduced-fat diet. *Am J Clin Nutr.* 1993;57:373–381.

[31]Schiller MR, Miller M, Moore C, Davis E, Dunn A, Mulligan K, Zeller P. Patients report positive nutrition counseling outcomes. *J Am Diet Assoc.* 1998;98:977–982.

[32]Barnard ND, Akhtar A, Nicholson A. Factors that influence compliance with low-fat diets. *Arch Fam Med.* 1995;4:153–158.

[33]Franz MJ, Splett PL, Monk A, Barry B, McClain K, Weaver T, Upham P, Bergenstal R, Mazze RS. Cost-effectiveness of medical nutrition therapy provided by dietitians for persons with non-insulin-dependent diabetes mellitus. *J Am Diet Assoc.* 1995;95:1018–1024.

[34]Monk A, Barry B, McClain K, Weaver T, Cooper N, Franz MJ. Practice guidelines for medical nutrition therapy provided by dietitians for persons with non-insulin-dependent diabetes mellitus. *J Am Diet Assoc.* 1995;95:999–1006.

[35]Caggiula AW, Watson, J E. Characteristics associated with compliance to cholesterol lowering eating patterns. *Patient Educ Counsel.* 1992;19:33–41.

[36]Berne E. *Principles of Group Treatment.* New York: Grove; 1966.

[37]Curry KR., Jaffe A. *Nutrition Counseling & Communication Skills.* Philadelphia: Saunders; 1998.

[38]Lorenz RA, Bubb J, Dans D, Jacobson A, Jannasch K, Kramer J, Lipps J, Schlundt D. Changing behavior: Practical lessons from the Diabetes Control and Complications Trial. *Diabetes Care.* 1996;19:648–655.

[39]Siminerio LM. Defining the role of the health education specialist in the United States. *Diabetes Spectrum.* 1999;12:152–157.

[40]Sigman-Grant M. Change strategies for dietary behaviors in pregnancy and lactation. *University of Minnesota Educational Videos from the 1999 National Maternal Nutrition Intensive Course.* Minneapolis: University of Minnesota School of Public Health; 1999.

[41]Glanz K, Greene G, Shield JE. Understanding behavior. In: American Dietetic Association. *Project Lean Resource Kit.* Chicago: American Dietetic Association; 1995:142–189.

[42]Prochaska JO, Norcross JC, DiClemente CC. *Changing for Good.* New York: Avon, 1994.

[43]Cormier S, Cormier B. *Interviewing Strategies for Helpers: Fundamental Skills and Cognitive Behavioral Interventions.* 4th ed. Pacific Grove, CA: Brooks/Cole; 1998.

[44]Lappalainen R, Koikkalainen M, Julkunen J, Saarinen T, Mykkanen H. Association of sociodemographic factors with barriers reported by patients receiving nutrition counseling as part of cardiac rehabilitation. *J Am Diet Assoc.* 1998;98:1026–1029.

[45]Iszler J, Crockett S, Lytle L, Elmer P, Finnegan J, Luepker R, Laing, B. Formative evaluation for planning a nutrition intervention: Results from focus groups. *J Nutr Ed.* 1995;27:127–132.

[46]Laquatra I, Danish SJ. Practitioner counseling skill in weight management. In: Dalton S. ed. *Overweight and Weight Management: The Health Professional's Guide to Understanding and Practice.* Gaithersburg, MD: Aspen; 1997:348–371.

Chapter

MAKING BEHAVIOR CHANGE LAST

Look to this day!
For it is life, the very life of life.
In its brief course
Lie all the verities and realities of your existence:
The bliss of growth;
The glory of action;
The splendor of achievement;
For yesterday is but a dream,
And tomorrow is only a vision;
But today, well lived, makes every yesterday
A dream of happiness,
And every tomorrow a vision of hope.
Look well, therefore, to this day!
—"Kalidasa," ancient Sanskrit poem

BEHAVIORAL OBJECTIVES:

- Explain significance of social support.
- Suggest ways significant others can support clients.
- Develop mental imagery and role-playing skills.
- Explain usefulness of social disclosure.
- Describe cognitive restructuring.
- Identify three components of dysfunctional thinking.
- Demonstrate basic cognitive restructuring counseling skills.
- Identify fourteen frequent cognitive pitfalls.
- Describe possible responses to stress.
- Identify common sources of stress.
- Distinguish three general categories of stress management strategies.
- Describe the basics of stress management counseling.
- List immediate determinants and covert antecedents of relapse.
- Describe the basic components of relapse prevention counseling.

KEY TERMS:

- **COGNITIONS:** what and how a person thinks and perceives based on life experiences
- **COGNITIVE RESTRUCTURING:** challenging destructive thoughts, beliefs, and internal self-talk and substituting self-enhancing cognitions
- **MENTAL IMAGERY:** a mental rehearsal of an anticipated experience
- **RELAPSE PREVENTION:** systematic approach to maintaining a behavior change; involves the identification of and preparation for high-risk situations
- **SOCIAL DISCLOSURE:** sharing information about self in order to enhance lifestyle change objectives
- **THOUGHT STOPPING:** a technique using the word *stop* to end destructive reoccurring thoughts

A number of factors can be considered to aid in maintaining dietary changes. Lifestyle change programs often address a client's social environment, cognition issues, as well as stress management and relapse prevention.

SOCIAL SUPPORT

Since the act of eating is often a social activity, either as an element of daily life or an integral part of special events, a client's social environment can have a significant impact on attempts to change eating behavior. The people closest to a person making dietary lifestyle changes will be affected by the new behavior patterns and in return can exert a powerful influence on your client. The changes often put a stress on the dynamics of a household. For example, a Friday night activity may have been to eat fried fish at a certain restaurant that only serves fried food. If a client needs to avoid this type of food, family members may resent the need for a change in the family pattern. In conditions where a client has many lifestyle change requirements, the feelings of resentment are likely to mount. One way of helping dispel negative feelings is by involving people close to your client in your counseling sessions.

Involving Family or Significant Others in Counseling

Encouraging significant others to provide a supportive environment has been shown to have a beneficial effect on dietary change objectives.[1] In one study in an acute inpatient hospital setting, a family or friend acted as a "care partner," resulting in increased patient and family satisfaction and adherence to the patient's medical regimen.[2] Also successful was an award-winning program designed to reduce hypertension through exercise.[3] In this program health care providers enlisted the support of a family or friend to be a designated "helper." One explanation for the positive effects of social support on health is that clients perceive a sense of support from others leading to a feeling of a more generalized sense of control.[3] A greater sense of control results in an increase in self-efficacy.

When meeting with family members or significant others, nutrition counselors can provide information, explore potential stress that the new food pattern could put on family interactions, and suggest ways in which family members can be supportive. Possible topics for discussion are presented in Exhibit 6.1.

EXHIBIT 6.1—Social Support Discussion Topics

- Inquire about any concerns or questions family members have about your client's dietary needs.
- If reasonable, invite family members to participate in the dietary changes and explore new tastes
- Ask whether there are times when the new diet pattern is likely to cause unusual stress.
- Explore willingness of family members to show support for your client.
- Suggest ways in which family members can support your client.

  Help keep undesirable foods out of sight to avoid tempting cues.

  Purchase foods to be avoided in varieties client does not like (e.g., an ice cream flavor client is unlikely to eat).

  Do not give foods to be avoided as gifts.

  Think of creative ways of celebrating a holiday or a birthday. For example, fresh fruit, gelatin, and fat-free sherbet can be quickly and attractively arranged on a plate for fat-free celebrations. Candles can even be placed in the composition.

  Give strokes.

  - Offer praise when desirable behavior is observed.
  - Express appreciation for the accomplishment of difficult tasks (such as taking blood sugar for the first time).
  - Provide preplanned or surprise rewards, such as, hugs, gifts, or back rubs.
  - Brag to others about positive behavior changes.

  Show patience for extra time needed to calculate or prepare foods.

  Offer help to plan ahead when visiting or going on trips.

  Avoid teasing or tempting with foods that need to be avoided.

  Avoid scolding, nagging, preaching, and embarrassing. Although such behavior may be well intentioned, the overall effect is destructive.

  Do not give criticism unless there have been three compliments.[4]

  Use positive statements about the new food pattern; avoid using the words *strange* or *different* to describe foods on the pattern. A motivated family member may be willing to keep a record of positive and negative statements regarding the new dietary pattern.

Sources: Raab C, Tillotson JL, eds., *Heart to Heart.* DHHS (PHS) publication 83-1528. Washington, DC: U.S. Department of Health and Human Services; 1983; Prochaska JO, Norcross JC, Diclemente CC, *Changing for Good.* New York: Avon; 1994; Snetselaar LG, *Nutrition Counseling Skills.* Gaithersburg, MD: Aspen; 1997.

Direct involvement with a client's social network is not always possible. In any case, clients should be encouraged to be vocal and request support for cooperation and assistance from their family and friends. Clients could cover any of the topics reviewed in Exhibit 6.1 directly with potential supporters. To keep a social environment conducive to change, counselors should remind clients to thank others for their involvement.

If your client does not have readily available support in their immediate environment, possible alternatives can be explored. The following list contains some suggestions:

- **Locate a distant support buddy.** Possibly a relative, associate, or friend in a distant location could be involved through telephone conversations, e-mail, or instant messaging.
- **Join clubs or organizations.** Clients may obtain direct or indirect support by taking part in organizations compatible with their lifestyle goals. This may include joining a walking club, gym, dance troop, or vegetarian society.
- **Locate a self-help group.** Self-help groups for most disorders can be explored in local medical centers as well as the Internet, where chat rooms may be available. Encourage your clients to become active in support groups by taking on responsibilities such as organizing a meeting, writing a newsletter, or volunteering for a committee. Active involvement increases commitment and expands the likelihood of making connections to others.
- **Take classes.** Classes related to your client's condition may be available in community education programs, supermarkets, or health centers. By taking part in these programs, clients enhance their skills and make contacts that can support their behavior change endeavors.

One of my clients was a pleasant, overweight teenage girl. She was very anxious to lose weight, but her mother seemed to sabotage her daughter's attempts by baking her favorite cakes, keeping the kitchen stocked with high-fat foods, and encouraging her daughter to eat. When her mother came to one of our sessions, she admitted that this was a problem. Although her mother loved her daughter very much, she voiced her fear that her daughter would become thin, be attractive to men, get married, and leave her. Eventually her mother agreed to go to a psychotherapist.

EXERCISE 6.1—Social Support Survey

Find two people who have made or attempted to make a lifestyle change who believe their social environment had an impact on their efforts. Ask them to describe the involvement. What was the impact of their social supporters or nonsupporters? Specifically what did they do to help or hinder the situation? Record these reactions in your journal.

Assertiveness Skills

A client's immediate social environment can sometimes exert a negative influence on eating behavior. If so, ways need to be sought to reduce the impact without causing undue social stress. Sometimes this involves assertive behavior, such as suggesting an alternative restaurant, calling a host ahead of time to discuss potential problems, or offering to bring a vegetable platter to a social function. Scenarios for dealing with difficult issues can be developed in counseling sessions through role playing, microanalysis of the scenario, or mental imagery.

Role Playing This strategy can effectively prepare clients for behavior change. For role playing to be an effective tool, a secure relationship should have been established with your client. Therefore, it is usually not advisable to use this technique during the first session.

Prepare for the Role Play.

- Analyze the concern, discuss possible scenarios to handling a situation, and decide on the best course of action.
- Explain the goals and objectives of role playing. Usually that means preparing for a difficult encounter.
- Assign roles.
- Set time limits. Generally you do not want role playing to go more than five minutes; in fact, an effective role play can be as little as two minutes. You want to leave time for processing the experience.

Enact the Role Play.

- Arrange chairs for appropriate interaction.
- As a counselor, you take on the role of one or more of the characters, and your client plays him- or herself. If you believe there is a benefit to modeling a certain behavior, the role play of a scenario could be done twice—once with you taking on the role of the client and a second time with the client playing him- or herself. See Exhibit 6.2 for a role play example.
- Stop the role play within five minutes or less.

EXHIBIT 6.2—Role Play Example

Problematic Scenario

Mother-in-law: You are coming for our beefsteak dinner on Sunday, right?

Client: I can't eat beefsteak—didn't John tell you?

Mother-in-law: Yes, but surely you can cheat once in a while. There is nothing like beefsteak dipped in melted butter.

Client: No, I can't! If you find it to be a problem that I can't eat beefsteak, then I will just stay home.

Mother-in-law: I don't understand why you can't just let this be a pleasant family get-together. It isn't like you are going to die if you eat it.

Effective Scenario

Setup: Client needs to establish control. Statements or arguments to be used must make sense to the mother-in-law. When dealing with difficult people, *blending* is an initial effective behavior. This involves agreeing on common ground.

Mother-in-law: You are coming for our beefsteak dinner on Sunday, right?

Client: That is wonderful of you to invite us, but I can't eat beefsteak. I have to try to get my cholesterol down.

Mother-in-law: Yes, but surely you can cheat once in a while. There is nothing like beefsteak dipped in melted butter.

Client: Well, beefsteak sure sounds delicious, and I guess I could cheat, but I really don't want to. I have three months to try to get my cholesterol down with my diet. I don't want to end up like Aunt Joan with all her heart problems. Or become a burden to John and the kids. Maybe I can bring a platter of grilled vegetables, and then I can come and enjoy?

Mother-in-law: Well, that sounds good, too. I must say you have more willpower than I do. You bring your grilled vegetables, and I can throw in some chicken, too.

Process the Experience.

- Ask your client these questions: What went well? What could have been done differently? How did you feel about the interaction? What may happen differently in the actual situation? Do you feel confident about how you will handle this situation when you encounter this problem?
- Provide feedback. Your comments should always be supportive and positive. For example, "It was really good that you . . ." If you have suggestions, they should be prefaced with tentative remarks, such as "You might try . . ." or "You could consider . . ."

Microanalysis of the Scenario This method is used to talk through an anticipated experience, identifying as many contingencies as possible and deciding on the best response. The following are some questions that could help lead the way in anticipation of a telephone call as illustrated in Exhibit 6.2:

- Where will you be when you make the phone call? What do the surroundings look like? What will be going through your head?
- What will you say to approach the topic of dinner on Sunday night?
- What will you say about cholesterol levels?
- What do you believe will be your mother-in-law's response?
- What do you want your mother-in-law to feel?

Mental Imagery A common technique for developing new behavior patterns, **mental imagery** involves a mental rehearsal of an anticipated experience. Clients imagine themselves thinking, feeling, and behaving in precisely the way they would like in the actual situation.[5] Counselors can help their clients reconstruct a past or potential scene in their mind and play out the scenario with a desirable ending. After microanalyzing the scenario, clients can be asked to close their eyes and play out the scene in their minds. After they have completed the exercise, ask whether any new concerns came to mind. Clients should be encouraged to do this activity several times before the actual encounter, thereby allowing them to practice their responses several times.

Social Disclosure

Closely related to social support is the concept of **social disclosure.** Disclosure of behavior records and progress in meeting goals to peers or professionals has been shown to exert a powerful influence in chang-

EXERCISE 6.2—Practice Using Microanalysis and Mental Imagery

Practice microanalysis and mental imagery with a colleague. Each should select an anticipated encounter, preferably an uncomfortable one, and take turns assuming the role of counselor. Record your experiences and impression of the technique in your journal.

ing behavior.[6] Even announcing the intent to engage in a new behavior can have a significant influence. Regular disclosure can be done formally in a weekly meeting with support buddies or informally during walks or while eating lunch with friends.

One of my clients was a busy professional who could not easily see how she could join a support group without putting additional stress in her life. However, she did feel that social disclosure would benefit her attempts to change her eating and exercise behavior. Her solution was to post her eating and exercise records each day in her office. Periodically throughout the day someone would inquire about the records, and this would precipitate supportive short conversations.

COGNITIVE RESTRUCTURING

The mind is its own place and in itself can make a heaven of hell, or a hell of heaven.
—John Milton

Another component of maintaining a behavior change is how the troubles encountered when attempting to continue with the new behaviors are perceived. Around 55 A.D., a Roman Stoic philosopher, Epictetus, maintained that difficulties related to problems are rooted in how problems are perceived rather than the actual troubles caused by the problems.[7] Today cognitive therapists embrace this concept of **cognitive restructuring**, focusing on identifying irrational thoughts and modifying them. The premise is that since **cognitions** (what and how a person thinks and perceives based on life experiences) are learned thinking behaviors, they can be relearned.[4] The objective is to change behavior patterns by changing destructive thinking patterns. Cognitive coping strategies, such as using positive self-talk, have been shown to effectively change lifestyle behaviors,[8] and in one study on smoking cessation these strategies outperformed behavioral methods.[9]

Thinking patterns have been categorized as *opportunity thinking* and *obstacle thinking*. Each mind-set can "influence our perceptions, the way we process information, and the choices we make in an almost automatic way."[10] A pattern of opportunity thinking allows finding constructive ways to deal with difficult situations. On the other hand, engaging in obstacle thinking leads to self-destructive behavior—making a difficult situation worse or giving up and retreating from problems. For example, an opportunity thinker diagnosed with high blood cholesterol is more likely to feel enthused by the challenge and may focus on the resources available to learn about new foods, cooking techniques, and support groups. In contrast, obstacle thinkers are likely to engage in self-pity and be bogged down in the difficulties of obtaining or preparing appropriate foods or adjusting their lives to take part in a support group.

Authorities have identified common cognitive distortions leading to obstacle thinking patterns that adversely affect attempts to change behavior.[11] They are listed in Exhibit 6.3 (p. 144) with examples of how they could be expressed in attempts to change lifestyle behaviors. Although the categories are presented as distinct entities, it is not unusual for several of them to manifest at one time. See Exhibit 6.4 on page 145 for an example.

The process of changing dysfunctional thinking addresses three factors:[9]

- **Internal dialogue.** All of us engage in an ever-constant dialogue that influences our feelings, self-esteem, behavior, and stress level.[9] By influencing this dialogue to provide self-enhancing messages, clients can better cope with difficult situations and are more likely to find the resources to take positive actions.
- **Mental images.** Athletes have used mental imagery to help produce a desired performance. By visualizing the accomplishment of an intended task, clients are more likely to attain an intended goal.
- **Beliefs and assumptions.** Core beliefs are deeply ingrained, leading to assumptions that trigger automatic thoughts. See Exercise 6.4 (p. 145) for an exercise to evaluate your core beliefs. For example, a core belief could be "If I do everything right, I will not have any health problems." The assumption is that everything right must be done, which could influence the development of some of the distorted cognitions listed in Exhibit 6.3.

Changing patterns of thinking that have been part of a person's makeup for many years can take a great deal of effort. The process of changing cognitions is a complex process, and there are psychotherapists that specialize in this type of therapy. Nutrition counselors, however, could incorporate some cognitive interventions into their sessions. The steps of this process are as follows:[10]

1. **Education.** Many individuals are not aware that thoughts are controllable. Your first step in cognitive restructuring is to educate your clients about this

EXHIBIT 6.3—Countering Negative Thinking

People who are attempting to make lifestyle changes need to guard against destructive negative thinking. For example, a person who says such statements to himself or herself as the following needs to find substitutions:

- *"A physical activity program is out for me. A woman at the gym said I should be ashamed of myself! She is right! I am going to ignore the people in my support group who say I should accept and love myself. Besides, I tried walking once, but I got a blister. That just goes to show that I wasn't made for exercise! Probably if I walked every day my blood pressure wouldn't come down anyway."* The talk exhibited here will certainly lead to defeat. This woman is focusing on negative feedback, generalizing that a single blister indicates she should stop walking, and assuming the worst outcome will happen. This talk could be transformed into "I am still searching for a way to make physical activity work for me."
- *"I did have an orange for a snack instead of my usual donut, but that didn't mean much because later I ate some potato chips. First, I just ate just one chip, and then I figured, 'this is absolutely awful so the diet is over and I might as well eat the whole bag of chips.' I guess I just don't have willpower. I am such a jerk!"* This person should give herself more credit for eating fruit for a snack, and focus on how and why that success happened. Identifying an episode as "awful" is not helpful since such a label can lead to a feeling that the situation is so bad that a solution is impossible. Using the word "absolutely" compounds the negativity of the phrase. The idea that once she started eating chips, there was no use stopping often happens when certain foods are considered off-limits. Also, blaming an indulgence on lack of willpower is always counterproductive since the characteristic is considered a personal failing so a change in lifestyle could not possibly occur. Instead, this woman should ask herself what she learned from the situation, and tell herself what she will do differently next time. In addition, all clients should be discouraged from using derogatory terms to describe themselves, such as "jerk." Once denigrated, a person is not likely to expect success in future attempts at lifestyle change. Instead, the person should remind himself or herself that he or she is learning so that better choices can be made in the future.
- *"I really do not want to go on the hayride because I must not have the cookies and hot chocolate afterwards. The other people in my group ought to be more considerate of the fact that I have diabetes. They should know better. I can't stand this!"* Sometimes people focus on one small difficulty and distort the total picture. Instead of searching for acceptable options, this obstacle thinker is caught-up criticizing others. This person uses words such as "should," "ought," and "must" to create an impossible standard, resulting in negative feelings and can lead to a relapse.

concept, reminding them that just because a thought pops into our head does not mean it is obligated to stay. In particular, we do not want to allow self-destructive thoughts and irrational messages to remain. They will influence our actions, and our behavior is likely to be counter to our lifestyle objectives. A leading psychologist is reported to have written, "One of the most significant findings in psychology in the last twenty years is that individuals can choose the way they think."[10]

Counselor: *Possibly your thoughts may be hampering your progress to make behavior changes. Some people are surprised to learn that we have control over what messages our brains deliver. It has been found that by directing our self-talk, we can improve the outcomes of our behavior change attempts.*

❷ **Identify dysfunctional thinking.** Analyze existing beliefs and assumptions, self-talk messages, and mental imagery patterns.

- Sometimes clients are well aware of their negative thought patterns but may have never thought they could be changed.
- **Show Exhibit 6.3** and ask your client whether he or she can identify with any of the common cognitive distortions.

EXHIBIT 6.4—Example of Destructive Self-Talk Using Cognitive Distortions

Mario is fifty years old with a high blood cholesterol level. He attended a holiday party after a stressful day at the office. At the party, Mario told himself it would be OK to eat some higher-fat foods since he didn't have lunch. However, once he stared eating and drinking alcohol, Mario ate some fruit and what he considered too many avoid foods, including several types of high-fat cheeses and cold cuts, various pastries, fatty snacks, and vegetables with dip. After leaving the party, Mario was annoyed and went home and ate some of his daughter's Halloween chocolate. He said to himself, "I deserve to have a heart attack the way I eat. This is awful. I am a terrible person. I should have prepared a lunch. I should have eaten something before I went to the party. I should have eaten more vegetables. I'm surprised I ate any fruits at all. How did that happen? I might as well continue to blow the diet and eat the chocolate. I'll never be able to eat right."

- **Review and analyze a situation your client identified as difficult.** The following are some questions to help explore cognitions:

Counselor: *What were your feelings before, during and after the event?*

What were you saying to yourself?

- **Keep a journal.** Recording thoughts and feelings before, during, and after the behavior change objective can help identify obstacle or self-enhancing thinking patterns. The Eating Behavior Journal, Lifestyle Management Form 5.1 (see Appendix G), can be used for this purpose.

3 Explore validity of self-destructive statements. Counselors can help their clients challenge their obstacle thinking patterns by exploring the validity of the internal messages.

- **Ask self-evaluating questions.** Probing questions rather than counselor evaluations aid clients in discovering inconsistencies.[12] This exploration provides a template for clients to challenge their irrational beliefs on their own. Possible questions include the following:[4,11,13,14]
  - **a.** Is the idea accurate?
  - **b.** What evidence exits that this idea is not correct?
  - **c.** Why is it so horrible and terrible that you ate that food?
  - **d.** Do people learn new behaviors by performing perfectly all the time?
  - **e.** Where is it written that you can not stand a situation?
  - **f.** Is there any factual evidence that supports this idea?
  - **g.** What are the worst things that could happen if what you must, should, or ought to do doesn't happen?
  - **h.** Are there good things that would occur if what should happen did not happen or what should not happen did happen?
  - **i.** What good does it do to focus on negative thoughts?
- **Use humor.** Corey[4] reports that humor is one of the most popular techniques of rational emotive behavior therapy (REBT) practitioners use to illustrate the absurdity of certain self-destructive ideas and to help clients not to take themselves so seriously. Greenberg and Jacobs[15] provide a humorous self-help approach to assist individuals in recognizing irrational thinking patterns.

Counselor: *Heavens! You should be boiled in oil. You ate the whole bag of chips!*

4 Stop destructive thoughts. Thought stopping is a technique that was developed to put an end to recurrent self-destructive thoughts and self-dialogue. It involves mentally saying the word *stop* and pushing away destructive automatic thoughts and substituting constructive thoughts.[16] To enhance the forcefulness of the word *stop,* a big red stop sign can be imagined or, if an individual is alone, a book can be slammed on a table or the back of a hand can be slapped. A constructive, affirming thought is then substituted.

5 Prepare constructive responses to substitute automatic dysfunctional cognitions. Once clients have recognized that they have obstacle-thinking patterns, a counselor should explore their openness to

EXERCISE 6.3—Evaluation of Destructive Self-Dialogue

In the example of self-dialogue in Exhibit 6.4, Mario has made several statements exhibiting cognitive distortions shown in Exhibit 6.3. Can you identify them? Write them in your journal.

EXERCISE 6.4—Core Belief Activity

Read the following statements and put a check next to the ones that apply to you.

- ❑ I need to have love and approval from peers, family, and friends to be worthwhile.
- ❑ I must not fail or make a mistake. I must be a success.
- ❑ Life should be easy, and I should not be frustrated. I can achieve happiness through passivity and inaction.
- ❑ I should always be in control of my emotions. I should be able to control negative feelings, never showing unhappiness or depression.
- ❑ I should never argue with someone I love.
- ❑ If I am alone, I will be miserable and not feel worthwhile.
- ❑ It is horrible when things or people are not as I expect them to be.
- ❑ All evil and wicked people should be punished.
- ❑ If someone criticizes me, something is wrong with me.
- ❑ I must live up to other people's expectations.
- ❑ I am ugly unless I have a perfect outward appearance.
- ❑ My worth depends on my achievements, intelligence, status, or attractiveness.
- ❑ If I do everything right, I will be successful.

Answers

If you checked six or more statements, you seem to view the world as all good or all bad. You are likely to be hard on yourself and engage in obstacle thinking when attempting to make lifestyle changes.

If you checked three to six statements, you are being too hard on yourself. Your tendency to be rigid may leave you feeling bad when you make mistakes or when things fall below your expectations.

If you checked fewer than three statements, you have positive views about life. You are more likely to set realistic goals and not to be discouraged when things do not go as you had planned.

In your journal describe your reaction to this activity. What did you learn about yourself? Is this an activity you would like to do with a client? Why or why not?

Source: Adapted from Wylie-Rosett J, Swencionis C, Caban A, Friedler AJ, Schaffer N, *The Complete Weight Loss Workbook*, pp. 155–156, with permission of the American Diabetes Association. © 1997.

preparing more effective thinking patterns. The following are some intervention techniques:

- **Identify and develop constructive thoughts to substitute for dysfunctional ones.** This can be done through using challenging self-evaluation questions, such as these:

Client: *Why must people treat me fairly?*

Is a hot dog the only reason for going to a ball game?

- Clients can also substitute opportunity thinking, such as this:

Client: *Rain makes walking outside inconvenient; it doesn't mean I need to stop my walking program. This shows I need to prepare for a rainy day. I will buy a good raincoat.*

- **Use imagery.** In this technique, an intense mental rehearsal is used to set new patterns of thinking. Ellis and Harper[13] describe this as an effective method involving clients imagining their feelings and self-talk in a worst-case scenario using previously established self-destructive thinking patterns and allowing negative feelings to emerge. Then a plan is made for a better response, and the imagined scenario is replayed using opportunity-thinking patterns. By repeating this exercise a number of times before encountering the activating event, clients will be better equipped to respond with nondestructive thinking when the event does occur. Events a nutrition counselor may visually imagine with a client could be ordering food at a fast-food restaurant, handling desserts at a holiday meal, or reducing cups of coffee or soft drinks consumed in a day.

❻ **Substitute constructive thoughts for destructive ones.** Replace destructive thoughts with previously prepared constructive thoughts—for example, "I learned that I shouldn't buy potato chips."

Rule Number 1: Don't sweat the small stuff.
Rule Number 2: It's all small stuff.
—Robert Eliot, M.D.

STRESS MANAGEMENT

Although stress is a normal part of life and can serve as a positive force to stimulate performance, too much stress will harm health and impair attempts to make lifestyle changes. Since food is often used to provide nurturing and stress reduction, finding alternative methods of coping with stress is very important. Also, stress has been found to be a major predictor of relapse, overeating, and dysfunctional eating patterns[17,18] and is linked to six leading causes of death—heart disease, cancer, lung ailments, accidents, cirrhosis of the liver, and suicide.[14]

Dr. Hans Selye, the scientist accredited with identifying the link between health and stress, defined *stress* as a nonspecific response of the body to threats or requirements for action or change.[20] The response is considered nonspecific because various physical, mental, and emotional factors could be affected. The physiological response stems from the stimulation of the hypothalamus from an imagined or real threat.[21] The hypothalamus activates the sympathetic nervous system to increase heart rate and blood pressure in order to deliver extra nutrients to muscles and the brain. Also, blood sugar and lipids rise to meet the anticipated increase in needs for energy, while breathing accelerates to supply extra oxygen for energy metabolism. Blood supply is diverted from the skin to large muscle groups. The release of stress hormones from the adrenal glands prepare the body to be on a heightened alert to make a quick response by shutting down tissue repair, digestion, reproduction, growth, and immune and inflammatory responses. The body prepares for fight or flight. The observable symptoms can include sweaty palms, rapid breathing, dilated pupils, dry mouth, nervous or shaky speech, crying, and a feeling of butterflies in the stomach, or heart in the throat.[20] If the stress response is continually being triggered, there is a negative effect on the mind, body, and quality of life, increasing the likelihood of developing one or more of the eight physical indicators of stress (see Exhibit 6.5). However, our bodies have a countering mechanism available that turns off the stress response and allows all systems to return to their normal state. Herbert Benson[22] refers to this natural restorative process as "the relaxation response."

EXHIBIT 6.5—Indicators of Stress

1. Increases in blood pressure
2. Suppressed immunity
3. Increased fat around the abdomen
4. Bone loss
5. Increases in blood sugar
6. Increased levels of cortisol
7. Weaker muscles
8. Increases in blood cholesterol levels

Source: McEwen BS, Protective and damaging effects of stress mediators. *New Engl J Med.* 1998;338:171–179.

Major life changes, whether positive or negative, are stressful events usually due to a series of new adjustments. Life-changing events include change in marital status, job status, financial status, birth or adoption of a child, death of a loved one, new residence, caring for a loved one with a debilitating illness, and diagnosis of a serious illness.[23] In addition to major life events, the National Institute of Mental Health has identified ten common sources of stress (see Exhibit 6.6). However, what is considered stressful for one person may not be the case for another since perceptions of events and conditions differ among individuals. This point is illustrated by stage-of-life data. American adults usually cite work, finances, and family issues as their top stressors, while older adults are likely to select social isolation.[19] Even within the same age category, differences exist as to appraisal of a situation as stressful. Stressed individuals are more likely to see difficulties as dangerous, difficult, or painful and are not likely to have the resources to cope with a problem.[21]

Stress management techniques have been successfully employed to enhance dietary lifestyle changes and to improve health.[24,25] Intervention strategies addressing these objectives can be divided into three general categories:

- **Strategies that focus on problem solving.** These strategies are problem-focused in that the goal is to

EXHIBIT 6.6—Ten Common Sources of Stress

1. Overscheduled daily calendars
2. Job stress/demands
3. Lack of play/downtime
4. Lack of time with family, friends, and significant other
5. Inequity in home responsibilities
6. Lack of time to explore own interests
7. Guilt (about everything)
8. In families: children's behavior and how to discipline
9. Lack of time
10. Lack of money

Source: Bradlye AC, Under pressure: Identifying and coping with stress. *American Fitness,* 1997;15:26–33. © 2000 Aerobics and Fitness Association of America.

remove or reduce exposure to specific stressors. They include cue management, time management, improving social relationships, improving health practices, assertiveness training, conflict management, communication skills, and engaging in distracting behaviors, such as doing puzzles.

Problem Solving

Glasgow et al.[26] describe a specific technique called STOP for systematically analyzing a problem and developing a solution. This problem solving method involves the following:

S—Specify the problem
T—Think of options
O—Opt for the best solution
P—Put the solution into action

- **Strategies that address the physiological response.** These include meditation, the relaxation response, visual imagery, soothing music, prayer, humor, emotion-focused coping, breathing exercises, exercise, and biofeedback.
- **Strategies that deal with cognitive coping skills.** These include cognitive restructuring and self-acceptance.

Some of these strategies have been covered in previous sections and chapters. Several excellent resources are available that explore these techniques in depth (see the resources listed at the end of the chapter for examples).

Laughter is inner jogging.
—Norman Cousins

Stress Management Counseling

As a nutrition counselor, you cannot possibly be an authority on all stress reduction methods. However, you can educate clients on the impact of stress on behavior change objectives, explore stress as an issue in your clients' lives, assist them in identifying stressors, provide information about possible stress reduction techniques, help clients locate stress reduction resources, and aid in developing stress-reducing behavior change goals. If your career path includes nutrition counseling, then you may consider obtaining training in some of the stress management techniques[27] or acquiring expertise in stress inoculation, a comprehensive approach to stress management.[28,29] The following are some general guidelines for implementing a stress management intervention:

Furnish Information about the Impact of Stress on Behavior Change Objectives

- **Explain reaction to stress.** Review two major components of stress: physiological arousal and internal monologue or thoughts that provoke anxiety, hostility, or pain.

Counselor: *When you feel stressed, two things are going on simultaneously. Physically your heart pounds, hands sweat, breathing rate increases, and you are likely to feel tightness in your muscles. Mentally your thoughts and self-talk are either intensifying the physical arousal with self-destructive statements, such as "I'm washed up," or they are providing tension reducing counsel, such as "I have learned a lot from this situation."*

- **Provide data about the impact of stress on food intake and health.**

Counselor: *Stress is an issue in many people's lives. It can severely affect health and impair attempts to change food behavior. We may be consciously or unconsciously looking for ways to calm down. Many of us naturally turn to food because we learned to associate it with nurturing.*

Investigate Whether Stress Is a Problem for Your Client and an Issue Your Client Wants to Address

- Review Symptoms of Stress, Lifestyle Management Form 6.1 (Appendix G). Possibly by reviewing a list, stress as a problem may be identified.
- Sometimes people are aware of feeling stressed but are not aware of the causes of their stress. If this is the case, authorities recommend using a stress log. See Stress Awareness Journal, Lifestyle Management Form 6.2.

Explore Possibilities for Reducing Stress

- Discuss methods and resources available for assisting individuals in reducing, minimizing, mastering, or tolerating stress. These include ways to problem-solve or reduce exposure to stressors, procedures to directly address physiological responses, and development of cognitive coping skills. A comprehensive stress management program would address all of them. You can help your client explore options by reviewing Tips to Reduce Stress, Lifestyle Management Form 6.3.
- Provide your client with community resource options. This requires that you have available a list of local referrals.

- Recommend books or other literature pertinent to your client's needs. Several excellent self-help books are available dealing with stress reduction. See the end of the chapter for a list of resources.

Set Behavior Change Goals That Address Stressors

- The procedure for goal setting is the same as that explained in Chapter 4.

There is no failure except in no longer trying.
—Elbert Hubbard

RELAPSE PREVENTION

There are many obstacles to initiating a lifestyle behavior change, but maintaining that change is a major challenge. Mark Twain once quipped, "It's not difficult to stop smoking—I've done it dozens of times." Relapse rates for dietary regimens are disconcerting, ranging from 50 to 100 percent.[12] However, these numbers could be misleading because they do not reflect the cumulative effect of multiple attempts to change dietary habits over time.[30] Described as a normal part of the change process in the transtheoretical model (see Chapter 1), relapse can in fact be part of a positive spiral that leads to enduring change. Also, relapse numbers do not account for self-changers, people who may have acquired skills in programs during previous attempts to change behavior but were not ready to follow through on their objectives at that time.

In response to this issue, Marlatt[31] developed a **relapse prevention** model that has been successfully employed for weight loss programs and addiction treatment interventions.[32,33] The premise of this model (see Figure 6.1) is to ascertain which factors are threats for relapsing and then to develop cognitive and behavioral strategies to prevent or limit relapse episodes.[34] There are two major categories of factors: *immediate determinants* and *covert antecedents.*[33]

EXERCISE 6.5—What Are Your Symptoms of Stress?

Look over the symptoms of stress listed in Lifestyle Management Form 6.1. Identify any that apply to you and record them in your journal. In the chapter assignment, you will compare these impressions to your stress record findings.

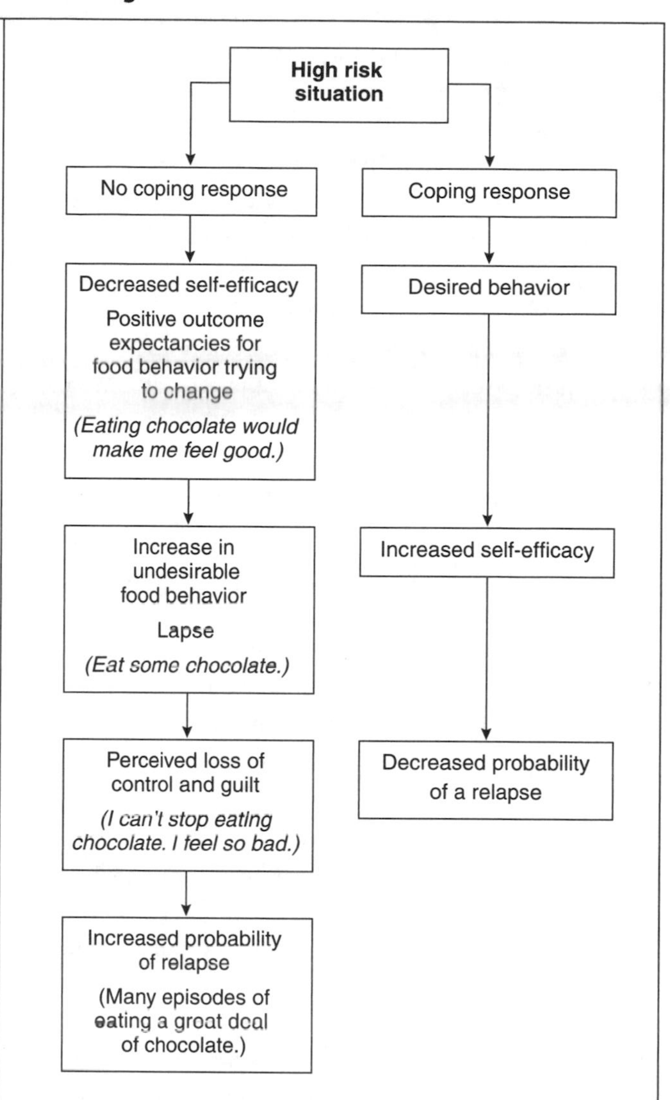

Figure 6.1 Cognitive-Behavioral Model of the Relapse Process
Sources: Adapted from Marlatt GA, Gordon JR, eds., *Relapse Prevention: Maintenance Strategies in the Treatment of Addictive Behaviors.* New York: Guilford; 1985, p. 38, by permission of Guilford Press.

Immediate Determinants

High-risk situations, a person's coping skills, positive outcome expectancies, and reaction to a lapse fall into this category.

High-Risk Situations Certain situations or events provide an alluring environment to revert to previously established behavior patterns. These high-risk conditions threaten a person's sense of control and frequently precipitate a relapse. The following are some of the most common high-risk situations:[33]

- **Negative emotional states.** Feelings of anger, anxiety, depression, or frustration as one may feel about an impending divorce or credit problem can precede a relapse. Also, feeling bored or lonely often leads to undesirable food intake.

- **Interpersonal conflict.** Conflicts with others, such as an argument with a friend, often trigger a relapse.
- **Social pressures.** Being around others who are eating foods that are to be avoided can result in direct and indirect pressure to relapse.
- **Positive emotional states.** Celebrations or events frequently serve as cues to eat certain foods and can provide a vehicle for a relapse.

Coping Skills Whether or not a high risk situation results in a relapse will depend on the ability and determination of the individual to cope with the difficulty.

Positive Outcome Expectancies Previous pleasurable experiences associated with particular foods and the significance of those perceptions can add to the desire to lapse in a stressful or high-risk situation.

Reaction to a Lapse A *lapse* is a single act, a slip, and a momentary return to previous behavior; and a *relapse* is a series of lapses, loss of control, a return to previous behavior. Lapses are a momentary indulgence that increases the likelihood of a full-blown relapse, but the progression is not inevitable.[33] Since no one is perfect, there should be preparation for handling a lapse. The way in which the slip is viewed is an important predictor of a relapse. If the lapse is considered a personal failure ("I am a failure. My wife is going to be so disappointed.") or due to a global attribute ("I have no willpower. I will always eat the wrong foods."), the probability of a relapse increases. On the other hand, if the slip is viewed as a learning experience providing the groundwork for developing more effective strategies for the future, relapse is less likely to occur.

Covert Antecedents

Some factors that precede a relapse are not obvious and are referred to as covert antecedents. These include apparently irrelevant decisions, lifestyle imbalances, and urges and cravings.

Apparently Irrelevant Decisions (AIDs) A series of seemingly harmless minidecisions can provide a conduit for a relapse. For example, buying a bag of potato chips for the "children" or a bag of cookies "in case guests stop by" creates conditions that can bring an individual to the brink of a relapse.[33]

Stress Level A person experiencing a high degree of stress is automatically generating negative emotional states, thereby creating high-risk situations (see Exhibit 6.7). In addition, there is an increased desire to counteract the emotional states with previously learned eating pleasures.[33]

Cognitions Cognitive factors such as rationalization and denial set the stage for a relapse. For example, "I deserve a whole batch of brownies after this rejection." Here rationalization that the indulgence is justified adds to the creation of a relapsing environment.

Urges and Cravings The desire for immediate gratification can take the forms of *urges,* sudden impulses to indulge, or *cravings,* subjective desires to experience the effects of an indulgence.[33]

Relapse Prevention Counseling

Relapse prevention programs were originally designed as a follow-up treatment to maintain gains made during an intervention. More recently, therapists report that they are integrating relapse prevention strategies throughout the course of interventions.[1,34,35] As one authority remarked, "Life is a chronic relapsing condition."[36]

Many of the strategies previously covered can be integrated in a relapse prevention program. Most versions of

EXHIBIT 6.7—Examples of Common High-Risk Situations

- Negative emotional states: anger, anxiety, depression, frustration, boredom, loneliness
- Positive emotional states: celebrations, events
- Conflict with others, arguments
- Social gatherings, parties
- Holidays
- Traveling and vacationing
- Eating out
- Snacking
- Lack of coping skills
- Negative self-talk
- Stress
- Hunger, urges, food cravings
- Fatigue
- Lack of social support

such a program incorporate three main categories of strategies: skills training, cognitive restructuring, and lifestyle balancing. Let's look at some of the specific components of these strategies.

Description and Introduction To begin, you may want to describe the relapse prevention model and introduce the concept of high-risk situations. Larimer et al.[33] suggest using the metaphor of a road journey to introduce the concept of anticipating high-risk situations and preparing a "toolbox" for dealing with them.

Counselor: *Behavior change can be compared to a highway journey that has easy and difficult stretches. The difficult portions can be handled effectively by being prepared (having a road map, a spare tire, a cell phone, and so forth), paying attention to road signs (warning signals), and utilizing driving skills learned previously for handling troublesome conditions. In your journey to change your food behavior, there are high-risk situations, too, and during our sessions together we will work on identifying them and the preceding warning signals. Also, we will focus on skills that you already have and incorporate them into coping strategies.*

Identification of High-Risk Situations To anticipate and prepare for high-risk situations, the conditions that precede a lapse must be identified. Attention should be paid to warning signs such as stress or apparently irrelevant decisions. The following list some methods of identifying situations where coping difficulties could arise:

- Investigate past lapse and relapse episodes.
- Journal activities, cognitions and eating behavior. See Lifestyle Management Form 5.1 or 6.2.
- Review a list of common high-risk conditions. See Exhibit 6.7.

Behavioral and Cognitive Strategies to Deal with High-Risk Situations The metaphor of a toolbox can be used to describe the availability of a variety of coping strategies. Since all strategies are not applicable to every temptation, clients need a repertoire from which to choose. Sometimes several strategies can be used at the same time. For example, behavioral strategies for dealing with temptations at a cocktail party could include eating a small snack before the event, bringing a vegetable platter or other dish with acceptable food choices, and making a contract that a glass of water will be drunk before tasting any party foods an individual is trying to avoid. In addition, positive self-talk strategies could be employed. A number of the strategies previously covered can also be utilized to deal with high-risk situations: tracking, cue management, countering, eating behavior interventions, reinforcement, contracting, goal setting, modeling, stress management, relaxation techniques, social disclosure, and cognitive restructuring.

Strategies to Minimize the Occurrence of High-Risk Situations By minimizing exposure to stressors, risk of encountering high-risk situations declines. This could include such activities as removing foods that one is attempting to avoid from the home, sitting in a different chair at the dining table or in front of the television, or taking a new route to work so as not to pass a particular bakery. This objective of relapse prevention incorporates cue management and countering techniques previously covered in Chapter 5.

Enhancement of Self-Efficacy One of the major objectives of the relapse prevention model is the enhancement of self-efficacy. Several ways of addressing this objective are possible:

- **Collaborative counseling style.** A collaborative counseling style is employed to encourage clients to take an objective view as well as ownership of their behaviors and goals. As a result, clients are more likely to feel a sense of accomplishment when goals are obtained, resulting in an increase in self-efficacy.
- **Set clear, modest, and achievable goals.** One of the objectives of setting goals is to encourage a behavior change to take place. Another objective is for clients to feel a sense of mastery over their problems when the goals are achieved. This increases the belief that additional or more complex goals can be achieved. The focus of goal setting, at least in the beginning of an intervention, should be on the smallest specific goal worth pursuing rather than on how much can be accomplished.[37] The overriding theme is that behavior change will occur as new skills are acquired rather than by employing will power.
- **Providing feedback.** Providing positive feedback about the accomplishment of new tasks, even if they were not related to the dietary objectives, can increase self-efficacy.

Lapse Management The objective of lapse management is to have a plan in place for handling a slip so that it does not escalate into a full-blown relapse.[31] Sometimes a written motivational set of instructions can be useful for a client to refer to in the case of a lapse. The following example contains two possible ways to introduce the topic:

Counselor: *Since no one is perfect, I thought we should talk about lapses and relapses. Lapses are momentary slips, such as eating a bag of potato chips. A relapse is a total abandonment of the food program objectives. Research has shown that if you handle the lapse effectively, there doesn't need to be a relapse. We could compare this to going on a trip. Let's assume you were driving to from New Jersey to Florida to go to Disney World. If you got lost in Baltimore, you wouldn't simply give up and go home. The same is true for a lapse. It should be treated as a stumbling block, not a reason to give up.*

It is a good thing that you had this lapse while we were still seeing each other. It gives us an opportunity to understand what happened and work out a plan to deal with the lapse. Actually lapses are a normal part of the change process. Let me show you a diagram of the process of change.

One of my clients stated that she did not believe she could make a food behavior change if there was no chocolate in her life. To deal with this craving and to keep her indulgence under control, she contracted to eat twelve chocolate chips every day. The chips were kept in her freezer. Each day she counted them out, put the chips into a small bowl, and immediately returned the package to the freezer. This plan worked well to satisfy her cravings, and the rest of her behavior change goals were also accomplished.

A good way to introduce the concept of relapse to your clients is to show them a visual representation of Prochasca's States of Change model and discuss the process of change (see Lifestyle Management 6.4 in Appendix G).

Cognitive Restructuring
Cognitive restructuring has been handled previously; however, self-talk related to lapses is particularly important in the relapse prevention model. Emphasis should be placed on lapses as a learning experience and the need to use or learn new skills. Clients should be warned not to fall into the trap of blaming themselves as a failure or having moral weakness and not to proclaim a prophecy that the slip means a total relapse.

Urge Management Techniques Clients need to realize that they are likely to experience an urge to return to previous eating styles. Ways of handling the desires should be addressed, such as the following:

- Contract to consume a certain amount of the food causing urges and cravings.

CASE STUDY 6—Amanda: The Busy Sales Rep

Amanda, age twenty-seven, is a sales representative for a large pharmaceutical company. She lives in a western suburb of Chicago, but her present territory involves the north side of Chicago. Her day is quite full. Typically she leaves her townhouse by 6:30 A.M. to make the drive into the city for an 8:00 call at a physician's office or a hospital. She loves her job, but the long hours and stressful lifestyle have taken a toll on her personal life as well as her health.

Despite efforts to eat well, Amanda cannot seem to control her weight. She is about forty pounds heavier than when she first started with the company despite her best efforts to diet. She believes that the nature of her job makes eating well impossible and has all but given up on any hope of losing weight permanently. She is out of town at least ten days a month and spends a good deal of time in airports or in her car. A shake and a burger do the trick, as she never quite knows when her next meal might be. Her many business meetings and associated social engagements have food as a central focus. When she finally gets home, she finds her cupboards are bare, including her refrigerator. She is never home long enough to use up any fresh produce or dairy.

Amanda's closest friend is Christine, who lives in her townhouse complex. She has known Christine for many years, and they have shared the efforts and woes of dieting. Christine is seventy-five pounds overweight and experiencing some depression at this time, mainly because of her weight. They socialize almost daily when Amanda is home and vacation together one to two times a year. Despite this great friendship, Amanda feels influenced by Christine's negative feelings toward life and is ambivalent about what she needs to do about the relationship to improve her own personal well-being.

EXERCISE 6.6—Case Study Review

What are possible sources of social support for Amanda? Suggest two changes in lifestyle that seem appropriate for her. Identify dysfunctional thinking patterns that could interfere with her attempts to change her behaviors. Identify a role-playing scenario that could be useful in a nutrition counseling session. How would you approach the topic of stress management with Amanda?

- Plan for a response when the urge arises. For example, the client could say to him- or herself that he or she can have the piece of cake after drinking a glass of water or eating an apple. Often the urge to eat the undesirable food has passed by the time the water or apple is consumed.
- Plan for a nonfood countering activity such as a relaxation response, knitting, jumping, and so on.
- Use an image technique called *urge surfing.* In this method, the client visualizes the urge or craving as a wave that crests and then washes onto a beach. Clients are told to imagine riding the wave on a surfboard rather than struggling against it or giving into the want.[33]

REVIEW QUESTIONS

1. Explain why dietary changes can put a stress on relationships.

2. Name ways social supporters can aid individuals in making dietary changes.

3. What explanation has been given to explain the benefits of having social support for making lifestyle changes?

4. Identify four ways an individual can search for social support if there is no one in the immediate environment readily available.

5. Explain the following techniques: role playing, microanalysis, and mental imagery.

6. Describe social disclosure.

7. Define cognitions.

8. Identify two categories of thinking patterns.

9. List fourteen examples of common cognitive pitfalls.

10. Name and explain three factors that can be addressed in the process of changing dysfunction thinking.

11. List six steps in a cognitive intervention.

12. Explain thought stopping.

13. Explain the physiological response to an anticipated stressor.

14. What countering mechanism is available to turn off the stress response?

15. Explain the STOP technique for problem solving.

16. Identify and explain four possible steps for a general stress management intervention.

17. Name two major components of stress.

18. What is the premise of the relapse prevention model?

19. What are four immediate determinants of a relapse?

20. Identify and explain four common high-risk situations for a relapse.

21. Name and explain four covert antecedents of a relapse.

22. What factors can lead a lapse to develop into a relapse?

ASSIGNMENT—Identifying Stress

The objective of this assignment is for you to become more aware of the stresses in your life. In Exercise 6.5 you identified your symptoms of stress. Use Lifestyle Management Form 6.2 (Appendix G) to record stressful activities, symptoms of stress, and internal self-talk at the time of the stressful event and at the time of the occurrence of the symptom. Note that the stressful event and the symptom may not occur simultaneously. In your journal answer the following questions:

1. Were the symptoms you identified in Exercise 6.5 congruent with your journal findings? Explain.

2. What were the main stressors in your life over the three days?

3. Review your internal dialogue entries. Was there evidence of cognitive distortions? Explain.

SUGGESTED READINGS, MATERIALS, AND INTERNET RESOURCES

Changing Self-Talk

Burns DD. *Feeling Good: The New Mood Therapy.* New York: Avon; 1999. A self-help presentation geared toward overcoming depression. An accompanying workbook, *The Feeling Good Handbook,* can be purchased as well. Burns also has a Web site, www.FeelingGood.com, with an abundance of information about therapies for both professionals and nonprofessionals. The tutorials on cognitive behavioral therapy are particularly interesting.

Ellis AE, Harper RA. *A Guide to Rational Living.* North Hollywood, CA: Wilshire; 1997. Helps readers deal more effectively with their lives by letting go of irrational beliefs and distorted ideas. The Albert Ellis Institute has a Web site, www.REBT.org, where additional information and resources can be found about rational emotive behavior therapy. The institute provides a variety of publications, tapes, training workshops, and several professional REBT certificates.

Investigating Assessment Forms

Thomas PR (Ed.) *Weighing the Options Criteria for Evaluating Weight-Management Programs.* Washington, DC: National Academy Press; 1995. Helpful for identifying and evaluating various assessment instruments related to self-esteem, body image, eating disorders, self-efficacy, dieting readiness, stress, social support, physical activity, and diet.

Guided Imagery

Guided Imagery Resource Center. Image Paths, Inc., 891 Moe Drive, Suite C, Akron, OH 44310-2538; (800) 800-8661; fax: (330) 633-3778; www.healthjourneys.com. Information about different types of imagery and procedures for ordering guided audiotapes for a variety of health conditions.

Humorous and Motivational Quotes

Be More Creative Web site. http://www.bemorecreative.com/quotesites.shtml. Links to more than one hundred quotation Web sites.

Stress Management Resources

Benson H, Stuart EM. *The Wellness Book: The Comprehensive Guide to Maintaining Health and Treating Stress-Related Illness.* New York: Simon & Schuster; 1992.

Davis M, Eshelman ER, McKay M. *The Relaxation & Stress Reduction Workbook.* 4th ed. Oakland, CA: New Harbinger; 1999.

Goliszek A. *60 Second Stress Management.* Far Hills, NJ: New Horizon; 1992.

Creative Arts Resources for Stress Reduction

MUSIC

- American Music Therapy Association, 8455 Colesville Road, Suite 1000, Silver Spring, MD 20910; (301) 589-3300; www.musictherapy.org.
- Campbell D. *The Mozart Effect.* New York: Avon; 1997.

DANCE

- American Dance Therapy Association, (410) 997-4040; www.adta.org.
- Halprin A. *Dance as a Healing Art: Returning to Health Through Movement and Imagery.* Mendocino, CA: LifeRhythm; 2000.

ART

- American Art Therapy Association, Inc. 1202 Allanson Road, Mundelein, IL 60060-3808; (847) 949-6064 or (888) 290-0878; www.arttherapy.org.
- Ganim B, Fox S. *Visual Journaling.* Wheaton, IL: Quest Books; 1999.
- Malchiodi C. *The Art Therapy Sourcebook.* Los Angeles: Lowell House; 1998.

POETRY

- Fox J. Poetic Medicine: *The Healing Art of Poem-Making.* New York: Tarcher/Putnam; 1997.
- Milosz C. *Book of Luminous Things: An International Anthology of Poetry.* New York: Harcourt Brace; 1996.
- Moyers B. *The Language of Life: A Festival of Poets.* New York: Doubleday; 1995.
- National Association for Poetry Therapy, #280, 5505 Connecticut Ave., NW Washington, DC 20015; (202) 966-2536; www.poetrytherapy.org.

Breathing to Reduce Stress

Weil, Andrew. *Breathing: The Master Key to Self Healing.* The well-known physician offers this package: $16 for two audiotapes or $21 for the CD version; call (888) 337-8345 to order.

REFERENCES

[1]Stevens VJ, Corrigan SA, Obarzanek E, Bernauer E, Cook NR, Herbert P, Mattfeldt-Beman M, Oberman A, Sugars C, Dalcin AT, Whelton PK. Weight loss intervention in phase 1 of the Trials of Hypertension Prevention. *Arch Intern Med.* 1993;153:849–858.

[2]Grieco AJ, Garnett SA, Glassman KS, Valoon PL, McClure ML. New York University Medical Center's Cooperative Care Unit: Patient education and family participation during hospitalization—the first ten years. *Patient Educ Counsel.* 1990;15:3–15.

[3]Fishman T. The 90-second intervention: A patient compliance mediated technique to improve and control hypertension. *Public Health Rep.* 1995;110:173–179.

[4]Prochaska JO, Norcross JC, DiClemente CC. *Changing for Good.* New York: Avon; 1994.

[5]Corey G. *Theory and Practice of Counseling and Psychotherapy.* 5th ed. Pacific Grove, CA: Brooks/Cole; 1996.

[6]Stevens VJ, Rossner J, Hyg MS, Greenlick M, Stevens N, Frankel HM, Craddick S. Freedom from fat: A contemporary multi-component weight loss program for the general population of obese adults. *J Am Diet Assoc.* 1989;89:1254–1258.

[7]Kiy AM. Cognitive-behavioral and psychoeducational counseling and therapy. In: Helm KK, Klawitter B, eds. *Nutrition Therapy Advanced Counseling.* Lake Dallas, TX: Helm Seminars; 1995:135–154.

[8]Tsoh JY, McClure JB, Skaar KL, Wetter DW, Cinciripini PM Prokhorov AV, Friedman K, Gritz E. Components of effective intervention. *Behav Med.* 1997;23:15–28.

[9]Shiffman S, Paty JA, Gnys M, Kassel JA, Hickcox M. First lapses to smoking: Within-subjects analysis of real-time reports. *J Consult Clin Psychol.* 1996;64:366–379.

[10]Neck CP, Barnard WH. Managing your mind: What are you telling yourself? *Educational Leadership.* 1996;53:24–27.

[11]Burns DD. Feeling Good: *The New Mood Therapy.* New York: Avon; 1999.

[12]Holli BB, Calabrese RJ. *Communication and Education Skills for Dietetics Professionals.* 3d ed. Baltimore, MD: Williams & Wilkins; 1998.

[13]Ellis A, Harper RA. *A Guide to Rational Living.* Hollywood, CA: Melvin Powers Wilshire; 1997.

[14]Wylie-Rosett J, Segal-Isaacson CJ. *The Complete Weight Loss Workbook: Proven Techniques for Controlling Weight-Related Health Problems, Leader's Guide.* Alexandria, VA: American Diabetes Association; 1999.

[15]Greenberg D, Jacobs M. *How to Make Yourself Miserable for the Rest of the Century.* New York: Random House; 1987.

[16]Glanz K, Greene G, Shield JE. Understanding behavior. In: American Dietetic Association. *Project Lean Resource Kit.* Chicago: American Dietetic Association; 1995:142–189.

[17]Foreyt JP, Goodrick GK. Attributes of successful approaches to weight loss and control. *Appl Prev Psychol.* 1994;3:209–215.

[18]Kayman S, Bruvold W, Stern JS. Maintenance and relapse after weight loss in women: Behavioral aspects. *Am J Clin Nutr.* 1990;52:800–807.

[19]American Psychological Association. Psychology at work. Available at: http://helping.apa.org/work/index.html. Accessed May 1, 2000.

[20]*The Vital Years—How to Grow Better and Older.* Nutley, NJ: Hoffman–La Roche, Department of Community Affairs; 1985.

[21]Davis M, Eshelman ER, McKay M. *The Relaxation & Stress Reduction Workbook.* Oakland, CA; 1995.

[22]Benson H. *The Relaxation Response.* New York: Morrow; 1975.

[23]D'Arrigo T. Stress & diabetes. *Diabetes Forecast.* 2000; 53:56–61.

[24]Spence JD, Barnett PA, Linden W, Ramsden V, Taenzer P. Recommendations on stress management. *Canadian Med Assoc J.* 1999;160(suppl):S46–S51.

[25]National Institutes of Health (NIH), National Heart, Lung, and Blood Institute. Clinical guidelines on the identification, evaluation, and treatment of overweight and obesity in adults—The evidence report. *Obes Res.* 1998;6(suppl 2):S51.

[26]Glasgow RE, Tooberi DJ, Mitchell DL, Donnelly JE, Calder D. Nutrition education and social learning interventions for type II diabetes. *Diabetes Care.* 1989;12:1105–1110.

[27]Warpeha A, Harris J. Combining traditional and nontraditional approaches to nutrition counseling. *J Am Diet Assoc.* 1993;93:797–800.

[28]Meichenbaum D. *Stress Inoculation Training.* New York: Pergamon; 1985.

[29]Meichenbaum D. Stress inoculation training: A 20-year update. In: Lehrer PM & Woolfolk RL, eds. *Principles and Practice of Stress Management.* 2d ed. New York: Guilford; 1993:373–406.

[30]Shattuck DK. Mindfulness and metaphor in relapse prevention: An interview with G. Alan Marlatt. *J Am Diet Assoc.* 1994;94:846–848.

[31]Marlatt GA. Relapse prevention: Theoretical rationale and overview of the model. In: Marlatt GA, Gordon JR, eds. *Relapse Prevention.* New York: Guilford; 1985.

[32]Foreyt JP, Poston WSC. The role of the behavioral counselor in obesity treatment. *J Am Diet Assoc.* 1998;98 (suppl):S27–S31.

[33]Larimer ME, Palmer RS, Marlatt GA. Relapse prevention: An overview of Marlatt's cognitive-behavioral model. *Alcohol Res Health.* 1999;23:151–160.

[34]Laws DR. Relapse prevention: The state of the art. *J Int Violence.* 1999;14:285–302.

[35]Gorski TT. Relapse prevention therapy and how it grew. *Behav Health Management.* 1999;19:33.

[36]Miller WR, Jackson KA. *Practical Psychology for Pastors.* 2d ed. Englewood Cliffs, NJ: Prentice Hall; 1995.

[37]Botman JA. Seven steps toward behavioral compliance. *Behav Health Treatment.* 1997;2:8.

Chapter

PHYSICAL ACTIVITY

An early morning walk
is a blessing for the whole day.
—Henry David Thoreau

BEHAVIORAL OBJECTIVES:

- List benefits of regular physical activity.
- Explain risks associated with exercise.
- Describe Healthy People 2010 physical activity goals.
- Differentiate between moderate and vigorous physical activity.
- Explain barriers to becoming physically active.
- Clarify the role of a nutrition counselor in physical activity counseling.
- Evaluate physical activity readiness using standard assessment tools.
- Demonstrate physical activity counseling approaches for four motivational levels.
- Identify issues of concern for physical activity goal setting and action planning.
- Describe the basics of an introductory walking program.

KEY TERMS:

- **AEROBIC ACTIVITY:** physical activity requiring oxygen; usually sustained longer than 3 minutes; required to develop cardiorespiratory fitness
- **ANAEROBIC ACTIVITY:** physical activity not requiring oxygen; utilized during high-intensity activities and at the beginning of sustained aerobic activities
- **CARDIORESPIRATORY FITNESS:** reported as maximum oxygen uptake (VO_2 max)
- **FLEXIBILITY:** full range of joint motion without discomfort
- **MAXIMUM HEART RATE:** roughly 220 beats per minute minus age
- **MODERATE PHYSICAL ACTIVITY:** use of large muscle groups that are at least equivalent to brisk walking
- **MUSCULAR ENDURANCE:** repetitive muscle contractions over a prolonged period
- **MUSCULAR STRENGTH:** ability to generate appropriate force
- **PHYSICAL FITNESS:** set of attributes relating to the ability to perform physical activity; often viewed as cardiorespiratory fitness; components include flexibility, suitable body composition, muscular strength, and muscular endurance
- **VIGOROUS PHYSICAL ACTIVITY:** repetitive activities using large muscle groups at 70 percent or more of maximum heart rate

Improving the health of Americans through physical activity and good nutrition must become a national priority.
—Martha N. Hill, R.N., Ph.D., past president, American Heart Association[1]

Although media coverage of the need for exercise gives the general appearance of widespread interest, the actual number of North Americans actively involved in regular physical activity remains rather low. Authorities in the United States and Canada have voiced concern about a growing lack of exercise,[2,3] and some have referred to the problem as an "epidemic of inactivity."[4] In 1997, only 15 percent of adults performed the recommended amount of physical activity, and 40 percent engaged in no leisure-time activity at all.[5] Inactivity increases with increasing age, and women are less likely to be active than men.

Several initiatives have been instituted to address this serious problem. To raise awareness that physical activity is needed for a healthier life and can be included in many everyday activities, the Center for Disease Control developed a national campaign called Ready, Set, It's Everywhere You Go (www.cdc.gov/nccdphp/dnpa/readyset/index). In addition, the National Association of Governor's Councils on Physical Fitness and Sports have instituted a program, Let's Get Physical (LGP) 2000 Challenge, to encourage individuals to participate in regular, moderate physical activity and make nutritious food choices (www.physicalfitness.org/lgpstart). Healthy People 2010 identified improving health, fitness, and quality of life through daily physical activity as one of the focus areas of the national initiative.[5]

ROLE OF A NUTRITION COUNSELOR IN PHYSICAL ACTIVITY COUNSELING

Since physical activity is an integral part of fitness and the combination of nutrition and physical activity are a primary strategy for reducing the risk of coronary heart disease, hypertension, diabetes, and osteoporosis, nutrition counselors need to be knowledgeable about the principles of exercise[6] and utilize these principles with clients.[7] To incorporate physical activity issues into nutrition counseling effectively, a counselor needs the following:

1. **A basic knowledge about the relationship between physical activity and health.** Such information can be obtained from course work, readings, and attendance at professional meetings. Most of the physical activity professional organizations, government health programs, and nonprofit health associations have educational Web sites and materials. Nutrition counselors could consider enhancing their physical activity counseling skills by obtaining additional training and/or credentials in the area of physical activity.
2. **Collaboration with physical activity professionals.** Many nutrition professionals utilize the services of physical activity professionals for obtaining information and guidance.[6] Opportunities for interaction are often available in fitness facilities and at physical activity professional meetings.
3. **Referral resources.** Nutrition counselors should keep a list of physical activity professionals, diabetes educators, professional organizations, fitness facilities, and fitness-knowledgeable physicians.
4. **Educational resources for clients.** Nutrition counselors should have educational materials on hand for their clients. Fact sheets and pamphlets are available from many professional organizations. See the resources at the end of the chapter for books geared to individuals beginning an exercise program.
5. **Medical approval.** Clients should receive a medical evaluation before beginning or increasing physical activity.

Specifically, a nutrition counselor can do the following:

- Provide advice and guidance regarding the health benefits of regular exercise.
- Provide referrals.
- Explore community physical activity resources with clients.
- Collaborate with clients to set physical activity goals and objectives.

A survey of registered dietitians regarding the extent of their involvement in physical activity counseling revealed the following:

- Most (64 percent) provide physical activity advice and guidance regarding the role of physical activity in a person's life.
- Most (65 percent) felt they know enough about physical activity to get their client started on a program of physical activity.

PHYSICAL ACTIVITY AND FITNESS

Differences Between Physical Activity and Physical Fitness

Physical fitness relates to a set of performance and health attributes connected to the ability to perform activity.[3] Performance-related attributes include agility, balance, coordination, power, and speed.[5] Health-related attributes include body composition, cardiorespiratory function, flexibility, **muscular strength**, and **muscular endurance.**[5] Sometimes physical fitness is viewed as specifically **cardiorespiratory fitness** and is expressed as maximum oxygen uptake (VO_2 max).[8] Physical activities need to be performed to achieve physical fitness. **Aerobic activities** involve large muscles over a sustained period, generally thirty minutes or more, and require oxygen to provide energy. **Anaerobic activities** do not need oxygen for energy as would occur during high-intensity exercise or at the beginning of sustained aerobic activities.

Benefits of Regular Physical Activity

Physical activity has a favorable effect on musculoskeletal, cardiovascular, respiratory, and endocrine systems.[3] Poor diet and lack of exercise have been identified as the second leading actual cause of death in the United States[9] (see Figure 7.1). Regular physical activity increases quality of life and decreases risk of developing or progression of many of the leading causes of illness and death in the United States. Regular physical activity can benefit health in the following ways:

- **Lower mortality.** Even moderately active individuals have a lower mortality than sedentary individuals. One study found that a daily two-mile walk could add years to a person's life.[10]
- **Improve cardiovascular health.** Physically active people have a substantially lower overall risk for coronary heart disease. This favorable effect is thought to be due to larger coronary arteries and increased heart size and pumping capacity.[11] As the heart becomes more efficient, heart rates decline, resulting in resting rates of about sixty beats per minute in trained athletes.[12] Also, regular physical activity decreases serum triglyceride and increases levels of HDL (a cholesterol-carrying blood protein) associated with a lower risk of cardiovascular disease.[13]
- **Positive effect on blood pressure.** Recurrent physical activity has been shown to prevent or delay the onset of high blood pressure, and regular exercise reduces blood pressure in people with hypertension.[3] This positive effect is thought to be due in part to a reduction in circulating low-density lipoprotein (LDL) cholesterol and fats, thereby increasing flexibility of blood vessel walls.[12]

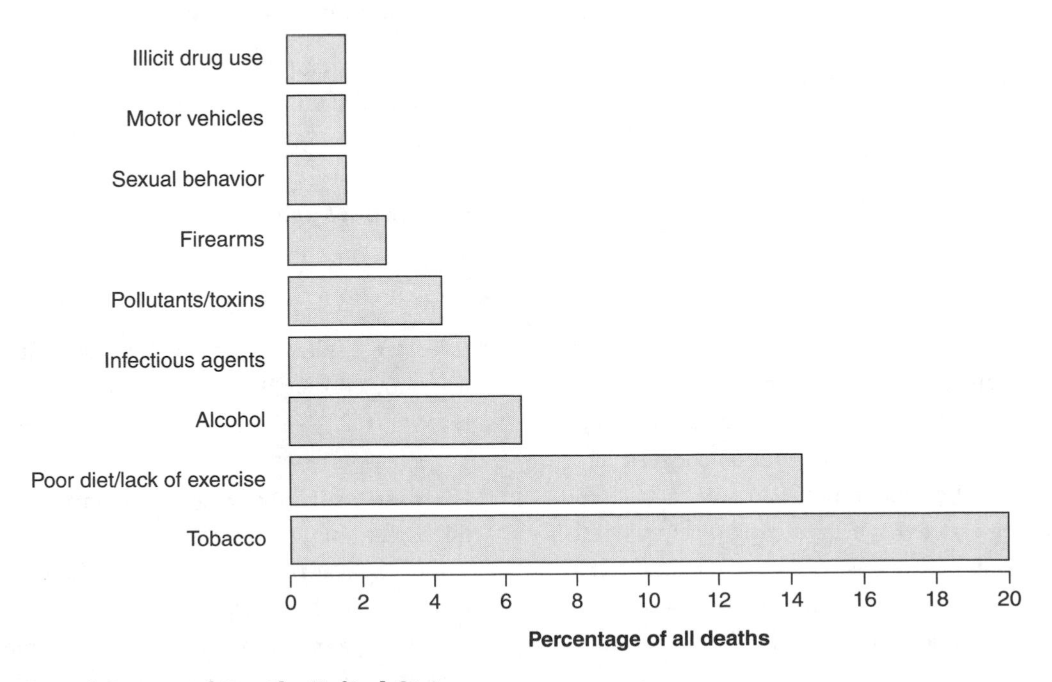

Figure 7.1 Actual Causes of Death, United States
Source: McGinnis JM, Foege WH, Actual causes of death in the United States. *J Am Med Assoc.* 1993;270:2207–2212. (1990 data).

- **Reduced risk of cancer.** In general, regular exercise is associated with lower rates of cancer.[12] There is a reduced risk of colon cancer, in particular, and possibly cancers of the prostate, endometrium (the lining of the uterus), and breast.[3]
- **Lower risk of developing diabetes.** Physically active muscles enhance insulin sensitivity, readily accept glucose, and lower blood glucose levels. As a result, regular physical activity lowers the risk of developing non-insulin-dependent diabetes mellitus and helps control blood sugar levels for those who have the condition.[3]
- **Weight control.** Since energy intake has not increased over the past few decades, several experts believe the expanding prevalence of obesity in North America is attributed to a progressive decrease in the amount of calories expended for work, transportation, and personal chores.[14] Regular physical activity is inversely related to rate of weight gain with age and plays a key role in long-term weight control and/or maintenance of weight loss.[14] Another advantage of regular physical activity is the favorable effect on body fat distribution away from the abdominal area, which is associated with the development of several chronic disease states.
- **Promotion of optimal bone density.** Physical activity, especially weight-bearing exercise, can contribute to building bone and help protect against osteoporosis.[15]
- **Enhanced psychological well-being.** Physical activity counters anxiety and depression and improves mood and the ability to cope with stress.[11] These benefits may be due to chemical alterations in concentration and/or activity of dopamine, norepinephrine, and serotonin. Also, the release of a morphine-like substance may contribute to the relief of pain and a feeling of euphoria.
- **Enhanced immune system.** Moderate physical activity is associated with resistance to colds and other infections; however, very strenuous activity can have an adverse effect on the immune system.[11]
- **Reduced risk of falls.** Falls among older individuals can be debilitating. Increasing evidence indicates that balance, strength, and flexibility training can reduce the risk of falling and allow maintenance of an independent living status.[3]

EXERCISE 7.1—Survey of Perceptions of Benefits of Physical Activity

Interview three people who do not exercise regularly. Ask each his or her beliefs about the benefits of exercise. Review the Benefits of Regular Moderate Physical Activity fact sheet, Lifestyle Management Form 7.1 (Appendix G), with the interviewee. Ask the individuals for their reactions. Record your observations in your journal.

Risks Associated with Exercise

Although there are numerous benefits to a physically active lifestyle, some risks are associated with exercise, especially at a high level of intensity.

- Sudden cardiac deaths, though extremely rare, are a serious concern. These occurrences are most often associated with primarily sedentary individuals who have preexisting coronary heart disease and engage in vigorous activity.[8] The risk of myocardial infarction or sudden death also slightly increases for well-trained athletes; however, the net effect of regular exercise is a lower risk than the risk of cardiac death associated with a sedentary lifestyle.[3] One study found the overall risk of cardiac arrest of habitually active men was only 40 percent of that of sedentary counterparts.[16]
- Musculoskeletal injuries are the most common harmful side effect of physical activity. Such injuries can generally be prevented by gradually working up to the desired intensity level and avoiding excessive physical activity.[3]

Note, too, that moderate physical activity is not associated with a significant risk of sudden cardiac death or musculoskeletal injury.[8]

Exercise Myths

Several myths about exercise persist that some clients may use to rationalize their inactivity.

- **Exercise causes arthritis.** Moderate physical activity is not associated with joint damage and in fact is recommended for individuals with arthritis during nonacute phases.[3]
- **Working out with weights is only for men.** Strength training tones muscles without making women appear overly muscular or "bulked up" like a professional male bodybuilder. There are numerous benefits for both men and women, including greater bone density, muscle strength, and balance.

- **It is dangerous for older people to start exercising.** All people can benefit from regular physical activity. Even people in a nursing home as old as ninety-eight have been able to improve their walking speed and ability to climb steps.[17]

PHYSICAL ACTIVITY GOALS

Physical activity goals have been shifting away from a focus on strenuous exercise to a broader range of health-enhancing physical movement. As evidence accumulated that **moderate physical activity** (use of large muscle groups equivalent to brisk walking) was a viable option for sedentary individuals, a number of national organizations (American College of Sports Medicine, American Heart Association, Centers for Disease Control, and National Institutes of Health) have released statements about the benefits of less vigorous exercise in addition to strenuous workouts. The Surgeon General's Report on Physical Activity and Health[3] concluded that moderate physical activity can significantly improve health and reduce substantially the risk of developing or dying from a variety of common ailments. However, greater health benefits can be attained by increasing the amount (duration, frequency, or intensity) of physical activity.[3]

Healthy People 2010[5] physical activity goals for the coming decade for United States citizens includes the following:

- Increase the proportion of adults who engage regularly, preferably daily, in moderate physical activity for at least thirty minutes per day.
- Increase the proportion of adults who engage in vigorous physical activity that promotes the development and maintenance of cardiorespiratory fitness three or more days per week for twenty or more minutes per occasion.
- Increase the proportion of adults who perform physical activities that enhance and maintain muscular strength and endurance.
- Increase the proportion of adults who perform physical activities that enhance and maintain flexibility.

Moderate Physical Activity

The minimum caloric expenditure able to produce health benefits is approximately 150 kilocalories per day, or 1,000 kilocalories per week. To achieve this goal, the general recommendation is to engage in at least thirty minutes of moderate physical activity a day. However, the nature of the activity alters the time requirement. The time needed for less intense activities needs to be longer to expend 150 kilocalories, and the time required to utilize this amount of energy during strenuous activities is reduced (see Exhibit 7.1). While sustained physical activity for at least thirty minutes is preferable, intermittent episodes of, say, ten minutes of exercise that add up to thirty minutes can be beneficial.

EXHIBIT 7.1—Examples of Moderate Amounts of Activity

A moderate amount of physical activity is roughly equivalent to physical activity that uses approximately 150 kilocalories of energy per day, or 1,000 kilocalories per week. Some activities can be performed at various intensities; the suggested durations correspond to expected intensity of effort.

Less Vigorous, More Time

Washing and waxing a car for 45–60 minutes
Washing windows or floors for 45–60 minutes
Playing volleyball for 45 minutes
Playing touch football for 30–45 minutes
Gardening for 30–45 minutes
Wheeling self in wheelchair for 30–40 minutes
Walking 1¾ miles in 35 minutes (20 minutes/mile)
Basketball (shooting baskets) for 30 minutes
Bicycling 5 miles in 30 minutes
Dancing fast (social) for 30 minutes
Pushing a stroller 1½ miles in 30 minutes
Raking leaves for 30 minutes
Walking 2 miles in 30 minutes (15 minutes/mile)
Water aerobics for 30 minutes
Swimming laps for 20 minutes
Wheelchair basketball for 20 minutes
Basketball (playing a game) for 15–20 minutes
Bicycling 4 miles in 15 minutes
Jumping rope for 15 minutes
Running 1½ miles in 15 minutes (10 minutes/mile)
Shoveling snow for 15 minutes
Stair walking for 15 minutes

More Vigorous, Less Time

Source: U.S. Department of Health and Human Services (USDHHS), *Physical Activity and Health: A Report of the Surgeon General.* Atlanta, GA: Department of Health and Human Services, Centers for Disease Control and Prevention, National Center for Chronic Disease Prevention and Health Promotion, 1996 (http://www.cdc.gov/nccdphp/sgr/sgr.htm).

Vigorous Physical Activity

While moderate activity is beneficial, there are cardiovascular advantages to engaging in vigorous physical activities involving large muscle groups. Healthy People 2010 documentation defines **vigorous physical activity** as that resulting in the individual reaching 70 percent of maximum heart rate; however, authorities caution sedentary, unfit individuals to aim for a lower level of 50 percent[18] or 55 to 64 percent.[19] These activities generally include running, fast walking, jump roping, swimming, bicycling, and stair walking.

Methods to Determine Level of Exertion

Perceived Exertion. A method used to determine if a vigorous level of exertion has been attained is the perception of an activity feeling "somewhat hard" to "hard" as indicated by breathlessness and/or a fatigue level.[20] Generally a physical effort should allow conversation, not singing, but if one is out of breath quickly or the activity must be stopped to catch breath, then the exertion is too much.

Heart Rate. Another common method used to determine level of exertion is to measure heart rate during or immediately after exercise and compare that rate to a target heart rate or a target zone. Your *target heart rate* is your recommended heart rate for exercising. Some authorities prefer to recommend a range of heart rates to aim for when exercising referred to as a target zone. If you are exercising above the target heart rate or zone, you are exercising too vigorously; if you are exercising below that level, you are not exercising strenuously enough.

Although various monitoring devices are available to measure heart rate, the usual procedure is to simply take a ten-second pulse at the base of the neck or the wrist. The following steps explain how to determine an appropriate target heart rate and/or target zone:

1. First ascertain the **maximum heart rate** (MHR). You can determine the MHR from a chart (see Table 7.1) or by direct calculations. The following is an easy, conservative formula for estimating maximum heart rate:

   220 − your age = MHR

   Note that some medications, such as beta-blockers, lower the MHR. If you have a client taking a beta-blocker, a physician should be contacted to determine whether the calculations need to be adjusted. To avoid the calculation difficulties for such people, perceived exertion is often the preferred method of monitoring intensity of physical activity.
2. Next, select a desired level of exertion. Less fit individuals should generally work at an intensity level of 50 to 70 percent of maximum heart rate, while physically fit individuals can aim for higher bouts of intensity of 70 to 85 percent of maximum heart rate.[8]
3. A heart rate that corresponds to a particular percentage of MHR can be calculated or read off a standardized chart to determine the target heart rate or the target zone. See Table 7.1 for a chart and Exhibit 7.2 for calculation instructions.

Muscular Strength

The American College of Sports Medicine recommends resistance training using standard equipment or free weights two to three times per week.[19] Recom-

EXHIBIT 7.2—Calculation of Target Heart Rate and Target Zone

Target Zone

Example: Calculation of the target zone of a less fit individual who is fifty-two years old

First, maximum heart rate (MHR) must be determined: 220 − 52 = 168 MHR. Less fit individuals should be working at 50 to 70 percent of MHR.

50% MHR = 168 × 0.50 = 84 beats per minute (10 seconds = 84 ÷ 6 = 14 beats)

70% MHR = 168 × 0.70 = 118 beats per minute (10 seconds = 118 ÷ 6 = 20 beats)

Target Zone = 84 to 118 beats per minute or 14 to 20 beats in 10 seconds

Target Heart Rate

A specific heart rate could be selected to aim for during exercise. The fifty-two-year-old may have been rather sedentary and should aim for a lower level. If 52 percent of MHR was selected, then the target heart rate can be calculated as follows:

52% MHR = 168 × 0.52 = 87 beats per minute (10 seconds = 87 ÷ 6 = 15 beats)

Target Heart Rate = 87 beats per minute or 15 beats in 10 seconds

TABLE 7.1—Maximum Heart Rate (MHR) and Target Heart Rate Zone

		TARGET ZONE 50%–70% MHR		TARGET ZONE 70%–85% MHR	
AGE	MHR	BPM*	10 S	BPM	10 S
20	200	100–140	17–23	140–170	23–28
30	190	95–133	16–22	133–162	22–27
40	180	90–126	15–21	126–153	21–26
50	170	85–119	14–20	119–145	20–24
60	160	80–112	13–19	112–136	19–23
70	150	75–105	13–18	105–128	18–21

*BPM = beats per minute; S = seconds.

mendations include at least eight to ten different exercise sets with eight to twelve repetitions each (arms, shoulders, chest, trunk, back, hips, and legs).[19] American Heart Association recommendations suggest moderate- to high-intensity (ten to fifteen pounds) strength training.[13] For weight training, clients should be referred to an exercise specialist to receive proper guidance.

All age groups can benefit from strength training; however the advantages to older individuals seem particularly significant. These include maintaining independence in performing activities of daily life, preserving bone, and reducing the risk of falling.[3]

Flexibility

Stretching exercises improve **flexibility** (range of motion), preventing the development of rigid joints by improving the elasticity of muscles, tendons, and ligaments.[5] Flexibility affects many aspects of life including walking, stooping, sitting, avoiding falls, and driving a vehicle.[5] Older adults particularly benefit from continued physical function contributing to independent living.

EXERCISE 7.2—Calculate a Target Heart Rate and Target Zone

Select a desired intensity level of exertion for yourself. Calculate the corresponding target heart rate and a target zone. and record the calculations in your journal.

EXERCISE 7.3—Record Your Physical Activity Patterns

Record your physical activities for seven days in the Physical Activity Log, Lifestyle Management Form 7.2 (see Appendix G). Compare your level of activity to the examples of moderate amounts of activity listed in Exhibit 7.2. Record your reaction to the activity and the evaluation in your journal. What did you learn from this assignment? Do you have any interest in making changes in your activity pattern?

Static stretching is often incorporated into warm-up and cool-down periods of aerobic activity. Popular longer programs that can greatly impact flexibility include yoga and T'ai Chi Chuan. Such stretching exercises are recommended at least two times a week.[19]

BARRIERS TO BECOMING PHYSICALLY ACTIVE

A major reason for the decline in physical activity in the past few decades is our high-tech society.[21] A great deal of time is spent behind a computer monitor or in front of a television instead of toiling on a farm or doing physical chores. The most common barriers to becoming physically active identified in the Healthy People 2010 report are lack of time, access to convenient exercise facilities, and

EXERCISE 7.4—Survey of Physical Activity Barriers and Benefits

Survey five people who do not engage in regular physical activity. Ask them why they are not physically active. Survey another five people who have a regular exercise program. Ask them how they have overcome barriers and what they believe motivates them personally. Record their responses in your journal. How did your findings compare to the most common physical activity barriers identified in the Healthy People 2010 report?

safe environments in which to be active.[5] Lifestyle Management Form 7.3, Physical Activity Options, provides some suggestions for overcoming barriers.

ASSESSMENT OF ACTIVITY LEVEL AND READINESS TO INCREASE PHYSICAL ACTIVITY

Some clients come to counseling who are already physically active and may even be participating in vigorous sports such as marathon running or dancing. If this is the case, the nutrition counselor's responsibility will be to address any special nutritional needs related to their physical activity. However, an assessment of their physical activity status may still be appropriate since they may not be meeting all the Healthy People 2010 physical activity goals.[5] Lifestyle Management Form 7.5 can be used as a quick self-assessment of physical activity status. Motivated clients who are not sure of their activity level may be willing to log their physical behaviors for a week. (See Lifestyle Management Form 7.2.)

Considering the low activity level of many people, nutrition counselors more often deal with an individual who is considering an increase in activity level. For these individuals, you have two readiness assessment responsibilities—medical and motivation. The medical assessment will guide you as to whether a referral to a physical activity professional is essential and the assessment of motivation allows you to tailor a counseling intervention approach.

Medical Assessment

Depending on motivation level, your role may be to help your client become aware of the importance of physical activity, or it may be to help guide them to actually make physical activity lifestyle changes. If the latter is the case, then you will need to assess whether a medical release or a referral is in order. The following provides some guidelines for handling this matter:

- **Physical Activity Readiness Assessment Questions.** The American Heart Association provides a set of ten questions based on the Physical Activity Readiness Questionnaire developed by the British Columbia Ministry of Health to guide individuals in deciding whether a medical assessment is necessary.[22] These questions are incorporated into the Physical Activity Medical Readiness Form, Lifestyle Management Form 7.4. If a client answers yes to any of the questions, then medical approval is required. However for your professional safety, a medical release is always advisable. The advantage of using these questions in your practice, even if a medical evaluation was a prerequisite, is to be sure that the medical professional is advised of any yes answers to provide a better evaluation of the situation.
- **Medical release form.** If you are using a medical release form for all clients to whom you are supplying nutrition counseling services, you may include a question asking whether there is any reason why an increase in moderate physical activity would be prohibited. See Lifestyle Management Form 7.6 for an example.
- **Delay physical activity.** Physical activity should be delayed in the following conditions:[8]
  - Women in the third trimester of pregnancy or experiencing a high-risk pregnancy
  - Clients experiencing acute symptoms (such as fever) during an illness or infectious disease
- **Referral to physical activity professional.** Referral to an exercise specialist is always useful, but it is essential in the following cases after receiving medical approval:
  - Clients with any symptoms of cardiovascular or metabolic disease
  - Clients recovering from coronary heart disease
  - Clients with severe bone or joint problems
  - Clients who begin exercise program and experience dizziness, chest pain, undue shortness of breath, difficulty breathing, or unusual discomfort
- **Referral to a diabetes educator.** Coordinating food intake, medication. and the demands of physical exertion can be complex. The integration can usually be done successfully but the guidance of a diabetes educator is advisable.

Motivation Level

As with counseling about food issues, physical activity counseling techniques need to be adjusted to a client's motivational level. See the physical activity counseling algorithm in Figure 7.2. Physical Activity Status, Lifestyle Management Form 7.5, can aid in determining motivation level as well as current endurance, flexibility, and strength activity patterns. The number a client circles on this form identifies one of four motivational levels: not ready, unsure, ready, and active.

ASSESSMENT FEEDBACK

The procedure for giving feedback about physical activity status is similar to the guidelines given in Chapter 3 for explaining dietary status to a client. A physical activity evaluation

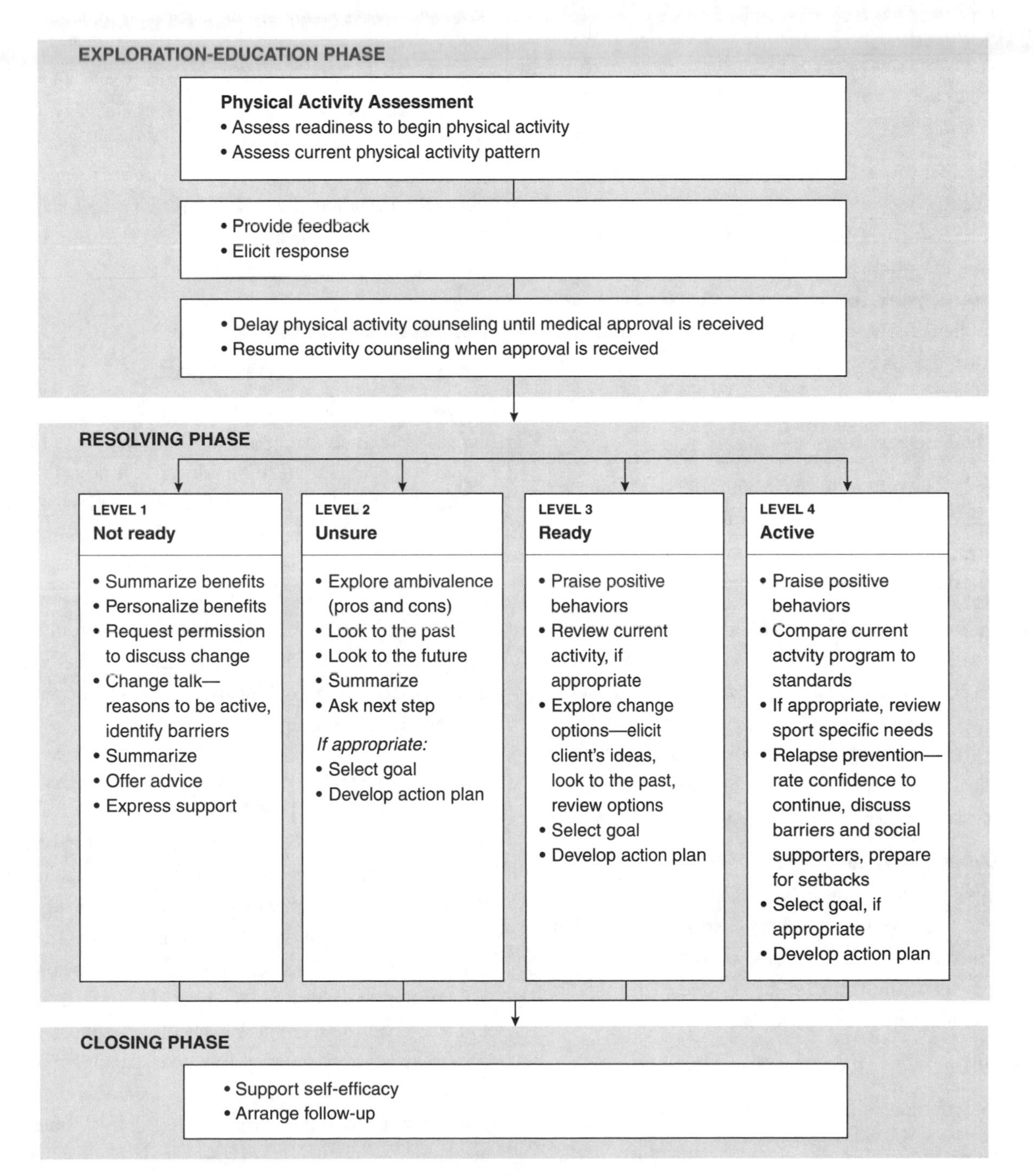

Figure 7.2 Physical Activity Motivational Counseling Algorithm

form to aid in that feedback is provided in the Physical Activity Feedback Form, Lifestyle Management Form 7.7 (see Appendix G). Healthy People 2010 goals and the American College of Sports Medicine position provide the national standards for assessing personal physical activity habits in this form.

While giving feedback, the objective is to create an environment of self-discovery in which clients can judge whether their present physical activity habits are congruent with where they would like to be. If an individual perceives a discrepancy, his or her motivation to change is likely to increase.[23] The procedure for giving feedback follows:

- **Give clear, concise nonjudgmental feedback.** The completed assessment form should not be simply handed to a client but gone over point by point with no hint of criticism.
- **If appropriate, elaborate on the assessment.** If your client shows concern or interest, certain components of the assessment may require elaboration. For example, clients may wish to have a description of the organizations that set the national standards or an explanation of the stages of change model. At this time it may be helpful to have on hand Lifestyle Management Form 7.1, Benefits of Regular Moderate Physical Activity, and Lifestyle Management Form 6.4, Prochaska and DiClemente's Spiral of Change.

Counselor: *The national standards used for this form were taken from the 1998 position statement of the American College of Sports Medicine, the premier international organization of fitness experts and sports medicine scientists, and from the Healthy People 2010 report, a document of the Department of Health and Human Services that identifies health related goals established by leading authorities across the United States.*

- **Elicit client response.** This is an opportunity for clients to make self-motivational statements. Counselors should not inform clients as to how they ought to feel but rather should encourage clients to express their thoughts and be allowed to form their own conclusions.

Counselor: *What do you think about this evaluation?*

Does this information surprise you?

- **Summarize.** At the end of an assessment feedback, Miller and Rollnick[23] suggest that a counselor summarize what transpired during this period. The summary should include (1) risks or problems that were identified; (2) client reactions, including self-motivational statements; and (3) a request for the client to add or correct your summary.

Counselor: *Daniel, the medical release form signed by your doctor did not indicate a need to avoid an increase in moderate physical activity, nor did your answers on the Medical Readiness Questionnaire. As for most people in North America, your present level of physical activity fell short of national standards for endurance, flexibility, and strength. Your motivation level for increasing physical activity was found to be unsure, which means you are thinking about a change but have not made a firm commitment to increase physical activity. You also said, "I guess I have a problem here." Does this seem to be where we are? Is there anything you would like to add or correct?*

PHYSICAL ACTIVITY COUNSELING PROTOCOLS

Counseling approaches for changing physical activity behaviors need to be tailored to a client's motivational level. A protocol developed for physicians[8] has been adapted here for nutrition counselors, and selected motivational interviewing strategies have been incorporated.[23,24] Refer again to the physical activity assessment and counseling algorithm in Figure 7.2. This algorithm incorporates four levels of readiness to change. Table 7.2 describes physical activity counseling approaches during the resolving phases for each level.

Implementation of Resolving Phase Counseling Approaches

In Chapter 3, the dietary counseling algorithm branched into three preaction counseling approaches addressing the needs of the majority of individuals who seek guidance for a nutritional lifestyle change. Although many clients are also in a preaction readiness stage for physical activity, the algorithm presented here, Figure 7.2, has a fourth branch to accommodate clients who are already physically active and may even be participating in vigorous sports. The counseling branches are presented as distinct approaches, but counselors need to be flexible to accommodate fluctuation in readiness to change that may occur during a counseling intervention.[23] Such cases will

TABLE 7.2—Resolving Phase: Physical Activity Counseling Approaches for Four Levels of Readiness

READINESS TO CHANGE		COUNSELING APPROACH
Not ready	**Goal:**	**Raise doubt about present level of physical activity.**
Score = 1	**Major task:**	**Inform and facilitate contemplation of change.**
	Approach:	• Summarize benefits of physical activity. • Personalize benefits to health status. • Request permission to discuss change. • Ask key open-ended questions to promote change talk. Elicit personal reasons to be active. Elicit identification of barriers to physical activity. • Summarize. • Offer professional advice, if appropriate. • Express support.
Unsure	**Goal:**	**Build motivation and confidence.**
Score = 2	**Major task:**	**Explore ambivalence.**
	Approach:	• Ask key open-ended questions to explore ambivalence. Client identifies advantages of not changing Client explores consequences of inactivity. Client identifies hoped-for benefits. • Look to the past. • Look to the future. • Summarize ambivalence. • Ask about next step.
Ready	**Goal:**	**Negotiate a specific plan.**
Score = 3	**Major task:**	**Resolve ambivalence and elicit a firm commitment.**
	Approach:	• Praise positive behaviors. • Review current activity program, if appropriate. • Explore change options. Elicit client's ideas for change. Look to the past. Review options that have worked for others, if needed. • Client selects an appropriate activity goal. • Develop an action plan.

(Continued)

require a cross-over of counseling strategies among the four approaches.

Level 1—Not Ready to Change (score = 1 on the Physical Activity Status Form, Lifestyle Management Form 7.5) The major objective of working with clients in this stage is to furnish a warm, nonjudgmental atmosphere while providing information and encouraging consideration of beginning an activity program.

Summarize Benefits of Physical Activity. David Satcher,[5] director of the Centers for Disease Control and Prevention, has stated, "Many Americans may be surprised at the extent and strength of the evidence linking physical activity to numerous health improvements."

TABLE 7.2—Resolving Phase: Physical Activity Counseling Approaches for Four Levels of Readiness (Continued)

Active	**Goal:**	**Continue the activity program.**
Score = 5	**Major task:**	**Prevent relapse.**
	Approach:	• Praise positive behaviors. • Review current activity program. • Review sport-specific nutrient needs. • Prevent relapse. Explain relapse prevention. Client rates confidence in ability to continue. Identify potential barriers. Explore solutions to barriers. Prepare for setbacks. Identify social supporters. • Set goal and develop an action plan, if appropriate.

Source: Adapted from Long B, Woolen W, Patrick K, Calfas K, Sharpe D, Sallis J, *Project PACE Phsyician Manual.* Atlanta, GA: Centers for Disease Control; 1992.

Lifestyle Management Form 7.1, a fact sheet addressing this concept, could be reviewed during your session to increase awareness.

Counselor: *Even though exercise seems to be a media buzzword, many people are not aware of the many ways in which even a moderate amount of physical activity can benefit health.*

Personalize Benefits to Health Status. Clients probably know that being physically active is a good idea, but they may feel with so many important things to do in life, it is hard to make each one a priority. Your clients may not have thought how increasing their activity level could be a benefit for them personally.

Counselor: *Mrs. Bernstein, physical activity could be particularly beneficial to you to help lower your blood pressure and aid in your effort to lose weight. Brisk walking just three times a week could produce results. You said you had some trouble sleeping at night. Increasing your activity could also help you to fall asleep more quickly and to sleep well.*

Ask Key Open-Ended Questions to Promote Change Talk. Helping clients think and talk about change can aid in the development of motivation to change.[24] However, since the client has already indicated no desire to change, it is generally best to begin this discussion with a tentative approach, requesting permission to discuss the issue.[25]

Counselor: *Would you be willing to continue our discussion and talk about the possibility of a change in your physical activity level?*

- **Elicit personal reasons to be active.** People are often more likely to be persuaded to change if they are presenting the arguments for change.[23]

Counselor: *We've looked at a list of reasons to have at least a moderate amount of physical activity in your life. If you were to start an exercise program, which of the benefits do you think would most likely apply to you?*

On a scale of 1 to 10, with number 1 being no increase in physical activity and 10 being "yes, I will start," what number would you select? That is interesting that you chose the number 2. How come not number 1?

- **Elicit identification of barriers to physical activity.** People at this motivational level may have given very little serious thought to why they do not become more involved in physical activity. Having clients identify reasons could lead to discussions of overcoming physical activity barriers. Physical Activity Options, Lifestyle Management Form 7.3, could be useful to review during this discussion. King[26] suggests assessing a client's psychological relationship to physical activity as a possible barrier. Negative feelings about exercise may have arisen from past embarrassments, punishments, or weight loss failures. Sometimes a negative physical experience, such as an injury or violation of the body, can hamper exercise attempts. If this is the case, then the issue should be explored with a psychotherapist. King also encourages the use of the term *joyful movement* for exercise to enhance a pleasurable perspective.

Counselor: *You chose number 2 on the scale. How come you didn't choose 4? What would have to change for you to select 4? Is there something I could do to help you get to 4?*

What would have to change for you to consider increasing the amount of joyful movement in your life?

Summarize. Summaries help reinforce what has been said, tying together various aspects of a discussion. They indicate to clients that you have been closely listening to them and prepare them to move on to a new topic. They encourage clients to rethink their position.

Give a summary of reasons not to change first and then give a summary of reasons to change. End your summary with any self-motivational statements your client may have made during the discussion and a "what's next" question.

Counselor: *Juanita, I'd like to review what we have talked about regarding physical activity. We looked at some of the benefits of being physically active, and you said you were surprised that health benefits could be attained with much less effort than you thought. You indicated that your biggest obstacles to beginning a physical activity program are finding time in your already busy schedule and that you have never found exercise to be enjoyable. I gave you some reading material to help you think about how to fit physical activity into your life, and you said you would look it over. You also said that increasing physical activity is something you should probably give some thought to. Was that a fair summary? Did I leave anything out? Where does all this leave us now?*

Be prepared to fully accept any answer to the "Where does all this leave us now?" question. Do not jump on a half-hearted ready-to-take action statement. In our eagerness to guide clients toward making a change, we may put too much weight on what was actually indicated. Before shifting gears and moving to action, probe the issue further with more open-ended questions.

Counselor: *When you came in today, you indicated that you were not ready for a change, but now it appears that you are having a change of heart. Is that correct? So we don't start going down the wrong path, I want to be sure that is what you really want to do. What can you say to convince me?*

Offer Professional Advice, If Appropriate. Well-timed and compassionate advice can aid in motivating behavior change.[23] In the ideal situation, a client asks for advice, but if that is not the case, then the counselor can ask permission to give advice.

Counselor: *Obviously you know my opinion that starting a physical activity program is what you should do, but of course only you really know what can work for you. Physical activity is important for everyone's health, and specifically for you I believe this would help you with your desires to lose weight and lower your blood pressure. A common procedure is to start walking ten minutes, three times a week, and to increase walking time gradually. What do you think?*

Express Support. Several authorities have documented that counselor expectations for a client's ability to change can produce a favorable outcome.[23]

Counselor: *Only you know what will work for you and when you will actually be ready to start an exercise program. When the time is right, I know you will be able to do it. I want you to know that I respect your decision and anytime you are ready, I'm available to help. If it is OK with you, I would like to bring up the topic periodically during our future sessions.*

Level 2—Unsure about Changing (score = 2 on the Physical Activity Status Form, Lifestyle Management Form 7.5) At this stage, clients have not made a firm commitment to change. The task at hand for clients and counselors is to build motivation and confidence to change.[23]

Ask Key Open-Ended Questions to Explore Ambivalence. Exploring ambivalence is very important to help clients tip the scales in favor of making a change.[23] This involves exploring the pros and cons of changing physical activity patterns. Barriers to increasing physical activity need to be discovered in order to explore ways to combat them. The fact sheet on physical activity barriers, Physical Activity Options (Lifestyle Management Form 7.3), could be useful to review during this discussion.

- **Client identifies disadvantages of changing.**

Counselor: *What do you believe are the reasons you haven't increased your level of activity?*

What problems would occur if you increased your level of physical activity?

What are some reasons why you would like things to stay just like they are?

- **Client explores consequences of inactivity.**

Counselor: *What concerns do you have about not increasing your level of physical activity?*

- **Client identifies hoped-for benefits.**

Counselor: *What are some reasons for increasing physical activity?*

What do you hope would be better for you if you increased your level of physical activity?

What are some good things that would happen if you were more physically active?

Look to the Past. There may be skills and resources that have worked for your client in the past that can be utilized in the present to build a more physically active lifestyle.

Counselor: *Can you think of physical activities you have done in the past that have been enjoyable?*

Look to the Future. By imagining the future, clients may be able to work out some ways physical activity could fit into their world.

Counselor: *I can see why you have some concerns about increasing your level of physical activity. Could we just stand back and imagine that you have an optimum amount of physical activity in your life? What would your life be like?*

Summarize Ambivalence and Reiterate Self-Motivational Statements. Periodic summaries throughout a counseling session can be useful for exploring ambivalence and for selectively emphasizing issues or self-motivational statements that could tip the balance in favor of making a change.

Counselor: *It sounds as if you have opposing feelings about exercise. On one hand, you have difficulty seeing how you can find time to exercise, and when you do have time you don't have the energy or the motivation to begin. On the other hand, your weight and blood pressure have been creeping up over the last ten years, and you know that some exercise would help both problems. Before you had children, you used to enjoy bike rides with a cycling group on the weekends. You said, "With all the ways that exercise could help me, I should be able to find a way to make an exercise program work." Did I miss anything? What are you thinking about this?*

Ask About Next Step. Ideally a client will be ready to select a goal and develop an action plan for increasing physical activity. However, you need to be ready to accept whatever direction your client wants go about planning a change. For your client to experience intrinsic motivation to change, he or she must perceive the freedom to choose a course of action.[23] The fact is that very few people like to be told what they must do. Feeling coerced into taking a certain course of action is likely to produce resistance, counterproductive to your counseling goals. If your client indicates that he or she would like to make a behavior change, then setting a goal and developing an action plan would be appropriate. See Chapter 4 for an explanation of the process.

Counselor: *What do you think your next step should be?*

Level 3—Ready to Change (score = 3 or 4 on the Physical Activity Status Form, Lifestyle Management Form 7.5) Individuals at this level of readiness have indicated that they are ready to make a change. When working with these clients, you should recognize that the objectives are to resolve any ambivalence, elicit a firm commitment to change, and develop specific goals and action plans.

Praise Positive Behaviors. Some clients who are classified in this category have begun limited amounts of activity, and those attempts should be praised.

Counselor: *It is really great that you have been doing some physical activities and are ready to increase your level of exercise.*

Review Current Activity Program, If Appropriate. Some people in the category have indicated that they are physically active on occasion. These activities may be expanded to develop a comprehensive program.

Counselor: *I see from your assessment form that you are physically active sometimes. What kinds of things do you do?*

Explore Change Options. While exploring viable options for increasing physical activity, Berg-Smith et al.[24] emphasize the need for a counselor to remain neutral while conveying the following messages to a client:

1. There are a number of physical activity options to choose among.
2. You are the best judge of what will work.
3. We will work together to review options.

- **Elicit client's ideas for change.**

Counselor: *I have a list of possible options for people who are initiating a physical activity program that we could go over, but you may already have definite ideas of what would work for you. After all, you are the best judge of what will work for you.*

- **Look to the past.**

Counselor: *When have you enjoyed being physically active in the past?*

- **Review options that have worked for others, if needed.** Reviewing the Physical Activities Options Fact Sheet, Lifestyle Management Form 7.3, could be useful.

Counselor: *Would you like to review the types of physical activities that have worked for others?*

Client Selects an Appropriate Physical Activity Goal. The goal-setting process covered in Chapter 4 can be applied to physical activity. Emphasis should be placed on activities that your client finds convenient and enjoyable.

Develop an Action Plan. The process of developing an action plan was also covered in Chapter 4 and should be applied to physical activity.

Level 4—Physically Active (score = 5 on the Physical Activity Status Form, Lifestyle Management Form 7.5) Some clients come to counseling who are already physically active and may even be participating in vigorous sports. When working with these clients, your major goal will be to have them continue with their program, and the major task will be to prevent relapse.

Praise Positive Behaviors. As discussed in the previous stage, positive behaviors should be praised to encourage continuation of a physically active lifestyle.

Counselor: *It is wonderful that you have such a physically active lifestyle.*

Review Current Activity Program. A physically active person may not be meeting all the guidelines for frequency, duration, intensity, flexibility, and strength training as set by national standards. Any problem areas should be identified and addressed. If Lifestyle Management Form 7.5 was not used for an assessment during the exploration-education phase of the counseling session, then you may wish to utilize it at this point to ensure that all aspects of fitness are covered.

Counselor: *You are doing so well with aerobic and flexibility activities; however, your strength training does not meet the national standards. What do you think about this?*

Review Sport-Specific Nutrient Needs. Clients may be engaging in a particular sport or at a level of intensity that gives rise to special nutritional needs.[27] These should be investigated and addressed with the client. Clients may also have questions about possible ergogenic aids (substance taken to improve physical performance) they have heard about. As in many areas of nutrition, there is a continual influx of new claims and supplements in the sports arena. One person cannot possibly keep up with every one of them. When these issues arise, you should tell your client that you are not familiar with the specific claim but you will use your professional resources to investigate the claim. Many professional organizations involved in physical activity have Web sites with educational reviews of new claims (see the end-of-chapter resources for some possible Web sites to visit). Listservs available through professional organizations such as the Society for Nutrition Education can also be helpful for posting a question.

Counselor: *Do you have any questions about foods or supplements related to your physical activity program?*

Prevent Relapse. A major task of counseling people at this level of motivation is relapse prevention. This can be addressed by helping your clients understand that perfection in a physical activity program should not be expected and setbacks occur, but they do not need to cause an abandonment of a physical activity program. Identifying potential barriers, preparing possible solutions for anticipated problems, and enlisting social support can also aid in relapse prevention.

- **Explain relapse prevention.** Often roadblocks that could interfere with continuation of a physical activity program can be anticipated, and preparations can be made to prevent the problem from resulting in a relapse.

Counselor: *You are doing really great in your physical activity program. The major task we have now in counseling is to work on making sure that your behavior continues. First let's look at how confident you are that you can keep up this activity pattern.*

- **Client rates confidence in being able to continue.** Confidence in ability to be physically active is an important component of a person's ability to maintain an active lifestyle.[5] This confidence can be assessed with the aid of the confidence assessment ruler, Lifestyle Management Form 3.1.

Counselor: *If we use the numbers on this picture of a ruler to represent how confident you are that you can maintain your present level of physical activity for the next three months, what number would you select? Number 1 indicates not at all, and number 12 denotes very confident.*

- **Identify potential barriers.** An important component to preventing relapse is identification of potential barriers to continuing a physical activity program. By planning for an effective response to an anticipated problem, the difficulty will not lead to a complete abandonment of the program. Key questions about the client's number selection on the continuum can identify what concerns may be interfering with his or her self-confidence.

Counselor: *How come you chose 10 instead of 12?*

What would have to change for you to select 12 instead of 10?

What has prevented you from keeping with your physical activity program in the past?

- **Explore solutions to barriers.** Once potential problems have been identified, your client should be encouraged to explore possible solutions.

Counselor: *You said vacations are a difficult time to maintain an exercise program. Do you have any ideas of how to overcome difficulties while traveling?*

- **Prepare for setbacks.** One important concept to explore with your client is that setbacks are common, are to be expected, and do not mean a total program should be abandoned.

Counselor: *In anyone's physical activity program, setbacks are to be expected. This could be because of illness, family responsibilities, work demands, house guests, or travel. Some times you can anticipate the difficulties and prepare for them, but other times continuing your physical activity program is simply is not feasible. If this is the case, what is important is to just start up again as soon as possible.*

Identify Social Supporters. Receiving social support has a key influence on physical activity levels.[5] Asking clients to identify social supporters and suggesting to clients to seek out their assistance is part of the successful PACE program.[8] Social support could come from loved ones giving words of encouragement or having someone to jog with on occasion. Social support can also come from organized groups, such as a canoe club. Counselors need to use their judgment about exploring this issue for someone who is heavily involved in a team sport such as soccer or one such as dance in which social support is intrinsic to the organized activity. Social support should be encouraged, but the point should be made that a physical activity program cannot rely on another person.

Counselor: *You are doing so well with your physical activity program, and I hope you continue to do well. Getting support from friends or relatives has been found to be helpful in maintaining a program. Can you think of someone who could be supportive? It would really be great if there was someone who was also interested in working on increasing or maintaining his or her level of exercise with whom you could discuss your progress.*

Set Goal, If Appropriate. Since some clients at this motivation level are heavily involved in a sport, goal planning may not be indicated. Follow Chapter 4 guidelines for goal setting, if necessary.

Develop an Action Plan, If Appropriate. Development of an action plan would only be appropriate if a goal was selected. Follow Chapter 4 guidelines for developing an action plan.

ISSUES PERTINENT TO PHYSICAL ACTIVITY GOAL SETTING AND ACTION PLAN DEVELOPMENT

The following are some factors to take into consideration that specifically apply to physical activity planning:

- **Initial goals should be modest.** To avoid soreness and injury and to maintain motivation, a sedentary individual contemplating an increase in physical activity should start with short sessions (five to ten minutes). The activity could be as simple as getting off a bus at an earlier stop or parking a car at a distant part of the parking lot and walking the extra distance.[3]
- **Increase gradually.** Injuries can be prevented by gradually building up to the desired amount of physical activity.
- **Take into consideration sustainable factors.** The physical activities chosen should meet the following criteria:

  Enjoyable—Clients should be encouraged to think creatively. Enjoyable activities could include folk dancing, bike touring, gardening, stair climbing, or kickboxing.

  Safe—Safe areas can include community parks, gyms, pools, malls, and health clubs. If a safe location cannot be identified, then areas of the home should be evaluated for providing space for exercise equipment, such as a stationary bike or treadmill.

  Convenient—For some people, increasing physical activity works best by including short activities throughout the day, such as using steps instead of an elevator or taking a short walk after lunch. If this procedure is chosen, then goal selection and monitoring may require special consideration. Some people have effectively used an odometer attached to a shoe or belt. Depending on the quality of the device, clients can set goals and monitor their progress in terms of miles covered in a day or number of steps taken.

  Affordable—Usually a variety of community programs and facilities are located in schools, community colleges, and universities. In the United States the Young Men's and Women's Christian Association (YMCA and YWCA) provide an array of exercise programs for a modest cost. Also, many malls are available to walkers early in the morning before stores open.
- **Plan for variety.** A physical activity program should include a variety of activities to maintain motivation and decrease the risk of injury due to over use of any particular muscle group.
- **Consider the daily routine.** A plan should have physical activity as part of a daily routine. Clients should be encouraged to use more physical activity when feasible, such as using stairs rather than an elevator or walking to a store rather than driving.
- **Plan for sustained activity.** To maintain the benefits of exercise from both endurance and resistance training, the activities must be continuous. Health benefits will decrease within two weeks if physical activity is considerably reduced, and if the inactivity is sustained, the gains will entirely disappear within two to eight months.[3]

EXERCISE 7.5—Incorporating Physical Activity

Read Case Study 7 on page 174 and explain how physical activity would benefit Bill. If you were Bill's dietitian, what role would you take in counseling him about physical activity? Describe a specific physical activity goal that would be appropriate for Bill to begin making a change in his lifestyle.

WALKING BASICS

Walking is frequently suggested as an introductory exercise for sedentary individuals.[28] This activity provides excellent cardiovascular and endurance advantages;[29] however, walking does not improve muscle power or reduce age-related muscle loss.[30] For most people, walking is inexpensive, accessible, safe, and enjoyable. The only expense is a pair of good walking shoes.

Several commercial and noncommercial organizations have developed guidelines for individuals embarking on a walking program. (See suggested resources at the end of the chapter.) The American College of Sports Medicine (ACSM) has three protocols for individuals beginning a walking program based on their fitness level; these are presented in Table 7.3.

TABLE 7.3—ACSM Walking Program

	DAILY WALKING TIMES*		
WEEK	**LEVEL 1 (NOT WALKED AEROBICALLY IN YEARS)**	**LEVEL 2 (SLIGHTLY MORE FIT THAN LEVEL 1)**	**LEVEL 3 (PARTICIPATED IN SOME AEROBIC WALKING)**
1	10	20	30
2	12	20	30
3	15	25	35
4	15	25	35
5	20	30	40
6	20	30	40

*Walking sessions should be preceded by two-minute, slow walking warm-ups. Level 1 warm-ups should be done walking in place.
Source: Adapted, by permission, from American College of Sports Medicine, 1991, *ACSM Fitness Book*, 2d ed. (Champaign, IL: Human Kinetics) 96–101.

CASE STUDY 7—Officer Bill

Bill Melia has been on the Los Angeles police force for twenty-six years. He has a good family life—married for thirty years with three grown children. He enjoys his job and is very happy with his career, although at times the stress can be a challenge. Overall Bill copes by planning great vacations with his wife Lola; they often go to Mexico to visit their parents.

Bill's recent physical exam revealed subclinical diabetes. The physician indicated that this is an early stage and if left unattended would evolve into Type 2 diabetes. His doctor warned Bill that he needed to make significant lifestyle changes, including weight loss, to control the situation. An appointment was made with the office dietitian to discuss lifestyle modifications that would be helpful in preventing progression of the disease.

This was Bill's first experience with the possibility of a major health problem, and the news was difficult for Bill to handle. He started feeling depressed about the possibility of being on a strict diet and taking insulin injections. Bill remembered his grandmother in Mexico taking shots every day, and the thought of this being part of his life greatly worried him.

As Bill drives to meet the dietitian, he thinks of all the changes he would have to make. Working balanced meals into his schedule as a police officer would not be easy. Since he is never quite sure when he will have time to eat, Bill has large meals when the opportunity arises. Any given call over his radio could result in hours without being able to eat, so the larger the meal, the better. He supposed his interest in returning to working out would be something positive to share with the dietitian. He used to love to lift weights for increased muscle strength, and surely this would be a positive activity toward controlling his diabetes. Yes, he thought, that would do it.

REVIEW QUESTIONS

1. Explain physical activity and physical fitness.

2. List and explain ten benefits of regular physical activity.

3. Identify the two most common risks associated with exercise and explain how to best prevent the harmful effects.

4. List and explain the Healthy People 2010 physical activity goals.

5. What is the general guideline for moderate physical activity?

6. Explain two ways that can be used to monitor level of physical exertion.

7. Explain how to determine maximum heart rate, target heart rate, and target zone.

8. Describe the American College of Sports Medicine recommendations for strength training.

9. Why do older adults particularly benefit from strength and flexibility training?

10. Identify three types of activities that can increase flexibility.

11. List the three most common barriers to engaging in physical activity identified in the Healthy People 2010 report.

12. Explain when a client should obtain a physician's approval before engaging in physical activity counseling.

13. When is referral to an exercise physiologist warranted for physical activity counseling?

14. When should physical activity be delayed?

15. Explain the physical activity counseling algorithm.

16. Identify physical activity counseling strategies for each of the four motivational levels discussed in the section on physical activity counseling protocols.

17. Identify six concerns that specifically apply to developing physical activity goals and an action plan.

18. Explain the basics of developing an introductory walking program for a sedentary individual.

ASSIGNMENT—Physical Activity Assessment and Counseling

The objectives of this assignment are to gain experience using the physical activity readiness assessment forms and the physical activity protocols. Pair off with a colleague and take turns counseling each other.

PART I. Use the following interview guide checklist to conduct the counseling session with your colleague. Possible counselor questions, statements, and responses are given in italics. Additional examples can be found in the chapter.

Preparation

- ❑ Review physical activity counseling algorithm in Figure 7.2 and protocols in this chapter, as well as goal setting and action plan development (see Chapter 4).
- ❑ You and your partner should each complete copies of physical activity assessment forms:
  - ❍ Client Assessment Questionnaire, Life Management Form 4.1
  - ❍ Physical Activity Medical Readiness Form, Life Management Form 7.4
  - ❍ Physical Activity Status, Life Management Form 7.5
- ❑ Exchange completed assessment forms and fill out a Physical Activity Feedback Form, Life Management Form 7.7, for your partner.
- ❑ Bring a blank card and copies of physical activity fact sheets:
  - ❍ Benefits of Regular Moderate Physical Activity, Life Management Form 7.1
  - ❍ Physical Activity Options, Life Management Form 7.3

Interview—Since the focus of this assignment is practicing physical activity counseling skills, not all phases of a counseling session will be addressed. One of you should take on the role of counselor and one the role of client. After completing the counseling experience, reverse the roles.

Provide feedback

- ❑ Point-by-point, clear, concise, nonjudgmental

- ❑ Elicit response
- ❑ Give summary
  - ❍ Identify problems
  - ❍ Reiterate any self-motivational statements—*I never thought much about weight training.*
  - ❍ Ask for additions or corrections
- ❑ Elicit response—*What do you think about this?*

Evaluate need for physician approval. This activity is a simulation. The physical activity counseling practice session should continue. However, if you or your partner require an evaluation from a physician, it should be understood that any goal setting or action plan discussed would not be implemented until receiving proper medical clearance.

Customize the counseling approach. Tailor your counseling approach to the motivational level of your volunteer client. The circled number on the Physical Activity Status, Life Management Form 7.5, indicates motivational level: 1 = not ready, 2 = unsure, 3 and 4 = ready, and 5 = active.

Not Ready (score = 1)

- ❑ Summarize benefits of physical activity—see Benefits of Regular Moderate Physical Activity fact sheet, Life Management Form 7.1.
- ❑ Personalize benefits—use the completed Client Assessment Questionnaire, Life Management Form 4.1, as an aid.
- ❑ Request permission to discuss the possibility of a change.
- ❑ Ask key open-ended questions—reasons to be active, barriers.
- ❑ Summarize.
- ❑ Offer advice if requested, or request permission to offer advice.
- ❑ Express support.
- ❑ Support self-efficacy.

Unsure (score = 2)

- ❑ Explore ambivalence—ask key open-ended questions.
  - ❍ Advantages of not changing
  - ❍ Consequences of not changing
  - ❍ Hoped-for benefits
- ❑ Look to the past—*Have you ever been physically active?*
- ❑ Look to the future—*What would your life be like?*
- ❑ Summarize ambivalence and reiterate self-motivational statements.
- ❑ Ask about next step.
- ❑ Set goal and develop an action plan, if appropriate. See Ready checklist below.
- ❑ Support self-efficacy.

Ready (score = 3–4)

- ❑ Praise positive behaviors.
- ❑ Review current activity program, if appropriate.
- ❑ Explore change options to develop a broadly stated goal.
  - ❍ Elicit client's thoughts.
  - ❍ Look to the past.
  - ❍ Go over list of possibilities if requested—see Physical Activity Options fact sheet, Life Management Form 7.3.
- ❑ Client selects an appropriate activity goal.
- ❑ Develop an action plan.
  - ❍ Investigate physical environment—*Do you have everything you need? Do you have walking shoes?*
  - ❍ Examine social support—*Is there anyone who could help you achieve your goal?*
  - ❍ Review cognitive environment—*What will you be saying to yourself if you miss a day that you planned to walk?*
  - ❍ Explain positive coping talk, if necessary.
  - ❍ Select tracking technique—chart, journal, and so forth.
  - ❍ Ask your client to verbalize goal.
  - ❍ Write down goal on a card and give it to your client.
  - ❍ Support self-efficacy.

Active (score = 5)

- ❑ Praise positive behaviors.
- ❑ Review current activity program; compare to standards.
- ❑ Review sport-specific nutrient needs, if necessary.
- ❑ Prevent relapse.
  - ❍ Explain need to discuss relapse.
  - ❍ Use Assessment Ruler, Life Management Form 3.1, to rate confidence to continue.
  - ❍ Identify potential barriers.
  - ❍ Explore solutions to barriers.
  - ❍ Explain that setbacks are common.
  - ❍ Identify social supporters.
- ❑ Set goal and develop an action plan, if appropriate. (See previous motivational level.)
- ❑ Support self-efficacy.

PART II. Answer the following questions in a formal typed report or in your journal. For formal reports, number and type each question, and put the answers in complete sentences under the question. For journal entries, number each answer.

1. Write a narration of the experience. There should be three titled sections in the narration: preparation, feedback, and counseling approach.

2. What counseling strategies had the greatest impact?

3. Explain the impact of the Lifestyle Management Forms and the picture of the ruler, if used.

4. Describe the experience of being counseled as compared to being a counselor.

5. What did you learn from the experience?

SUGGESTED READINGS, MATERIALS, AND INTERNET RESOURCES

Nutritional Needs for Exercise

Berning JR, Steen SN. *Nutrition for Sport and Exercise.* 2d ed. Gaithersburg, MD: Aspen; 1998.

Rosenbloom CA, ed. *Sports Nutrition: A Guide for the Professional Working with Active People.* 3d ed. Chicago: American Dietetic Association; 2000. This book contains sports-specific nutrition recommendations.

Guiding Clients in Designing a Fitness Program

ACSM Fitness Book. 2d ed. American College of Sports Medicine. Champaign, IL: Human Kinetics; 1992. This basic skills book can assist in designing both anaerobic and aerobic activities. Included are assessments for aerobic, strength, and flexibility and a step-by-step fitness program.

American Heart Association. *Fitting in Fitness: Hundreds of Simple Ways to Put More Physical Activity into Your Life.* New York: Time-Life Books; 1997; $4.99. This book provides an abundance of suggestions for putting physical activity into daily life.

Kratina K, King NL, Hayes D. *Moving Away from Diets—New Ways to Heal Eating Problems & Exercise Resistance.* Lake Dallas, TX: Helm Seminars. Expands on the concepts of exercise resistance and joyful movement.

Physical Activity Counseling Manual

Project PACE (Physicians-Based Assessment and Counseling for Exercise). Address inquiries to PACE Project, San Diego Center for Health Interventions, San Diego State University, Student Health Service, 5500 Campanile Drive, San Diego, CA 92182-4720; http://www.paceproject.org/. This practical manual was developed for physicians to improve physical activity counseling skills for clients at three stages of readiness.

Reproducible Handouts

Parlay International Catalog. Parlay makes a number of handouts related to various topics. They are available through www.parlay.com, or call 800-457-2752.

Walking

Walking protocols for various levels of fitness can be obtained from the following:

- *Pep Up Your Life: A Fitness Book for Mid-Life and Older Persons* is a booklet available from the American Association of Retired Persons, 601 E. Street, N.W., Washington, DC 20049. It is also available on the Web at www.aarp.org; search for Pep Up Your Life.
- *The SURESTART Walking Program* was developed for Cheerios and is available on the Web. Go to www.cheerios.com and search for SURESTART.
- *The Healthy Heart Walking Book: The American Heart Association Walking Program.* New York: Macmillan; 1995, $14.95.

Internet Resources

Several health organizations have Internet resources that provide practical guidelines for developing and maintaining a physical activity program.

American Council on Exercise

www.acefitness.com/

American Heart Association

www.justmove.org/

American Medical Association

www.ama-assn.org/consumer/gnrl.htm

Centers for Disease Control and Prevention

www.cdc.gov/nccdphp/dnpa/dnpaaag.htm

Shape Up America

www.shapeup.org/sua

Gatorade Sports Science Institute

www.gssiweb.com

Surgeon General's Report on Physical Activity and Health

www.cdc.gov/nccdphp/sgr/sgr.htm

REFERENCES

[1]U.S. Department of Health and Human Services (USDHHS). *Physical Activity and Good Nutrition: Essential Elements for Good Health.* Atlanta, GA: Department of Health and Human Services, Centers for Disease Control and Prevention, National Center for Chronic Disease Prevention and Health Promotion; 1999.

[2]Federal, Provincial, and Territorial Advisory Committee on Population Health. *Report on the Health of Canadians.* Minister of Supply and Services Canada; 1996.

[3]U.S. Department of Health and Human Services (USDHHS). *Physical Activity and Health: A Report of the Surgeon General.* Atlanta, GA: Department of Health and Human Services, Centers for Disease Control and Prevention, National Center for Chronic Disease Prevention and Health Promotion; 1996 (http://www.cdc.gov/nccdphp/sgr/sgr.htm).

[4]Rippe JM. The role of physical activity in the prevention and management of obesity. *J Am Diet Assoc.* 1998;98 (suppl 2):S31–S38.

[5]U.S. Department of Health and Human Services (USDHHS). *Healthy People 2010.* Atlanta, GA: Department of Health and Human Services, Office of Disease Prevention and Promotion; 2000.

[6]American College of Sports Medicine, American Dietetic Association, International Food Information Council. For a healthful lifestyle: Promoting cooperation among nutrition professionals and physical activity professionals. *J Am Diet Assoc.* 1999;99:994–997.

[7]Kong JC. Guidelines for the dietitian in counseling the obese patient on walking as an exercise prescription. *Top Clin Nutr.* 1994;9:59–66.

[8]Long B, Woolen W, Patrick K, Calfas K, Sharpe D, Sallis J. *Project PACE Physician Manual.* Atlanta, GA: Centers for Disease Control; 1992.

[9]McGinnis JM, Foege WH. Actual causes of death in the United States. *J Am Med Assoc.* 1993;270:2207–2212.

[10]Hakim AA, Petrovitch H, Burchfiel CM, Ross GW, Rodriguez BL, White LR, Yano K, Curb JD, Abbott RD. Effects of walking on mortality among non-smoking retired men. *N Engl J Med.* 1998;338:94–99.

[11]Davis JM. What can a physically active lifestyle promise? In: Wardlaw GM. *Perspectives in Nutrition.* 4th ed. Boston WCB McGraw-Hill; 1999:314–315.

[12]American Medical Association Fitness Basics. In: *AMA Health Insight.* http://www.ama-assn.org/insight/gen_hlth/fitness/fitnes2.htm; 1995–1999.

[13]Fletcher GF, Balady G, Blair SN, Blumenthal J, Caspersen C, Chaitman B, Epstein S, Froelicher DSS, Froelicher VF, Pina IL, Pollock ML. Statement on exercise: Benefits and recommendations for physical activity programs for all Americans. *Circulation.* 1996;94:857–862.

[14]National Institutes of Health, National Heart, Lung, and Blood Institute. Clinical guidelines on the identification, evaluation, and treatment of overweight and obesity in adults—the evidence report. *Obes Res.* 1998;6(suppl 2):S51.

[15]American College of Sports Medicine. *ACSM's Guidelines for Exercise Testing and Prescription.* 5th ed. Baltimore, MD: Williams & Wilkins; 1995.

[16]Siscovick DS, Weiss NS, Fletcher RH, Lasky T. The incidence of primary cardiac arrest during vigorous exercise. *N Engl J Med.* 1984;311:874–877.

[17]Fiatarone MA, O'Neill EF, Ryan ND, Clements KM, Solares GR, Nelson ME, Roberts SB, Kehayias JJ, Lipsitz LA, Evans WJ. Exercise training and nutritional supplementation for physical frailty in very elderly people. *N Engl J Med.* 1994;330:1769–1775.

[18]American Heart Association. *Physical Activity and Cardiovascular Health: Fact Sheet.* www.justmove.org/; 1999.

[19]American College of Sports Medicine (ACSM) Position Stand. The recommended quantity and quality of exercise for developing and maintaining cardiorespiratory and muscular fitness, and flexibility in healthy adults. *Med Sci Sports Exerc.* 1998;30:975–991.

[20]Borg GA. Psychophysical bases of perceived exertion. *Med Sci Sports Exerc.* 1982;14:377–381.

[21]American Dietetic Association. *Project Lean Resource Kit.* Chicago: American Dietetic Association; 1995.

[22]American Heart Association. *Fitting in Fitness.* New York: Times Books; 1997.

[23]Miller WR, Rollnick S. *Motivational Interviewing Preparing People to Change Addictive Behavior.* New York: Guilford; 1991.

[24]Berg-Smith SM, Stevens VJ, Brown KM, Van Horn L, Gernhofer N, Peters E, Greenberg R, Snetselaar L, Ahrens L, Smith K. A brief motivational intervention to improve dietary adherence in adolescents. *Health Education Research.* 1999;14:399–410.

[25]Snetselaar L. Counseling for Change. In: Mahan LK, Escott-Stump S, eds. *Krause's Food, Nutrition, and Diet Therapy.* 10th ed. Philadelphia: Saunders; 2000:451–462.

[26]King NL. *Counseling for Health and Fitness.* Eureka, CA: Nutrition Dimension; 1999.

[27]Berning JR, Steen SN. *Nutrition for Sport and Exercise.* 2d ed. Gaithersburg, MD: Aspen; 1998.

[28]Fletcher GF. How to implement physical activity in primary and secondary prevention. *Circulation.* 1997;96:355–357.

[29]Clark KL. Promoting a healthful lifestyle through exercise. In: Kris-Etherton P, Burns JH. *Cardiovascular Nutrition: Strategies and Tools for Disease Management and Prevention.* Chicago: American Dietetic Association; 1998:127–134.

[30]Evans WJ, Spokas D. *FITness from 50 forward: A Manual Describing How to Begin and Continue an Exercise Program.* Chicago: American Dietetic Association; 1998.

Chapter

8

PROFESSIONALISM AND FINAL ISSUES

Being a professional is doing all the things you love to do on the days when you don't feel like doing them.
—Julius Erving

BEHAVIORAL OBJECTIVES:

- Describe procedures for providing a smooth transition to ending a counseling relationship.
- Evaluate counseling effectiveness.
- Describe issues of importance in a professional setting.
- Describe the Corey model for ethical decision making.
- Utilize ADA Code of Ethics and Standards of Professional Practice to evaluate counseling interactions.
- Identify boundaries between psychotherapy and nutrition counseling.
- Describe strategies for dealing with client behaviors counselors find difficult.
- Identify ways to approach clients with low literacy skills.
- Describe the basics for implementing a group counseling program.

KEY TERMS:

- **CODE OF ETHICS:** published ethical standards by a professional organization
- **GOAL ATTAINMENT SCALE:** a rating system with a range of values used to assess success in meeting goals
- **STANDARDS OF PROFESSIONAL PRACTICE:** components of high-quality dietetic practice

Nothing so difficult as a beginning
In poesy, unless perhaps the end.
—Lord Byron, *Don Juan*

ENDING THE COUNSELING RELATIONSHIP

Life is a journey of change. Nutrition counseling, either a brief intervention or an intensive program, is part of that journey and at some point needs to come to an end. For brief encounters, the transition to terminating the counseling experience generally would not pose any special issues. However, for clients with whom there has been long-term involvement, several considerations should be taken into account.

Reasons for termination vary. Sometimes insurance coverage or program protocols impose a time limit. In other cases, the counselor, the client, or both will feel that counseling goals have been obtained or at least reasonable progress in attaining them has been made. On occasion a counselor may believe that a client should be referred to a psychotherapist before resuming work on nutritional concerns. Referrals would also be in order when problems emerge, such as bulimia, that could be best handled by a specialist. At times a counseling relationship needs to end because the counselor or the client is preparing to move away or take a new job. The following sections offer suggestions for ensuring a smooth transition to the end of your counseling support.

Preparation for a Wrap-Up Session

First, provide transition time. Final meeting dates should be agreed on before the terminating session. For time-limited programs, counselors can remind clients of the upcoming final meeting. Care should be taken, if at all possible, not to spring a decision to end a counseling relationship at the last session. Counselors and clients may be surprised at the intensity of the emotional response to the end of the interaction.

Discuss reasons for ending the counseling relationship. Share your perception of your clients' progress, and request your clients to voice their views.

Wrap-Up Session

Consider taking these steps for a smooth finale with your clients:

- **Review beginnings.** Discuss the issues that brought your client to you in the first place. You might review initial assessments.
- **Discuss progress.** Identify goals and progress in meeting them. Having your progress notes handy could be useful.
- **Emphasize success.** A review of accomplishments can be a source of much pleasure as clients remember how certain tasks were anticipated with dread (for example, taking glucose readings) or how stuck they felt in the midst of indecision. Miller and Jackson[1] encourage a discussion of the skills a client used to bring about change. Research indicates that clients who feel a major responsibility for a transformation will continue to experience success rather than those who attribute fairing well because of an extrinsic factor, such as a therapist. A counselor could ask, "What do you believe you did to bring about this change?"
- **Summarize current status.** Highlight current biochemical and physical parameters, coping skills, social support and/or challenges, and physical and/or environmental issues.
- **Explore the future.** How will the changes that have been made be maintained? How will old challenges be addressed and what new challenges are likely to be encountered? How will those difficulties be handled? Are there additional changes to be made? Who will offer support in the future? This is also the time to discuss a referral if one is being made.
- **Discuss future involvement.** Follow-up meetings are advantageous to both clients and counselors. Such meetings provide opportunities for reassessment of nutritional status and evaluation of goals and a time to reinforce previously set behavior changes. A periodic check-in arrangement could be negotiated that involves personal meetings, phone calls, or e-mail. Sometimes recommendations are made for clients to initiate contact if changes in clinical parameters (such as a five-pound weight gain) or a significant life event occurs (divorce, marriage, or death of a close family member). These interactions help counselors document progress to implement long-term evaluation of counseling effectiveness.
- **Provide and elicit feedback concerning the significance of the relationship.** Allow time to express what the meaning of the relationship has meant to yourself and to your client. This generally means

expressing appreciation for each other. Often clients express appreciation by saying counselors did so much for them. In this case, thank your client for the compliment, adding the reminder, "The reason you did so well was because of your hard work." Possible lead-ins to relationship discussions include the following:

Counselor: *It has been such a pleasure working with you. I want you to know that I will miss our weekly meetings.*

I want you to know that I really appreciate what you have taught me about Cuban culture.

- **Consider holding a ceremony and exchanging symbols.** You may think about a special location or activity for the last meeting. Some possibilities include a walk in the park, a lunch, or a different meeting room. Sharing a particular food, especially if it had a significant connection to your client's diet, can make a profound impression. Exchange of gifts will be based on program or facility policy; however, it is generally considered appropriate to give and/or accept an inexpensive gift as a symbol of completion. Mementos symbolic of work done together can be particularly meaningful, such as a wooden apple for a client who made a major effort to increase intake of apples. One counselor regularly writes an individualized letter of support and encouragement to be given to clients on the final meeting day or to be mailed after the last meeting.
- **Final good-bye.** Generally you would expect to acknowledge the end, shake hands, and walk with your client to the door, waiting room, or usual exit. Murphy and Dillion[2] suggest a final parting by telling your clients you will be picturing them doing something they have longed for or intend to accomplish, such as running a five-mile race, taking a cooking class, or completing a degree. Such statements reinforce that your client's future welfare is important to you and that you will always be rooting in his or her corner.

Handling Abrupt Endings

Unfortunately, all closures are not tidy. Sometimes clients simply do not show up for a session, or they cancel future appointments. If there is an abrupt ending to an involvement, consideration should be given to sending a termination letter or e-mail as an attempt to have a closure experience. In your communication, you may wish to reinforce achievements and leave the door open to a future association. See Exhibit 8.1 for an example of such correspondence.

EVALUATION

Evaluation is an important component of counseling for your growth as a counselor. After an intervention, counselors should take time to assess the quality of their skills and contemplate what could have been handled differently. In addition, there is increasing pressure from the managed care industry for health professionals to produce outcome measurements. To meet new mandates, nutrition counselors need to incorporate brief, efficient, and inexpensive assessment procedures into their counseling programs. Large facilities are likely to have a tracking system in place for evaluation. Licensing regulations differ around the country, but they often require what and how data need to be reported. For small facilities and for counselors in private practice, nutrition counselors will likely find the need to develop their own procedures for producing outcome data.

Evaluation of Client Progress

Several of the factors necessary for an effective evaluation of client progress have already been covered throughout this book. Parameters used for evaluation can include the following types of data: behavioral (food diaries, exercise records), physical (body weights, blood pressure), biochemical (cholesterol and glucose levels), and functional (length of hospital stay). The initial as-

EXHIBIT 8.1—Example of a Termination Letter

Dear Mary,

Since I haven't heard from you after our meeting on December 2, I wanted to touch base with you. I hope you are doing well. On two occasions I called and left a message but did not receive a reply. I am assuming that you wish to stop working together at this time.

If you wish to resume lifestyle counseling at a future date, I would be happy to work with you again. You made commendable changes in your exercise pattern as well as an increase in your fruit, vegetable, and fiber intake.

It was a pleasure to work with you. I hope you have continued success in your goals to lower your blood pressure.

Sincerely,
Sally Frank

sessment establishes a baseline of client behaviors and problems that will be used for future comparisons. Standards, such as Recommended Dietary Intakes, established by national organizations provide a yardstick to determine normalcy. In addition, the American Dietetic Association has developed medical nutrition therapy protocols for several health problems that can be used as a standard to assess outcomes.[3,4]

The key to a client evaluation is having well-defined, measurable goals and a charting process that tracks implementation of strategies and goal attainment, such as the Client Progress Report, Lifestyle Management Form 4.8 (see Appendix G). If this form is used properly, counselors can continually assess whether goals were achieved and what strategies were successful.

Goal Attainment Scale Cormier and Cormier[5] describe a **goal attainment scale** (GAS) rating system that a counselor and client can work together collaboratively to establish. A range of values are assigned to possible results, from a score of +2 for a most favorable result to a −2 for a least favorable outcome. A score of 0 represents the anticipated level of performance. This type of rating system is particularly useful for providing outcome results to funding agencies or for supplying useable numerical scores to determine levels of change for statistical analysis. The graduated level of desired outcomes has the added advantage of allowing success at several levels of performance. This is useful when there is a tendency to set goals that are too ambitious. See Table 8.1 for an example.

EXERCISE 8.1—Design a Goal Attainment Scale

Choose a new goal or one on which you have worked previously and design a goal attainment scale. This scale can be developed for a long- or short-term goal based on behavior or biochemical readings, such as blood glucose levels.

Final Client Evaluation At the end of treatment, a final assessment should be conducted to determine degree of attainment of final goals. If a follow-up interaction was arranged at the termination, a posttreatment evaluation can be conducted to determine whether the benefits of counseling have been maintained. Possible implementation of these evaluations include an in-person follow-up interview, a questionnaire sent via e-mail or postal delivery, or a telephone conversation. The follow-up has the added advantage of indicating to clients that the counselor continues to be interested in their welfare.

Evaluation of Counseling Effectiveness and Skills

Counselors also need to evaluate their effectiveness. A number of methodologies can be utilized for this purpose.

Assessing Client's Nonverbal Behavior The nonverbal behavior of a client represents a key to his or her emotional state and can indicate how the counseling session is going. Any discrepancies between a client's nonverbal behavior and verbal messages should be noted.

TABLE 8.1—Goal Attainment Scale for Exercise for One Week

VALUE	DESCRIPTION OF VALUE	BEHAVIOR
−2	Most unfavorable outcome thought likely	Do not walk at all.
−1	Less than expected success with performance	Walk less than 60 minutes during the week.
0	Anticipated level of performance	Walk for 30 minutes two times.
+1	More than expected success with performance	Walk for at least 30 minutes, four times.
+2	Best expected level of performance	Walk for at least 30 minutes, more than three times.

Checking Checking, periodic summaries, is a technique covered in Chapter 2. This method allows counselors to evaluate whether they are on target during a counseling session.

Counseling Checklists (Interview Guides)
Counseling checklists, such as those provided in Chapter 9, can serve as a rudimentary assessment tool, even though their primary function is to help organize a counseling session. At the end of a session, a counselor can review the form to assess whether all planned counseling interventions were addressed.

Charting Charting can be a valuable tool to evaluate counseling effectiveness. The assumption can be made that if counseling goals have been met, then the counselor demonstrated effective skills. However, additional sources of evaluation are required since client ability or inability to meet counseling objectives does not always reflect back on the counselor. The following questions should be considered if general counseling goals were not met:

- Was the assessment adequate? Were major problem areas clearly identified?
- Were goals realistic and clearly defined?
- If specific goals were not achieved, was there an adequate assessment of why not?
- What behavior change strategies were attempted? How effective were they?
- What could have been done differently?

Videotape, Audiotape, or Observation Evaluations Counselors can conduct self-evaluations of their skills by using a video- or audiotape. An alternative would be to have a colleague or mentor conduct an assessment. Generally it is a good idea to use an assessment instrument to guide the evaluation. The assignment at the end of this chapter provides two assessment instruments. The Interview Checklist, Lifestyle Management Form 8.1, addresses general counseling effectiveness including an evaluation of the flow and organization of the interview, application of interpersonal skills, and quality of client responses. This form was originally developed at Brown University School of Medicine to assess medical student interviewing skills and was modified to meet the needs of a nutrition counselor. The second assessment instrument, Counseling Responses Competency Assessment, Lifestyle Management Form 8.2 (see Appendix G), was developed to increase understanding and awareness of basic counseling responses. Either form can be used as a self–rating instrument, without frequency tabulation, following a session to upgrade counseling skills.

PROFESSIONALISM IN THE HELPING RELATIONSHIP

To provide high-quality service, nutrition counselors should be familiar with the general tenets of professionalism and pay attention to the unique concerns related to nutrition counseling. By adhering to professional standards, nutrition counselors are likely to receive both tangible rewards, such as a higher salary, and intangible rewards, such as greater social status. This will contribute to satisfaction with job, profession, and quality of life. The components of professionalism reviewed in this section include standards of practice, ethics, clients rights, the boundary between nutrition counseling and psychotherapy, referrals, and proper dress attire.

Overview of Professionalism

According to Todd et al.,[6] professionalism is characterized by autonomy, altruism, community of equals, and quality of service. Let's take a closer look at each of these terms in the context of nutrition counseling.

Autonomy is the degree to which nutrition professionals are making their own choices and governing their own practices. This implies self-regulation in that only one's peers have the degree of specialization in nutrition to be qualified to evaluate another's performance. The nutrition profession decides on the body of knowledge as well as continuing education standards. Counselors practicing in private settings have greater autonomy than dietitians practicing in a hospital setting. Hospitals have a corporate structure and institutional guidelines that need to be followed thereby taking away some choices.

Altruism implies that professionals have their clients' best interests at hand. Since clients are not often in the best position to evaluate services of a professional, they depend on the integrity of the professional to make the

EXERCISE 8.2—Evaluate Counseling Effectiveness and Professional Behavior

Read the following scenario, and identify what the counselor could have done or said differently to have a more effective professional encounter. Record your ideas in your journal, and discuss them with your colleagues.

Anita: Hi, Nancy. Please come in. Have a seat and relax. I am just finishing up a few things.

(Four minutes later . . .)

Anita: OK. Let's see if we can get to the root of your problem here. How long have you had a weight problem?

Nancy: Well, when I was pregnant in 1987, I gained fifty-three pounds. I had only lost about twenty pounds when I realized that I was pregnant again. Unfortunately, I gained another thirty-five pounds with the second pregnancy.

Anita: Has your doctor said anything to you about it?

Nancy: Not really.

Anita: OK. Tell me about yourself.

Nancy: Well, I am forty-two years old, married, with two children. I work at a nursing home from 11 P.M. to 7 A.M. as the charge nurse. I am an LPN. I have a lot of stress in this job because I am the only nurse on duty. Whenever anything goes wrong, I have to make the decision as to what to do all by myself. Do I call the doctor and wake him up; do I send the patient straight to the emergency room; do I wait until morning; what if they die by then? To cope with my anxieties during my night shift, I eat.

Anita: What types of things do you eat?

Nancy: Usually families are trying to be nice and they bring in cookies, donuts, or chocolates. These things are always at the nurse's station.

Anita: Do you eat any meals?

Nancy: Yes, the kitchen sets us up with a hot meal, which they leave in the refrigerator, and we microwave it when we are ready to eat. It is the same dinner the patients had to eat.

Anita: Do you eat when you go home?

Nancy: Usually I stop at one of the fast-food restaurants on Route 5. I have a croissant sandwich or a biscuit with an egg and Taylor ham.

Anita: Well, we will have to change that!

Nancy: I have that and then in about an hour or so I go to bed. I sleep most of the day, which is a problem because of my daughter.

Anita: What about your daughter?

Nancy: She is in high school, and because I usually sleep until 6 or 7:00 P.M., she goes unsupervised after school. She has gotten to be a handful.

Anita: What about your husband—isn't he available to watch her?

Nancy: My husband works from 3:00 P.M. to 1:00 A.M. But he really doesn't care anymore. He feels that Marie's problems are a direct result of my inability to control her and blames me for everything. He has little respect for me, especially with me being so overweight.

Anita: Well, from the Client Assessment Questionnaire you completed, it seems to me that you are about a hundred pounds overweight. Therefore, we are talking about a long-term lifestyle change for you to get to your goal weight of 120 pounds. You need first of all to get your husband to be supportive of you! He should be happy with your efforts at improving yourself.

Nancy: I guess I can try.

Anita: Is there any chance you could get another job? You are so sedentary in what you are doing.

Nancy: I have a pretty good pension and seniority. I really can't change that part of my life.

Anita: Well, OK, lets see if we can design a diet for you that will work. But I must tell you that to be successful you are going to have to join a gym.

right decision on their behalf. Even money for services should be secondary to the quality of services.

Community of equals reflects equality in status within a peer group of professionals. This is exemplified in professional organizations and credentials that require particular standards be achieved. Various types of certificates are available through professional groups. The American Dietetic Association provides the R.D. (registered dietitian), but other organizations furnish certifications in diabetes, nutrition support, and pediatrics. To achieve and maintain the standards of these organizations, practitioners need to improve and update their knowledge and skills continually through reading, attending educational programs, and possibly using mentors. The use of mentors is relatively new in the field of nutrition counseling[7] and has been found to be extremely valuable.[8] Due to the diverse and rapidly changing nature of the nutrition profession, practitioners frequently join several professional organizations. Active involvement in the organizations provides a conduit for enhancing professional abilities and building a network of colleagues that can be called upon when needed.

Quality of service is of utmost importance in the professional setting. It typically involves application of skills and knowledge based on set standards that are good for the client. The professional, however, has the option of using professional judgment in individual situations. Of recent concern for nutrition counselors has been the increase in workload regarding the number of clients/patients that are expected to be consulted on a daily basis. Increased numbers decrease the amount of time available for each client, potentially compromising the quality of service. Health care reform requires nutrition professionals to rise to the challenge and to develop innovative nutrition intervention programs. As with most health care practitioners, new counselors should investigate the options and purchase malpractice insurance.

Standards of Professional Practice

To state clearly the components of high-quality dietetic practice, the American Dietetic Association, the largest organization of food and nutrition professionals, develops and publishes **Standards of Professional Practice.** See Exhibit 8.2 for an overview of the standards, and visit www.eatright.org and search for "standards" for a complete list. They are periodically reviewed and revised as the needs of the dietetic profession changes. The standards are an aid to individual practitioners to evaluate their professional practices and maintain competence.

The *Journal of the American Dietetic Association* publishes guidelines for practicing for a variety of nutrition positions as well as position papers on numerous topics in dietetics. Clinicians are encouraged to research those that apply to their practice. Go to www.eatright.org and search for positions.

Ethics

All counseling relationships need to be handled in an ethical way so as to promote and protect clients' welfare.[5] A counselor's value system is the guiding force in beliefs regarding ethical behavior. To help influence the manifestation of the best ethical behavior, professional organizations develop and publish ethical standards. The American Dietetic Association, for example, publishes the Code of Ethics for the Profession of Dietetics. The code, found in Exhibit 8.3, "provides guidance to dietetics practitioners in their professional practice and conduct."[9] See Exhibit 8.4 on page 188 for a procedural model for ethical decision making as described by Corey.[10]

Client Rights

Clients are entitled to know their rights and options during the course of an ongoing counseling relationship. Cormier and Cormier[5] point out that ethical counselors should provide their clients with enough information about the counseling process to enable them to make informed choices (also known as empowered consent). This includes (1) confidentiality, (2) the procedures and goals of counseling, and (3) the counselor's qualifications and practices. In addition, long-term care residents and acute care patients have established rights that allow for personal choices to be addressed.

EXHIBIT 8.2—Overview of the American Dietetic Association's Standards of Professional Practice

1. **Provision of services.** Provides quality service based on client expectations and needs.
2. **Application of research.** Effectively applies, participates in or generates research to enhance practice.
3. **Communication and application of knowledge.** Effectively applies knowledge and communicates with others.
4. **Utilization and management of resources.** Uses resources effectively and efficiently in practice.
5. **Quality in practice.** Systematically evaluates the quality and effectiveness of practice and revises practices needed to incorporate the results of evaluation.
6. **Continued competence and professional accountability.** Engages in lifelong self-development to improve knowledge and enhance professional competence.

EXHIBIT 8.3—Code of Ethics for the Profession of Dietetics

The American Dietetic Association and its credentialing agency, the Commission on Dietetic Registration, believe it is in the best interest of the profession and the public it serves to have a Code of Ethics in place that provides guidance to dietetics practitioners in their professional practice and conduct. Dietetics practitioners have voluntarily adopted a Code of Ethics to reflect the values and ethical principles guiding the dietetics profession and to outline commitments and obligations of the dietetics practitioner to client, society, self, and the profession.

The Ethics Code applies in its entirety to members of The American Dietetic Association who are Registered Dietitians (RDs) or Dietetic Technicians, Registered (DTRs). Except for sections solely dealing with the credential, the Code applies to all members of The American Dietetic Association who are not RDs or DTRs. Except for aspects solely dealing with membership, the Code applies to all RDs and DTRs who are not members of The American Dietetic Association. All of the aforementioned are referred to in the Code as "dietetics practitioners." By accepting membership in The American Dietetic Association and/or accepting and maintaining Commission on Dietetic Registration credentials, members of The American Dietetic Association and Commission on Dietetic Registration credentialed dietetics practitioners agree to abide by the Code.

Principles

1. The dietetics practitioner conducts himself/herself with honesty, integrity, and fairness.
2. The dietetics practitioner practices dietetics based on scientific principles and current information.
3. The dietetics practitioner presents substantiated information and interprets controversial information without personal bias, recognizing that legitimate differences of opinion exist.
4. The dietetics practitioner assumes responsibility and accountability for personal competence in practice, continually striving to increase professional knowledge and skills and to apply them in practice.
5. The dietetics practitioner recognizes and exercises professional judgment within the limits of his/her qualifications and collaborates with others, seeks counsel, or makes referrals as appropriate.
6. The dietetics practitioner provides sufficient information to enable clients and others to make their own informed decisions.
7. The dietetics practitioner protects confidential information and makes full disclosure about any limitations on his/her ability to guarantee full confidentiality.
8. The dietetics practitioner provides professional services with objectivity and with respect for the unique needs and values of individuals.
9. The dietetics practitioner provides professional services in a manner that is sensitive to cultural differences and does not discriminate against others on the basis of race, ethnicity, creed, religion, disability, sex, age, sexual orientation, or national origin.
10. The dietetics practitioner does not engage in sexual harassment in connection with professional practice.
11. The dietetics practitioner provides objective evaluations of performance for employees and coworkers, candidates for employment, students, professional association memberships, awards, or scholarships. The dietetics practitioner makes all reasonable effort to avoid bias in any kind of professional evaluation of others.
12. The dietetics practitioner is alert to situations that might cause a conflict of interest or have the appearance of a conflict. The dietetics practitioner provides full disclosure when a real or potential conflict of interest arises.
13. The dietetics practitioner who wishes to inform the public and colleagues of his/her services does so by using factual information. The dietetics practitioner does not advertise in a false or misleading manner.
14. The dietetics practitioner promotes or endorses products in a manner that is neither false nor misleading.
15. The dietetics practitioner permits the use of his/her name for the purpose of certifying that dietetics services have been rendered only if he/she has provided or supervised the provision of those services.
16. The dietetics practitioner accurately presents professional qualifications and credentials.
    a. The dietetics practitioner uses Commission on Dietetic Registration awarded credentials ("RD" or "Registered Dietitian"; "DTR" or "Dietetic Technician, Registered"; "CSP" or "Certified Specialist in Pediatric Nutrition"; "CSR" or "Certified Specialist in Renal Nutrition"; and "FADA" or "Fellow of The American Dietetic Association") only when the credential is current and authorized by the Commission on Dietetic Registration. The dietetics practitioner provides accurate information and complies with all requirements of the Commission on Dietetic Registration program in which he/she is seeking initial or continued credentials from the Commission on Dietetic Registration.
    b. The dietetics practitioner is subject to disciplinary action for aiding another person in violating any Commission on Dietetic Registration requirements or aiding another person in representing himself/herself as Commission on Dietetic Registration credentialed when he/she is not.
17. The dietetics practitioner withdraws from professional practice under the following circumstances:
    a. The dietetics practitioner has engaged in any substance abuse that could affect his/her practice;
    b. The dietetics practitioner has been adjudged by a court to be mentally incompetent;
    c. The dietetics practitioner has an emotional or mental disability that affects his/her practice in a manner that could harm the client or others.

(Continued)

EXHIBIT 8.3—Code of Ethics for the Profession of Dietetics (Continued)

18. The dietetics practitioner complies with all applicable laws and regulations concerning the profession and is subject to disciplinary action under the following circumstances:

- **a.** The dietetics practitioner has been convicted of a crime under the laws of the United States which is a felony or a misdemeanor, an essential element of which is dishonesty, and which is related to the practice of the profession.
- **b.** The dietetics practitioner has been disciplined by a state, and at least one of the grounds for the discipline is the same or substantially equivalent to these principles.
- **c.** The dietetics practitioner has committed an act of misfeasance or malfeasance which is directly related to the practice of the profession as determined by a court of competent jurisdiction, a licensing board, or an agency of a governmental body.

19. The dietetics practitioner supports and promotes high standards of professional practice. The dietetics practitioner accepts the obligation to protect clients, the public, and the profession by upholding the Code of Ethics for the Profession of Dietetics and by reporting alleged violations of the Code through the defined review process of The American Dietetic Association and its credentialing agency, the Commission on Dietetic Registration.

Source: Copyright The American Dietetic Association. Reprinted by permission from the *Journal of the American Dietetic Association*, Vol. 99, page 109.

Confidentiality A breach of confidentiality can do irreparable harm to a counseling relationship since it undermines the essential component of trust. Generally, counselors are not free to disclose information about their clients unless they first receive written permission.[5] A discussion regarding the confidential nature of the sessions should be included in the first counseling session with your client. The American Dietetic Association Professional Code of Ethics[9] states, "The dietetics practitioner protects confidential information and makes full disclosure about any limitations on his/her ability to guarantee full confidentiality." However, it is generally considered acceptable to discuss client cases with colleagues or for educational purposes (lectures, in-service training, and so forth) as long as the identity of the client remains anonymous.[7]

EXHIBIT 8.4—Steps for Making Ethical Decisions

1. Identify the problem or dilemma. Gather information that will shed light on the nature of the problem. This will help you decide whether the problem is mainly ethical, legal, or moral.
2. Identify the potential issues. Evaluate the rights, responsibilities, and welfare of all those who are involved in the situation.
3. Look at the relevant ethical codes for general guidance on the matter. Consider whether your own values and ethics are consistent with or in conflict with the relevant guidelines.
4. Seek consultation from more than one source to obtain various perspectives on the dilemma.
5. Brainstorm various possible courses of action. Continue discussing options with other professionals.
6. Enumerate the consequences of various decisions, and reflect on the implications of each course of action for your client. Again, consider the five ethical principles of autonomy, beneficence, nonmaleficence, justice, and fidelity as a framework for evaluation of the consequences of a given course of action.
7. Decide on what appears to be the best possible course of action. Recognize that the more subtle the ethical dilemma, the more difficult the decision-making process will be. Be aware that in reasoning through any ethical issue, different practitioners will make a variety of decisions.

Source: Corey G, *Theory and Practice of Counseling and Psychotherapy*. 5th ed. Pacific Grove, CA: Brooks/Cole; 1996:57. Used with permission.

Procedures and Goals of Counseling The procedures and goals of the counseling program should be discussed with your client during the first session, including a clarification of any fees and a time frame for payment. Depending on the setting of the counseling intervention, there may be an institutional form that clients will be asked to sign. If you are working in a doctor's office or a private practice office, you should design an appropriate form. Appendix G includes a sample agreement form for a student working with a volunteer. See Lifestyle Management Form 3.3, Student Nutrition Counseling Assignment—Lifestyle Management Agreement.

Qualifications and Practices of the Counselor It is important for your clients to know what you can and cannot do for them. For example, clients may be coming to you hoping to obtain a diagnosis for their condition. The counseling agreement form should contain information about your credentials and something about the scope of your practice. During your first session, it is important to ask your clients what they are expecting to get out of the counseling intervention with you. At that point any discrepancies should be clarified.

EXERCISE 8.3—Personal Inventory of Attitudes Relating to Ethical Issues

This inventory is designed to assess your attitudes and beliefs on specific ethical issues common to all counselors or particularly relevant to nutrition counselors. Select the response that comes closest to your position, or write your own response in *e*. There are no right or wrong answers. Discuss your selections with your colleagues.

1. A counselor's primary responsibility is to
a. the client.
b. the counselor's agency.
c. society.
d. the client's family.
e. ____________________

2. Regarding confidentiality, my position is that
a. it is never ethical to disclose anything a client tells me under any circumstances.
b. it is ethical to break a confidence when the counselor deems that the client might do harm to himself or herself or to others.
c. confidences can be shared with the parents of the client if the parents request information.
d. it applies only to licensed therapists.
e. ____________________

3. Concerning the issue of physically touching clients, my position is that
a. touching is an important part of a helping relationship.
b. touching a client is not wise.
c. touching a client is ethical when the client initiates physical closeness with the counselor.
d. it should be done only when the counselor feels like doing so.
e. ____________________

4. The way I can best determine my level of competence in working with a given type of client is
a. by having training, supervision, and experience in the areas in which I am practicing.
b. by asking my clients whether they feel they are being helped.
c. by possessing an advanced degree and a license.
d. by relying on reactions and judgments from colleagues who are familiar with my work.
e. ____________________

5. Regarding the ethics of social and personal relationships with clients, it is my position that
a. it is never wise to see or to get involved with clients on a social basis.
b. it is an acceptable practice to strike up a social relationship once the counseling has ended if both want to do so.
c. with some clients a personal and social relationship might well enhance the therapeutic relationship by building trust.
d. it is ethical to combine a social and counseling relationship if both parties agree.
e. ____________________

6. If I am counseling individuals who are engaging in a cultural practice that is morally repugnant to me (for example, the sacrifice of a dog to achieve healing may be repugnant to many Westerners), I believe it is my responsibility to
a. learn about their values and not impose mine on them.
b. encourage them to accept the values of the dominant culture for survival purposes.
c. modify my counseling procedures to fit their cultural values.
d. end the counseling relationship since I cannot accept their values.
e. ____________________

7. When working with an overweight client who has a long history of losing weight and gaining back more weight than lost, the focus of my counseling should be to
a. encourage the client to accept their present weight.
b. encourage the client to join a self-help group.
c. put the client on a strict calorie controlled diet.
d. encourage the client to set weekly goals to improve the quality of their diet.
e. ____________________

8. For clients who have minimal financial resources,
a. it is acceptable to file false claims for services.
b. refuse to take them as clients.
c. it would be appropriate for me to charge no fee or less than I charge other clients.
d. refer clients who can not pay to self-help groups.
e. ____________________

(Continued)

EXERCISE 8.3—Personal Inventory of Attitudes Relating to Ethical Issues (Continued)

9. When working with salespeople,
  a. it would be appropriate to accept expensive perks, such as a trip to a resort.
  b. it would be acceptable to accept modest perks, such as a fruit basket.
  c. it would not be acceptable to accept any presents.
  d. it would not be appropriate for me to continue working with a sales person who offered me perks.
  e. ______________________________

10. When counseling a client referred to me by a doctor who has given incorrect nutrition advice (such as "Don't eat fruit to a pregnant woman—it holds water"), I should
  a. tell the client it is OK to eat fruit.
  b. tell the client I will talk to her doctor about that and get back to her.
  c. avoid the topic and talk about what other foods she should eat.
  d. agree with the statement.
  e. ______________________________

Source: Adapted from *Manual for Theory and Practice of Counseling and Psychotherapy* by Gerald Corey.

Boundary Between Nutrition Counseling and Psychotherapy

According to the American Dietetic Association Professional Code of Ethics,[9] "The dietetic practitioner recognizes and exercises professional judgment within the limits of his/her qualifications and collaborates with others, seeks counsel, or makes referrals as appropriate." After reviewing various theoretical approaches to counseling, a beginning nutrition counselor may well feel overwhelmed and wonder what the boundaries are between counseling and psychotherapy. Expanding your skills in psychotherapy can in fact be useful if you decide to specialize in nutrition counseling. You may decide to obtain additional education in areas such as addiction or family counseling. However, several authorities have addressed the boundary issue.

One time when I was counseling a middle-aged man with high blood pressure and elevated serum cholesterol, he came to our weekly session distraught over a decision of his unmarried teenage daughter to have an abortion. It was against his moral beliefs, and he was having trouble functioning. He said I was a counselor and maybe I could give him some advice. I told him I sympathized with his dilemma, I understood that being a parent is sometimes a heart-wrenching task, and I wished I could change things for him. However since there did not seem to be anything he could do about the abortion, I suggested that he consider individual or family counseling with a psychotherapist so at least he could better cope with the issue.

Saloff-Coste, Hamburg, and Herzog[11] cover boundaries between psychotherapy and nutrition counseling in their article on eating disorders. They believe the nutrition counselor's territory includes almost any issue related to food, weight, eating patterns, and body image. King[8] emphasizes the need to maintain healthy boundaries in a counseling relationship; otherwise, clients become confused as to what you can do with them and for them. The counselor does this by setting an example in his or her discussions of what can be handled in the nutrition counseling session. The nutrition counselor needs to be clear that the psychotherapist's work is based on feelings and the nutrition counselor's work is based on food. If feeling issues do arise that do not fall into a nutritionist's scope of practice, the nutrition counselor should be cautious about giving advice. It would be appropriate to acknowledge those feelings, provide any nutrition information related to the issue, and suggest the client receive assistance from a therapist qualified to deal with his or her problem.[8]

Referrals

At times you will want to suggest to your clients that they seek additional assistance from a person qualified to work with them regarding their feelings. In some clinical situations, you will be working with a medical team, and your client will already be receiving help from a psychotherapist. If this is the case, then you can suggest your client discuss the issue in question with his or her therapist. Helm[12] states that referrals to a mental health profes-

EXERCISE 8.4—Working with Standards and Ethics

Reread the case study about John in Chapter 1 and answer the following questions:

1. Review the American Dietetic Association Code of Ethics in Exhibit 8.3. Identify by number and explain specific ethical issues that are of concern in the Chapter 1 case study.
2. If you have access to the Internet, consider investigating the American Dietetic Association's Standards of Practice in relation to this case study. Go to the American Dietetic Association Web site (www.eatright.org/). Search for standards to locate the complete Standards of Practice and identify, and discuss which standards of practice are applicable to the Chapter 1 case study.

sional should be made when a client discloses information such as suicidal tendencies, physical abuse, severe marital difficulties, feelings of depression, past unresolved sexual abuse, recurring self-destructive behaviors, eating disorders, and other severe problems that are beyond the scope of nutrition practice.

Referrals are in order whenever the needs of the client are outside the scope of a particular counselor's expertise. It is useful to have on hand a list of professionals, such as social workers, physical therapists, psychologists, or physical trainers.

I had a client who appeared to be playing unusual games with food. She was forever "finding" food hidden deep in her closet or buried under her bed. After several weeks of counseling, it appeared to me that her eating problems had deep-rooted emotional origins, and I suggested that she seek the help of a mental health professional.

Proper Dress Attire

Proper dress attire is a component of professionalism. Dressing for a successful counseling intervention means to dress in a manner that does not create discomfort for the clientele. Clean, neat, and modest clothing and jewelry are appropriate for most counseling sessions. Expensive suits and jewelry would not be appropriate for counseling in poverty programs. Tight-fitting, revealing clothes are probably not suitable for any setting but would be particularly inappropriate in an obesity clinic. Strong scents should be avoided since the odor can be irritating, and some people are allergic to them. Dangling jewelry or any clothing item that could be distracting should not be worn. Often nutrition counselors in health care settings wear a white lab coat with a nametag over professional attire.

SELECTED COUNSELING ISSUES

From the beginning of this text, the reader was guided through the basics of the counseling process based on the counseling model presented in Figure 3.1. This included preparation, implementation of a counseling intervention, the end of the relationship, and evaluation. Having completed this process, there remain a few issues that merit discussion for the novice counselor.

Counseling Strategies for Dealing with Difficult Behaviors

Clients sometimes exhibit behavior that nutrition counselors can find challenging. Salas and Hall[9] have suggested strategies for dealing with five behaviors often encountered in nutrition counseling. See Table 8.2 and suggested resources at the end of the chapter describing a video illustrating the strategies. Their focus is to encourage clients to rethink thinking patterns and to suggest alternative viewpoints. Developing the skills to effectively implement these strategies will not come quickly for most beginning counselors; however, an awareness of these strategies can begin the process.

Rambling

Sometimes clients tend to ramble, getting into topics unrelated to their condition. To keep the focus on relevant issues, the counselor will need to intercede. This may require periodically reviewing the established agenda or interrupting to redirect the conversation.

EXHIBIT 8.5—Characteristics That Help You Strive for Success

dependability	organization skills
commitment	computer literacy
critical thinking	socially responsible
entrepreneurship	respect for diversity
collaboration	compassion
team player	leadership
communication skills	initiative
networking	open-mindedness
multiskilling	assertive

Source: Adapted from Covey SR, *The 7 Habits of Highly Effective People.* New York: Fireside; 1990.

TABLE 8.2—Counseling Strategies for Handling Difficult Behaviors

CATEGORY	DESCRIPTION	STRATEGIES
Denial	A person who has a serious problem appears unconcerned in order to protect self from emotional pain.	• Explain the function of denial. • Use analogies to clarify the concept.
Chronic complaining	Complaining can be the outcome of fear, frustration, or inadequate information. Counselors may not be able to fully alleviate the difficulty.	• Respond empathetically. • Focus on what the client has done well.
Dependency	Some people do not take responsibility for their problems, look for someone else to solve them, and blame others when things go wrong.	• Examine expectations. • Help client understand thinking pattern. • Encourage client to shed the victim role.
Fatalistic thinking	A client has drawn a very negative conclusion.	• Evaluate the validity of the conclusion. • Recognize client's theory, but remain neutral. • Show respect for client's reasoning capabilities. • Ask permission to provide additional information.
Compulsive talking	Some people talk excessively to control a situation and as a result are not open to taking in new information.	• Respond empathetically. • Agree when possible. • Paraphrase when possible. • Explain behavior to client.

Source: Adapted from Salas C, Hall DH, *What's Going on Here?* Flushing, MI: DGH Productions; © 1995. Used with permission.

Counselor: *I think you said something really important that I would like to discuss in more detail.*

This is really interesting, but I'm afraid we are getting off the topic, and we won't have enough time to talk about your food issues.

Can we go back to something you said earlier about . . . ?

Counseling Low-Literacy Populations

Another final issue to consider for the beginning nutrition counselor is the low level of literacy among a significant portion of adults. One of five adults in the United States reads at the fifth-grade level or below.[10] At the present time, no rapid assessment tool exists to assess health-related reading ability.[11] Many times people who have very limited literacy skills are too embarrassed to reveal their condition, so counselors need to be attuned to indicators of literacy difficulties. Possible clues include the following:

- Forms have not been completed.
- Clients call to have instructions repeated.

EXERCISE 8.5—Apply Characteristics That Help You Strive for Success in Nutrition Counseling

Review the list of characteristics in Exhibit 8.5 that can aid in the pursuit of a successful career. In your journal, indicate how you believe they apply to a career in nutrition counseling; discuss your answers with your colleagues.

- An excuse is given for not being able to complete a form, such as "I'm not good with forms," "I don't care for forms," "I have visual problems," "I forgot my glasses," or "I have a headache."
- Clients state that they will bring their forms home for a spouse to complete.
- Family members, including children, serve as surrogate readers and complete forms.
- A long time is taken to form letters.
- Clients sign with an *X* or cannot sign their name.

For clients who have literacy problems, dietary instructions can be given on an audio- or videotape. Pictures can be drawn and crossed out as food is consumed. Color-coded stickers can be used to identify types of food, such as yellow for fat-free and red for whole milk.

Nutrition counselors need to keep a variety of easy-to-read educational materials on hand. Limited-literacy focus groups indicated that the best format for learning to improve eating habits include discussion groups with hands-on activities that are culturally appropriate and enjoyable.[14,16] The major barriers to improving eating habits identified in these groups were extra time and money needed to purchase and prepare healthful foods, food preferences of family members, ease of eating unhealthful food from fast-food restaurants or vending machines, lack of interest and skills in cooking, and insufficient knowledge about which foods are healthful.

GROUP COUNSELING

The focus of this text has been on individual counseling. Nutrition counselors may well be asked, however, to conduct a group counseling program at some time in their career; therefore, an overview of group counseling is presented here. In addition, using group counseling as a substitute for or in combination with personal counseling is likely to gain popularity as resources dwindle and administrators search for affordable avenues to deliver services.

Group counseling run by nutritionists is not intended to be a form of therapy but provides a format to find solutions to dietary problems.[12] In this section, the basics of group counseling are reviewed. See the resources at the end of the chapter for further help in developing skills in this area.

Traditionally, nutrition counselors conducted group sessions with an educational focus. However, more recently the view of education has included many of the approaches and goals of counseling by incorporating behavior change methods into programs. In one evaluation, group counseling was found to be more effective than individual counseling for those seeking weight loss.[13] A recent review of twenty years of diabetes education literature found advantages of combining one-on-one and group counseling.[14]

Advantages of Group Counseling

Group counseling affords many advantages:

- **Emotional support.** Groups help clients feel as if they are not alone in dealing with their nutritional concerns. Sharing experiences with others who really know what it is like can provide a great deal of emotional support. A cohesive group helps participants feel accepted and special.
- **Group problem solving.** Participants motivate each other to change as they share coping strategies and problem-solve together. As illustrated in the DOVE activity in Chapter 1, two heads are better than one. Sharing supplies additional ideas and generates suggestions that neither person would have thought of individually (that is, a synergy effect ensues).
- **Modeling effect.** Participants learn from each other by observing the accomplishments of others with similar problems.
- **Attitudinal and belief examples.** As participants describe their attitudes and beliefs regarding health behavior challenges and perceived failures, other group members tend to reevaluate their own belief systems.

Disadvantages of Group Counseling

Unfortunately, group counseling also holds some potential drawbacks:

- **Individual responsiveness.** Some people do not easily share in a group setting, and as a result their issues may never be addressed.
- **Group member personalities.** The dynamics of a group are heavily influenced by members' individual personalities. The ability of leaders to handle domineering, demoralizing, or needy individuals who may

tend to monopolize time will impact on the counseling environment for all group members.

- **Possibility of poor role models.** Poor role models can create additional burdens for a counselor to counteract.
- **Meeting the needs of all group members.** It may be difficult to organize a group with similar issues and health concerns. Meeting the needs of participants who widely differ in age, gender, ethnic background, and specific health problems can be a challenge.[19]

Group Process

To run effective groups, it is necessary to build on skills developed for conducting individual counseling sessions. Most of the discussion in this section pertains to running *closed groups*—that is, groups that do not accept new members after the first or second session. Generally closed groups allow for greater bonding to take place, providing a more suitable environment for behavior change to take place.

In the ideal cohesive group, each member feels a sense of belonging and acceptance. The following six steps have been identified as important for the development of cohesive, well-functioning groups.[12,15]

Step 1: Establish an open, warm environment and productive leader-participant relationships. The same rules for establishing rapport in one-on-one counseling apply to group counseling. Facilitators need to show empathy, appear warm and genuine, and use relationship-building responses, attentive behavior, and effective body language. The counselor is a model of trusting behavior for all group members to emulate to promote openness and interpersonal communication. Holli and Calabrese[16] suggest that facilitators start each meeting with an expression of intent to create an environment conducive to acceptance and open expression.

EXERCISE 8.6—Evaluate Past Group Experiences

Consider your best experience as the member of a small group. Describe the setting and the function of the group. What made the experience go well? What was your role in the group? Is there anything in that encounter that you would like to emulate in your work as a group facilitator? Explain.

The first session is crucial because a group's personality evolves early and is difficult to change at a later time. Therefore, interactive and fun activities should be planned according to the participants' maturity level and interests. The principal objective is to address participants' primary concern of feeling accepted and being acknowledged as worthy.[21] Possible get acquainted activities include a human treasure hunt (Exhibit 8.6), having participants interview each other and report findings back to the group, or using a bean bag to toss to group members to determine the next speaker to introduce self. Openings could include humor (possibly a cartoon on an overhead projector), an open-ended question, or an interesting story. As an opening to a subsequent session, you may wish to lead the group through a systematic relaxation exercise. (See the resources at the end of Chapter 6.)

Ground rules for the group counseling program should be established during the first session. This can be an informal discussion or a written copy of guidelines that each participant signs. Leaders can ask the group to formulate ground rules with the leader making informal suggestions, or to save time the facilitator can provide a preset list of guidelines for the group to comment on and modify. Guidelines generally include listening respectfully to others, responding nonjudgmentally about the behavior of others, respecting confidentiality, encouraging all members to share, and attending every session.

Step 2: Balance facilitator-generated and group-generated information. The challenge for nutrition counselors is to cover a preset curriculum and integrate client needs and experiences and allow the group to generate solutions and problem solve. Often counselors have a list of tasks identified as essential for clients to understand, but the facilitator-generated information will fall on deaf ears if group members have other concerns on their minds. For example, a diabetic who is worried about amputations due to complications may have trouble focusing on other issues such as glucose monitoring if the complication issue is not addressed first.[17] One way to handle this potential problem is to ask participants in the first session to identify their pressing concerns. Then cover the most pressing problems first.

Step 3: Design problem-solving strategies. Many opportunities for group problem solving should be provided rather than having the counselor tell partic-

EXHIBIT 8.6—Human Treasure Hunt Guidelines

This activity is a good icebreaker but probably would not work well for groups with fewer than eight members. Besides helping participants become acquainted with each other, it also encourages the process of sharing experiences and coping strategies.

1. Before your meeting, find out something interesting or special about each participant. Preferably identify a fact related to the group concern. For example, a fact could be that a person enjoys eating oatmeal each day in a group of heart attack survivors.
2. Compose a human treasure hunt sheet by writing a list of the facts on a sheet without names.
3. At the beginning of the meeting, hand out the form and ask group members to search for the member who meets that description.
4. Close the activity by reading the facts and identifying the person who corresponds to each fact.

ipants what they need to do. Participation in a group discussion is two to ten times more effective than a lecture format.[21] Specific guidelines for group problem solving can be found in Exhibit 8.7.

As covered in Chapter 5, social disclosure is a powerful force for behavior change. Group counseling provides an ideal setting for coupling group problem solving with this process. Participants design goals with feedback from the group and disclose their intent to accomplish certain tasks before the next group session. Depending on the size of the group, you may wish to break up into smaller groups for designing individual goals. The following session each participant reports back to the group his or her progress in meeting those goals.

Many of the interactive activities identified in Exhibit 5.2 can be applied in a group setting and lend themselves to group problem-solving discussions. In fact, some of the activities and demonstrations are more practical for a group setting. For example, groups can list advantages and disadvantages of maintaining blood glucose levels, select a low sodium meal from printed restaurant menus, or role-play placing special orders at a restaurant. An effective way to use role playing is to verbalize probable or possible self-talk while making selections from a holiday buffet or after making poor selections from a smorgasbord of tempting avoid foods.

Step 4: Provide the opportunity for group members to practice new skills. In step 3, the group worked as a whole to problem-solve or develop new strategies. Opportunities for each member to rehearse the skill can occur if members are divided into groups of two or three. For example, participants could each practice measuring portion sizes, analyzing blood glucose records of previous clients, or jointly modify a recipe. The new skill should be something clients can use before the next group session so members can report on their experiences using the skill in "real life."

Step 5: Use positive role models and pacing to keep the group motivated. Spending time reviewing and understanding the successes of group members can provide a model for other participants to make alterations in their lifestyle. Successful members inspire others to follow their example and stay in the group. However, clients who are having difficulty and appear to be monopolizing the group time with their problems can be frustrating for the rest of the group. In such cases, the facilitator should acknowledge the difficulty and tell the client to stay after the session to receive personal attention.

Step 6: Ask for evaluation and feedback. Throughout the counseling process and after trying out a new activity or strategy, the facilitator should ask the

EXHIBIT 8.7—Group Problem-Solving Guidelines

1. *Identify* a problem of one or more group members
2. *Assess* the conditions that contribute to the problem and identify factors that promote healthful practices and alleviate the problem. Evaluate the following:
   - *Physical environment*—aspects of the external environment that cue poor eating habits, as well as aspects that remind the person to eat appropriately
   - *Social environment*—social situations that support poor eating habits and identify people who could support good eating habits
   - *Cognitive environment*—thoughts and feelings that get in the way and positive ideas that can be used to promote positive habits
3. *Brainstorm solutions.* The objective of brainstorming is to generate as many ideas as possible for consideration. There is one major rule: no censorship. No idea is rejected, no matter how silly or useless it may appear initially.
4. *Select a solution.* Once all ideas are listed, the person selects one and plans the details of implementation and evaluation.

Source: Raab C, Tillotson JL, *Heart to Heart: A Manual on Nutrition Counseling for the Reduction of Cardiovascular Disease Risk Factors.* NIH Publication No. 83-1528. Washington, DC: U.S. Department of Health and Human Services; 1983.

group for their opinions. For example, "Since we began meeting, what did you find particularly useful?" or "Did you find analyzing glucose records of previous clients useful?"

Practical Considerations for Successful Groups

The following is an overview of practical matters that must be handled when organizing a group:[12,20]

- **Allow adequate time for organization.** Generally six to eight weeks are needed to arrange for a meeting location, publicize, and develop curriculum.
- **Select a comfortable meeting room and location.** The room should be large enough to accommodate all participants to sit around a table or in a U-shaped arrangement. A circular table provides a better environment for exchange among all group members rather than a rectangular table, at which participants are more likely to limit interaction to those directly across from them. Additional space to allow participants to break up into pairs or small groups would also be an advantage. If very large people are likely to attend, be sure the room has sturdy chairs of adequate size. The room should have satisfactory lighting, temperature control, and ventilation. Generally people feel more comfortable talking in a room that has a closed door. In addition make sure that the location is easily accessible via public transportation and/or car.
- **Ideal group size for closed groups is six to twelve.** This appears to be a good size for the development of a group identity. If the group gets too large, there will not be a free flow of conversation, and giving individual attention will be difficult. The tendency will be to revert to a lecture format. If the size is too small, the dynamics of the group will be severely impaired if one or two people are missing.
- **Contemplate collecting fees or refundable deposits before the first meeting.** Fees encourage better attendance. If a periodic pay schedule is used, fees should be paid before attendance at sessions. Some programs use refundable deposits that are gradually returned as participants attend meetings. In some cases, the refunds are graduated so that the final payment is the largest.
- **Appraise target group needs for selecting a meeting time.** If most members are working, then late afternoon, evening, or a weekend day should be considered. If possible, a survey could be taken of interested parties. A daytime meeting may work best for retired individuals.
- **Consider composition issues of the potential group.** Groups with common health concerns, age, and gender are likely to share similar needs and goals. However, some diversity can provide valuable perspectives. If the decision is to mix the composition, care should be taken to keep numbers somewhat balanced so no one feels left out.
- **Interview prospective group members.** Interviews could be handled over the phone, but it would be better to meet individuals in person. By assessing prospective participants, you can determine suitability for their participation and design a program that better meets their needs. Also, you can receive assurance that the people understand the specifics about the group purposes and procedures.
- **Group leaders should remain the same.** Group leaders should not be considered interchangeable. New leaders break continuity and cohesion and create disruption.
- **Be responsible.** Start meetings on time and always be well prepared. Test-run activities, such as measuring portion sizes. Make a list of equipment and supplies that will be needed. Arrive early to be sure the room is arranged properly and equipment is working. Always follow through on promises to locate information or resources or to make contact with members between sessions.
- **Plan sessions carefully.** Make an outline of topics, activities, and the amount of time expected to spend on each.
- **Consider refreshments.** Generally sharing of food encourages bonding. The selections could provide exposure to foods promoted for the group. It is not unusual for participants to offer to bring in samples of foods. Group members could sign up to act as host or hostess to bring in refreshments and to make reminder phone calls to other group members. A restaurant meal or a potluck could be considered for the last meeting.
- **Call members who miss meetings.** If a participant misses a meeting, encourage group members to call to review what was covered and to encourage attendance at the next meeting. You may even design a buddy system. Be sure to request permission before distributing phone numbers.

Specific Group Counseling Skills

Many of the responses and skills a group counselor utilizes to develop productive relationships are similar to those necessary to successfully conduct individual counseling sessions. However, there are certain group leadership skills that require special emphasis.

Blocking If counterproductive behaviors emerge, the counselor needs to block them from disrupting the group process. Disruptive behaviors include scapegoating, personal attacks, aside jokes, unrelated stories, and gossiping. The focus of blocking should be on the behavior, not on the person as a whole.[18] Specific techniques for handling difficult clients can be found in Table 8.3.

EXERCISE 8.7—Interview a Group Counselor

Interview an individual, either on the telephone or in person, who runs group counseling programs, preferably nutrition programs. Ask the following questions, and record the answers in your journal.

1. What type of groups do you facilitate, and how often do you run them?
2. How do you view your role in the group process?
3. What skills do you find particularly useful for facilitating groups?
4. Do you establish ground rules? If so, how are they set?
5. How do you set the agenda for the sessions?
6. How do you evaluate the effectiveness of the groups?
7. What advice do you have for a novice group counselor?
8. What did you learn from this interview?

Record the name of the person you interviewed and the date and time of the interview.

CASE STUDY 8—Interactive Personal Case Study

The objective of this experience is to utilize theories, strategies, and skills learned throughout the course of your nutrition counseling studies and apply them to yourself as if you were counseling you.

PART I. Think of a time in your life when you wanted to make a lifestyle change. (You could also choose a change you are currently attempting to establish. In that case, all the following questions would be posed in the present tense, rather than the past.) For this activity, it does not matter to what degree you feel your efforts were successful. You will analyze and reflect on the experience in as much detail as possible in a case study format. Number and answer each of the following questions:

1. Describe the issue:
   a. What was the problem and why were you inspired to try to change?
   b. Label the feelings you experienced at the time.
2. Provide background and depth into the case:
   a. How did the issue impact your life and the life of those close to you? What difficulties or potential problems were related to the problem.
3. Identify your motivation level:
   a. Identify your motivational level at the time, including readiness, degree of importance, and confidence to succeed using the Assessment Ruler, Lifestyle Management Form 3.1 (see Appendix G).
4. Identify your goal or goals.
5. Analyze your approach/intervention:
   a. What approaches/interventions did you utilize?
   b. Were they appropriate for your motivational level?
6. Describe the difficulties you encountered:
   a. What were your cues?
   b. What were your barriers to making a behavior change?
   c. What cognitive distortions may have hampered your progress?
   d. What prompted lapses or a relapse?

PART II. Now imagine yourself as a counselor working with a person just like yourself at the time you were attempting the behavior change. Considering what you have learned during the course of your counseling studies, answer the following questions:

1. Which theory or theories in Chapter 1 would you want to guide your counseling approach? Explain.
2. How would you view your role as a counselor?
3. What behavior change strategies do you believe would be effective? Explain.
4. How would you assist your client in preparing for a lapse?

TABLE 8.3—Suggestions for Dealing with Difficult Group Participants

PARTICIPANT BEHAVIOR	PROBLEM OR POSSIBLE MOTIVE	ACTION
Participant statement is definitely wrong.	• Making an obviously incorrect comment	• Must be handled delicately. Say, "That's great that it worked for you, but others have found . . ." or "That's one way of looking at it, but there are authorities that believe . . ." or "I see your point, but let's try looking at it this way."
Searching for leader's opinion	• Trying to put you on the spot • Trying to have you support one view • May be simply seeking your advice	• Generally, you should avoid solving problems. However, there are times you should give a direct answer. Before you do so, you may want to open up the discussion to other participants by asking, "Let us get some other opinions about this issue. How do you view this point?" (Direct your question to a particular person.)
Won't talk	• Bored • Indifferent • Feels superior • Timid	• Your action will depend on what you believe is motivating the participant. • Arouse participant's interest by asking for his or her opinion. Draw out the person next to the participant. • Ask the quiet person for an opinion of what was said. • If the participant is seated near you, ask a question so that the person is talking to you, not the group. Ask an easy question. Then say, "Nice answer—thank you."
Griper	• Has a pet peeve • Professional griper • Has legitimate complaint	• Point out we can't change policy here and that we must operate as best we can under the system. • Say you'll discuss problems with person privately later. • Ask group, "How do the rest of you feel?"
Side conversations	• May be related to the subject • May be personal • Distracting to the group and you	• Don't embarrass participants. • Call one participant by name, ask an easy question, or restate last opinion expressed and ask for an opinion. If, during the meeting you, you are in the habit of moving around the room, stroll over and stand casually behind or next to the participants who are talking.
Inarticulate	• Lacks ability to put thought in order	• Say, "Thank you—let me repeat that." Then put the idea in better language.

Source: © 1995, The American Dietetic Association. "Cardiovascular Nutrition." Used with permission.

Raczynski[19] makes the following general suggestions for dealing with disruptions:

- Use reflective listening skills.

Counselor: *Miguel, it appears that you want to change the topic when we talk about monitoring food intake.*

- Use assertiveness skills.

Counselor: *Let's get down to business.*

Grace, I'd like to stay with our discussion of self-talk and hear from others in the group before we cover new ground.

Hank, you are really telling an interesting story about your job, but let's plan to finish the story at the end of the session so we can review how everyone did with their goals this week.

- Enlist the support of the group to refocus.

Counselor: *Now that we analyzed the president's proclivities, we should get back to our discussion of handling the upcoming holiday.*

Silence As the facilitator of a group, a novice leader can feel pressure to fill silence. Often silence occurs early in a session when no one wants to initiate conversation. However, the silence can be a signal that the facilitator does not intend to dominate or force opinions.[21] Ordinarily, if the silence is long enough, someone will take the initiative for beginning a discussion.

Facilitating As a facilitator, a counselor acts as a guide on a journey of self-discovery and problem solving. Corey[10] and Snetsalaar[20] list some ways in which a leader can help the process:

- Encourage members to express their feelings and expectations openly.
- Teach members to focus on themselves and their feelings.
- Teach members to talk directly and plainly to one another.
- Foster member-to-member rather than member-to-leader interaction.
- Encourage quiet members to participate by using eye contact.

REVIEW QUESTIONS

1. Describe nine topics that could be addressed in a final session with a client.

2. Identify methods that can be used to evaluate client progress.

3. Explain ways in which nutrition counselors can assess their own effectiveness.

4. Describe a goal attainment scale.

5. Name and explain four factors that characterize professionalism.

6. Explain boundaries between nutrition counseling and psychotherapy.

7. Identify and explain three factors clients have a right to know before engaging in a counseling relationship.

8. Why do professional organizations develop codes of ethics?

9. Why do professional organizations develop standards of practice?

10. Identify four advantages of group counseling.

11. Identify four disadvantages of group counseling.

12. Explain the six steps of the group process.

13. What is the ideal size for a closed group?

14. Explain how a group counselor can effectively use blocking, silence, or facilitating skills.

ASSIGNMENT—Evaluate Your Counseling Effectiveness

Audio- or videotape a counseling session with a volunteer client following the counseling format outlined in Chapters 3 and 4, or use one of the checklists for a counseling session found in Chapter 9. Review the tape very carefully, and complete the Interview Checklist, Lifestyle Management Form 8.1, and the Counseling Responses Competency Assessment, Lifestyle Management Form 8.2 (Appendix G). Answer the following questions:

1. Explain what materials you reviewed and what educational materials or activities you prepared before meeting your client.

2. What verbal facilitation techniques and relationship-building responses appeared to work most effectively with your client? Explain why.

3. What behavior change techniques did you utilize? How effective were these methods?

4. Are there cues or messages that you missed or interpret differently after listening to the tape?

5. Do you believe you focused on the main lifestyle issues that concerned your client?

6. Did you keep the focus of the session on your client's main issues? If yes, what did you do to keep your client on track? If not, what could you have done differently?

7. What were your impressions of the emotions expressed by your client while you listened to the tape? Were these impressions different than what you understood at the time of the counseling session?

8. What was your emotional state at the time of the session? How do you believe this impacted the course of the session?

9. What is your overall impression of how your client responded? Explain.

10. If you could redo the counseling experience, what would you do differently?

11. What did you learn from this experience?

Additional Considerations of Videotape Observations

To focus on and assess the visual aspects of the counseling experience, watch your video with the sound turned off.

1. Describe your body language during the course of the counseling session. What messages do you believe your body behavior conveyed to your client?

2. Describe your client's body language during the course of the counseling session.

3. After analyzing the nonverbal behavior, did you change any of your impressions regarding the counseling encounter?

SUGGESTED READINGS, MATERIALS, AND INTERNET RESOURCES

Nutrition Counseling

DGH Productions, 3156 Pierson Place, Flushing, MI 48433; (810) 732-8100. Offers two very useful videos that demonstrate nutrition counseling strategies. Basic skills are covered in *Gaining Collaboration in Nutrition Counseling;* $49. Strategies for handling difficult behaviors addressed in Table 8.2 are covered in *What's Going On Here? Nutrition Counseling at the Next Level;* $79.

Nutrition Dimension, P.O. Box 6488, CA 95502; www.NutritionDimension.com; (888) 781-5388. Sells an audiotape series addressing a variety of nutrition counseling skills done by Kathy L. King, called *Counseling for Health & Fitness;* $79.

Low-Literacy Materials

Oregon State University Extension Web site, http://osu.orst.edu/dept/ehe/. An abundance of materials related to low literacy, including the development of easy-to-read materials and guidelines of how to get your message across when working with this population group.

Shield JE, Mullen MC. *Developing Health Education Materials for Special Audiences: Low-Literate Adults.* Chicago: American Dietetic Association; 1992. This audiotape and guide provides educators direction in planning, developing, and evaluating health education materials for low-literate adults. Also included are guidelines for adapting existing health education materials to meet clients' needs.

Evaluation

Raab D, Tillotson JL. *Heart to Heart: A Manual on Nutrition Counseling for the Reduction of Cardiovascular Disease Risk Factors.* NIH Publication No. 83-1528. Washington, DC: U.S. Department of Health and Human Services; 1983. This book contains a four-page nutrition counseling competency checklist.

Group Counseling

American Group Psychotherapy Association (AGPA). A professional organization that offers a number of training opportunities, certification, and a yearly conference. More information can be obtained from the Web site: http://www.groupsinc.org/

Association for Specialists in Group Work (ASGW). A member organization of the American Counseling Association (ACA), ASGW is a professional organization dedicated to promote quality in group work training, practice, and research. This organization publishes a quarterly journal, a newsletter, and a monthly newspaper. Students are able to join at a reduced rate by writing to American Counseling Association, c/o Accounting Dept., 999 Stevenson Avenue, Alexandria, VA 22304-3300, or calling (800) 347-6647. More information can be obtained from the Web site: http://coe.colstate.edu/asgw/

Corey G. *Theory and Practice of Group Counseling.* 5th ed. Belmont, CA: Wadsworth/Thomson Learning; 2000. Covers the basic elements of group process and reviews ten theoretical approaches to group counseling. There is a student manual available to reinforce concepts covered in this book.

REFERENCES

[1]Miller WR, Jackson KA. *Practical Psychology for Pastors.* 2d ed. Englewood Cliffs, NJ: Prentice Hall; 1995.

[2]Murphy BC, Dillon C. *Interviewing in Action: Process and Practice.* Pacific Grove, CA: Brooks/Cole; 1998.

[3]American Dietetic Association. *Medical Nutrition Therapy across the Continuum of Care: Supplement 1.* Chicago: American Dietetic Association; 1997.

[4]American Dietetic Association. *Medical Nutrition Therapy Across the Continuum of Care.* 2d ed. Chicago: American Dietetic Association; 1998.

[5]Cormier S, Cormier B. *Interviewing Strategies for Helpers: Fundamental Skills and Cognitive Behavioral Interventions.* 4th ed. Pacific Grove, CA: Brooks/Cole; 1998.

[6]Todd KR, Conklin MT, Guthrie L. *The Profession of Home Economics: A Casebook.* Upper Montclair, NJ: Department of Human Ecology; 1997.

[7]Curry KR, Jaffe A. *Nutrition Counseling and Communication Skills.* Philadelphia: Saunders; 1998.

[8]King NL. *Counseling for Health & Fitness.* Eureka, CA: Nutrition Dimension; 1999.

[9]American Dietetic Association. Code of ethics for the profession of dietetics. *J Am Diet Assoc.* 1999;99:109–113.

[10]Corey G. *Theory and Practice of Counseling and Psychotherapy.* 5th ed. Pacific Grove, CA: Brooks/Cole; 1996.

[11]Saloff-Coste C, Hamburg P, Herzog D. Nutrition and psychotherapy: Collaborative treatment of patients with eating disorders. *Bulletin of the Menninger Clinic.* 1993;57(4):504–516.

[12]Helm KK. Group process. In: Helm KK, Klawitter B. eds. *Nutrition Therapy Advanced Counseling Skills.* Lake Dallas, TX: Helm Seminars; 1995:207–213.

[13]Salas C, Hall DH. *What's Going on Here?* Flushing, MI: DGH Productions; 1995.

[14]Macario E, Emmons KM, Sorensen G, Hunt MK, Rudd RE. Factors influencing nutrition education for patients with low literacy skills. *J Am Diet Assoc.* 1998;98:559–564.

[15]Davis TC, Crouch MA, Long SW, Jackson RH, Bates P, George RB, Bairnsfather LE. Rapid assessment of literacy levels of adult primary care patients. *Fam Med.* 1991;23:433–435.

[16]Hartman TJ, McCarthy PR, Park RJ, Schuster E, Kushi LH. Focus group responses of potential participants in a nutrition education program for individuals with limited literacy skills. *J Am Diet Assoc.* 1994;94:744–748.

[17]Raab D, Tillotson JL. *Heart to Heart: A Manual on Nutrition Counseling for the Reduction of Cardiovascular Disease Risk Factors.* NIH Publication No. 83-1528. Washington, DC: U.S. Department of Health and Human Services; 1983.

[18]Brownell K. The psychology and physiology of obesity: Implications for screen and treatment. *J Am Diet Assoc.* 1984;4:406–413.

[19]Brown SA. Interventions to promote diabetes self-management: State of the science. *Diabetes Educator.* 1999;6(suppl):S52–S61.

[20]Klein L, Axelrod B. Using the workbook in a group setting. In: Wylie-Rosett J, Segal-Isaacson CJ, eds. *Leaders Guide: The Complete Weight Loss Workbook.* Alexandria, VA. American Diabetes Association; 1999:19–22.

[21]Holli BB, Calabrese RJ. *Communication and Education Skills for Dietetics Professionals.* 3d ed. Baltimore, MD: Williams & Wilkins; 1998.

[22]Funnell MM, Anderson RM. Putting Humpty Dumpty back together again: Reintegrating the clinical and behavioral components in diabetes care and education. *Diabetes Spectrum.* 1999;12:19–22.

[23]Corey G. *Theory and Practice of Group Counseling.* 5th ed. Belmont, CA: Wadsworth/Thomson Learning; 2000.

[24]Raczynski J. *The Latest in Nutrition Counseling.* Presentation at ADA Annual Meeting, Orlando, FL; October 20, 1994.

[25]Snetselaar LG. *Nutrition Counseling Skills for Medical Nutrition Therapy.* Gaithersburg, MD: Aspen; 1997.

Chapter

9

GUIDED COUNSELING EXPERIENCE

Before everything else,
getting ready is the secret of success.
—Henry Ford

BEHAVIORAL OBJECTIVES:

- Utilize standard counseling procedures.
- Employ interpersonal skills.
- Demonstrate use of basic assessment tools.
- Employ goal-setting processes.
- Tailor educational interventions.
- Utilize standard documentation procedures.
- Evaluate counseling effectiveness.

KEY TERMS:

- **COUNSELING CHECKLISTS:** step-by-step counseling guides
- **INFORMED CONSENT:** Sufficient information was supplied to make a decision regarding a course of action

This chapter provides a guided approach for conducting four one-hour individual counseling sessions with a volunteer adult client who would like to lose weight and/or improve the quality of his or her diet. It integrates the material covered in all previous eight chapters. This guide follows the counseling protocol and the motivational counseling algorithm, Figure 3.2, presented in Chapter 3, and contains the basic elements of the counseling process generally accepted as standard. The application of commonly used nutrition assessment and counseling tools are integrated into the guide. This guided experience was designed to introduce basic nutrition counseling procedures to a college student with some knowledge of nutrition. It can be applied to other settings but was not intended to cover the spectrum of all nutrition counseling experiences.

DEVELOPING A COUNSELING STYLE

The anticipation of meeting with your client for the first time can be exciting as well as stressful. Interviewing and counseling skills take time to master, but your competence will improve with each session. There is no one perfect counseling method; however, a novice nutrition counselor needs to start somewhere. We recommend that this structured program be used as a springboard on which to build and modify individual counseling styles that mesh with your disposition and personality.

Also, you will find that adjustments are necessary to meet your clients' specific needs. For example, in this guidebook we suggest setting very explicit goals. This has been found to work well for most clients. However, our experience has shown us that occasionally a client will rebel against this structure and prefer to set more general goals.

Your counseling strategies and style will evolve over the course of your career as you gain experience, take continuing education courses, and read educational materials. If nutrition counseling becomes a major career goal, then you may consider obtaining additional counseling credentials and/or establishing a relationship with a mentor.

FINDING VOLUNTEER CLIENTS

Volunteer clients should be relatively healthy, and any expectations of weight loss should be modest. No client should be accepted into the program with severe medical problems. Any individual with a complicated medical condition should be reviewed with your instructor before proceeding.

This program can be used with a friend or relative, but the impact of the learning experience is likely to be greater if you do not know the client. Established relationships can have patterns and issues that could interfere with the counseling relationship. The process of establishing a counseling relationship is a valuable learning experience that would be lost if the counselor is closely connected to his or her client. An additional concern is the strain such an experience can put on a relationship if a friend or family member is having difficulty meeting his or her obligations to take part in the program. An associate that has been coerced into the program may be a reluctant client and not have the motivation to make lifestyle changes.

In our clinic, we have been successful at finding volunteers by advertising on our university campus through flyers, school newspaper stories, school radio advertisements, e-mail listservs, and classroom announcements. Student counselors sign up for a counseling room for four weekly one-hour sessions. Volunteers enroll in the program during a designated registration time in our student center or with the instructor of the counseling course. Most volunteers follow through on their commitment to complete the program. However, for those that do not, we keep a record of all those potential volunteers who we were not able to accommodate during registration. Usually we are able to fill a vacancy from the record. As a last resort, a counseling student may need to secure a friend or relative to be his or her volunteer.

GOALS OF THE GUIDED COUNSELING EXPERIENCE

The overall goals of the guided counseling experience can be divided into skill goals and attitudinal goals.

Skill Goals

Skill goals as you complete this program include the following:

- Conduct a four-session counseling program with an adult volunteer client.
- Demonstrate use of standard nutrition assessment tools: client assessment questionnaire, food frequency

form, usual diet form, and computerized analysis of three-day log.

- Utilize appropriate interpersonal skills for facilitation and relationship building.
- Demonstrate use of basic counseling responses.
- Tailor education approaches and intervention strategies to a client's motivational level and needs.
- Employ effective goal setting during counseling sessions.
- Effectively evaluate, document, and plan after each session.
- Assess self for counseling effectiveness.

Attitude Goals

Attitudinal goals for this intervention are in accordance with those defined by Dubé et al.[1] for medical interviewing:

- An unconditional positive regard for clients
- A respectful and nonjudgmental approach to clients
- A willingness to work with clients as partners
- An openness to work with and learn from clients with diverse backgrounds and personal styles

THE FOUR COUNSELING SESSIONS

The following describes the procedures and goals for four one-hour counseling sessions. Each session has a detailed checklist to aid in the flow and pacing of a counseling session. Time frames are given for the phases to help with pacing. You may need to modify some segments of the protocol to keep the sessions moving.

The need and desire for the amount of structure in the checklists vary. Some students prefer to use the checklists as a reference and then develop their own interview guide. Many welcome the well-defined guidelines and take the checklists into their counseling sessions. If this is the case, be careful not to let the checklists interfere with the development of a counseling relationship. Before meeting with your client, you should read the checklists carefully so that all you need to do is glance at the papers to help you keep on track. All components of the checklists are addressed in more detail in other chapters; however, those that require additional explanation for the intervention are highlighted here.

Preparation for Session 1

Certain preliminaries should be addressed before your first session. These include completing a registration form, getting your client's consent, and giving him or her a welcome packet containing a client assessment questionnaire and a short food frequency form to complete before the first session. Lifestyle Management Form 3.3 is the agreement form for the program (see Appendix G). You need to thoroughly explain the content of the form. This is your client's informed consent. Adhere to the following protocol:*

1. Give the potential client a copy of the nutrition counseling–lifestyle management agreement form to read along as you read the content aloud.
2. Look up at the potential client often for any signs of confusion or misunderstanding, and check for understanding.
3. After reading through the entire document, ask whether the potential client has any questions or concerns. If there are any, address them.
4. Ask the potential client to summarize his or her understanding of what is being signed. You should correct misunderstandings and provide clarification.
5. You should stress that signing the form is voluntary; if the potential client *hesitates, you should offer the person the option of not signing the document.* If the person refuses to sign, thank the person for his or her interest in the program and assure that another person will be found.
6. Obtain a voluntary signature from the person on both copies of the form (client copy and clinic copy).
7. Give a copy of the form to the client, and submit the clinic copy to your instructor.

If your client did not formally register for the program before your first meeting, you should call him or her to arrange a registration meeting and to give the future client a welcome packet. If you were not able to meet with your client for registration before the first session, your first counseling session will probably require an extra twenty minutes.

*This protocol is based on one developed for a practicum for medical interviewing students at Brown University School of Medicine.[1]

If your client formally went through the registration process with someone else, you should call your client to verify your meeting time and place. Your telephone conversation could go something like this:

Hello, Mrs. Jones. This is [give your full name]. I am a student at University X taking a nutrition counseling course. I understand that you have been assigned to my time slot for nutrition counseling. I called to verify our meeting on [give day of the week and date] at [give time] in [give location]. The session should take approximately one hour. Is this OK? Have you been able to look over the forms? Do you have any questions? Please try to fill them out to the best of your ability. We will be going over them during our session. Do you know where the [name location] is? Please give me a call if you have any problems keeping our meeting time. Do you have my phone number? The best time to reach me is at [give time of day]. I look forward to meeting you.

Note that the session 1 checklist indicates that you should bring copies of the forms contained in your client's welcome packet in case the forms have been forgotten.

Session 1

The involving phase for session 1 is more extensive than for subsequent sessions. During the greeting, the tone of your voice and your body language should convey the message that you are happy to meet your client. Be sure to stand and smile during the welcome and invite your client to sit down. The following are examples of possible greetings:

Good morning, Mr. Gray. I am very happy to meet you. I am Sally Mason. I hope you managed to find my office easily. May I call you Jim?

Good afternoon, Mrs. Jones! I am Sally Mason. Please come in. It is great to finally meet you. How do you prefer to be addressed? Is calling you Mrs. Jones OK?

The amount of time you will need to devote to explaining the program will depend on how much of this topic was covered during the registration process. The following are some aspects of the program you may wish to emphasize.

Overview *The objective of the four sessions is for us to work in partnership to discover ways for you to make lifestyle changes so as to improve the nutritional quality of your diet and to achieve or maintain a desirable weight. This program is designed to achieve slow, steady changes.*

Nutritional Assessment *To analyze the nutritional quality of your diet and make recommendations, it will be necessary to complete some forms. You have completed two of them already, client assessment questionnaire and the food frequency evaluation. They were in your welcome packet. We will go over them today. I will also ask you to describe what you eat on a usual day. In a few weeks I will ask you to record your food intake for three days in order to analyze your diet with the aid of a computer program.*

Educational Component *Every session will include an educational component geared to your needs and interests. This could include short videos or reading materials.*

Goal Setting *Each session, we will evaluate the possibility of setting goals or modifying old ones based on your food habit problems and strengths.*

Food Management Options *You have a choice in how you would like to proceed with aid in making dietary changes. On your client assessment questionnaire, you indicated how much structure you believe would work for you.*

Audiotape *Possibly I will be asking your permission to audiotape a session to evaluate my counseling skills.*

To develop a better understanding of your client's issues, a typical day strategy and usual diet method have been combined during the exploration phase. You will need to jot down notes on the usual diet form while your client is talking. Care should be taken not to let this process interfere with the flow of conversation. For example, do not ask your client to repeat in order to correctly record what is usually eaten for lunch. Instead, ask clarifying questions at the conclusion of your client's story.

The resolving phase intervention is geared toward your client's degree of readiness to change. For most clients, the sequencing of the questions and topics should work well. However, you should not feel as if you are tied to the protocol. If you believe certain questions or topics would be useful to cover with your client, cross-over is appropriate.

The closing part of the interview is just as important as the opening. Each session should end on a positive note.

The postsession activities guide you to evaluate your client's needs, assess your intervention effectiveness, and prepare for the next session. To gain experience with two documentation methods, both IAO (Client Progress Report) and SOAP have been assigned. In subsequent sessions, only the Client Progress Report will be used for charting. Since the next session addresses physical activity, the checklist guides you to complete preparation tasks in this area.

You will also need to prepare an education intervention for your client that should last ten minutes or less.

Your guide to selecting a topic will be your discussions during session 1 as well as the areas of interest indicated in the client assessment questionnaire. In addition, one of the postsession activities is to prepare a preliminary copy of the food management tool (detailed food plan, exchange system, food group plan, or only goal setting) selected by your client in the client assessment questionnaire. If your client indicated a desire to just set goals each week, then a food management tool does not need to be developed.

Session 2

After the opening, one of your first objectives will be to review your client's progress since the first session, and then provide the tailored educational intervention you prepared for your client. The next counseling activity will be to review the preliminary food management tool you developed based on your session 1 discussions and your client's response to the inquiry on the client assessment questionnaire. Discuss with your client how the tool should be modified and how the tool should be used. Motivated clients may wish to implement the food plan immediately; others may want to use the pattern as a picture of what the client is attempting to achieve through small steps. For those who wish immediate implementation, you will need to provide instructions on the use of the tool.

Another issue to be addressed is the need to keep a food diary for at least three days to complete a computerized diet analysis. Review with your client procedures for keeping a food diary. The final part of the counseling session will be devoted to the protocol for physical activity counseling.

Session 3

After the greeting and reviewing progress in meeting previously set goals, go over your client's three-day food record and verify portion sizes and preparation methods. If your client did not bring his or her records, then conduct a twenty-four-hour recall. In that case, you will do a computer analysis based on this recall rather than a three-day food record. Then proceed with the tailored education intervention you prepared for your client.

Your next activity will be to review your client's food management tool, if appropriate. Whether or not this topic needs to be addressed will depend on the outcome of your food management tool discussion during the previous session. The rest of the exploration-education phase will cover the role of behavior management and the importance of a support system for making a behavior change. The checklist contains three behavior management strategies to discuss with your client; however, you may wish to address alternative strategies identified in Chapter 5 or 6 based on your client's needs.

Session 4

This is your final session. After greeting your client and setting the agenda, investigate your client's thoughts about his or her progress since the last session. Then assess all goals your client is currently pursuing and consider altering or continuing them. Present the computerized analysis to your client and review it point by point. Ask your client to give his or her thoughts about the feedback and then summarize the discussion regarding the analysis.

Next, present the tailored education intervention that you prepared for your client. The last activity of the exploration-education phase is to explore some aspects of relapse prevention counseling. Specifically explain the spiral of change in Lifestyle Management Form 6.4 and investigate high-risk situations that apply to your client. Discuss the concept of apparent irrelevant decisions, and identify ones that could apply to your client's lifestyle change goals.

Continuing with relapse prevention counseling into the resolving phase, discuss coping strategies to deal with the high-risk situations and apparent irrelevant decisions that your client identified. The final counseling objective is to address cognitive restructuring. First discuss the concept with your client and investigate possible dysfunctional thinking patterns your client may exhibit. You may wish to show Lifestyle Management Form 6.5 to your client to help in the identification. If this is an issue for your client, ask him or her to consider the validity of their destructive self-talk and prepare some coping strategies, such as thought stopping and alternative responses. Imagery can be used to rehearse the use of the strategies.

Ending the relationship in a meaningful manner will have a significant impact on the counseling encounter. The checklist contains a sequence of culminating points to address. At the end of the session, express your appreciation for your client's willingness to participate in the clinic, and present your client with a certificate of appreciation.

REFERENCE

[1]Dubé C, Novack D, Goldstein M. Brown University School of Medicine Student Syllabus, BI 371 Medical Interviewing. Providence, RI: Brown University; 1999.

PREPARATION FOR SESSION 1 CHECKLIST

REGISTRATION

- ❑ Complete registration form—LMF* 9.1 in duplicate.
  - ❍ Give one copy to the client and one copy to the counselor.
- ❑ Complete the agreement form—LMF 3.3—in duplicate.
  - ❍ Adhere to the protocol for obtaining a consent in preparation for session 1 guidelines.
  - ❍ Both you and your client should sign each copy of the form.
  - ❍ Give a copy to your client.
- ❑ Give or send the client a welcome packet containing the following:
  - ❍ LMF 4.1, Client Assessment Questionnaire
  - ❍ LMF 4.4, Food Frequency Questionnaire
  - ❍ LMF 7.6, Medical Release Form

PHONE CALL

- ❑ Verify date, time, and place of counseling session.
- ❑ Remind client about forms and inquire if there were any questions.
- ❑ Verify how to get in touch with each other if the meeting needs to be postponed.
- ❑ Express desire to meet your client.

PREPARE FOR SESSION 1

- ❑ Review the following procedures/guidelines:
  - ❍ Analysis and flow of a counseling interview/counseling session (Chapter 3)
  - ❍ Nutrition Counseling Motivational Algorithm (Figure 3.2 in Chapter 3)
  - ❍ Session 1 guidelines
  - ❍ "A typical day strategy" (Exhibit 3.2 in Chapter 3)
  - ❍ Goal-setting process (Chapter 4)
  - ❍ Protocol for a diet history interview (Exhibit 4.1 in Chapter 4)
  - ❍ Daily Food Guide—Food Guide Pyramid (Appendix A) and DASH Diet (Appendix B)
  - ❍ Review session 1 checklist.
- ❑ Bring copies of the following forms:
  - ❍ LMF 3.1, Assessment Ruler
  - ❍ LMF 4.1, Client Assessment Questionnaire
  - ❍ LMF 4.3, 24-Hour Recall/Usual Diet Form
  - ❍ LMF 4.4, Food Frequency Questionnaire
  - ❍ LMF 4.5, Food Group Feedback Form
  - ❍ LMF 7.4, Physical Activity Medical Readiness Form
  - ❍ LMF 7.5, Physical Activity Status
  - ❍ Copy of the Food Guide Pyramid (Appendix A)
- ❑ Bring visuals to estimate portion size.
- ❑ Minimize distractions.
- ❑ Remind yourself of the six relationship-building responses: attending, reflection, legitimation, support, partnership, and respect.

*LMF = Lifestyle Management Form.

SESSION 1 CHECKLIST	
TASK	**POSSIBLE DIALOGUE**
Involving Phase (10–15 minutes)	
❑ Greeting ❍ Verbal greeting ❍ Shake hands. ❍ Introduce yourself. ❍ Resolve how to address each other.	*Good morning. I'm very happy to meet you.* *How would you like to be addressed?*
❑ Small talk	*How did you hear about our program?*
❑ Investigate client's long-term objectives, expectations, needs, and concerns. *If appropriate, use* ❍ reflection statements. ❍ legitimation statements. ❍ respect statements. ❑ Summarize.	*Have you ever worked with a nutrition counselor before?* *What do you want to achieve in this program?*
❑ Explain program and counseling process. ❍ Use partnership statement.	*I'd like to tell you about the design of this program.* *My hope is that we will work together to build on skills you already have to make dietary changes.*
❑ Review weight loss expectations, if appropriate.	
❑ Discuss monitoring of weight, if appropriate.	
❑ Set agenda. ❍ Review Client Assessment Questionnaire. ❍ Ask about a typical day. ❍ Review Food Frequency Form. ❍ Discuss Food Guide Pyramid or DASH Diet. ❍ Set one goal, if appropriate.	*What we will do this session is . . .*
Transition ❑ Transition statement	*Now that we have gone over the basics of the program, we will explore your needs in greater detail.*
Exploration-Education Phase (25 minutes)	
❑ Review completed Client Assessment Questionnaire, LMF 4.1. ❍ Clarify any highlights on the form.	*I am wondering what came to your mind as you were completing this form.* *What topics in this form do you think have particular importance for your food issues?*
❑ A Typical Day	*Can you take me through a typical day so I can understand more fully what happens and tell me where eating fits into the picture?*

Exploration-Education Phase (25 minutes) (Continued)	
❍ Interrupt as little as possible. ❍ Do not impose your ideas. ❍ Speed up the pace, if necessary. ❍ Summarize.	*Start with when you get up.* *Is there anything else you would like to add?*
❑ Complete Usual Diet Form, LMF 4.3. ❍ Fill in during A Typical Day review. ❍ Clarify; ask questions.	*How was the chicken cooked?*
❑ Review completed Food Frequency Questionnaire, LMF 4.4. ❍ Clarify portion sizes; use visuals. ❍ Clarify preparation methods.	*Thank you for completing this questionnaire.* *What came to your mind as you were filling it out?* *Did you feel a need to clarify or expand on anything?*
❑ Complete Food Group Feedback Form, LMF 4.5.	
❑ Provide Feedback and Education ❍ Review feedback form, LMF 4.5, point by point, nonjudgmentally. ❍ Clarify when needed. ❍ Explain Food Guide Pyramid and DASH diet. ❍ Ask opinion of comparison. ❍ Give your opinion, if requested. ❍ Summarize.	*What do you think about the comparison?*
❑ What's next?	*How would you like to proceed?*
❑ Check readiness with Assessment Ruler, LMF 3.1. ❍ Check importance. ❍ Check confidence. 0–4 = not ready, 4–8 = unsure, 8–12 = ready	*To get a better idea of how ready you are to make a change, we will use this picture of a ruler. If 0 is not ready all and 12 is ready, where are you?* *Using the same scale, how do you feel right now about how important this change is for you? How confident are you that you can make this change?*
Transition to Resolving Phase ❑ Transition Statement	*Now I'd like to talk more about the possibility of changing your food patterns.*
Resolving Phase (15 minutes)	
Level 1—Not Ready (0–4 on the ruler)	
❑ Raise awareness. ❑ Personalize benefits. ❑ Request permission to discuss. ❑ Ask key open-ended questions.	*Summarize benefits of following the pyramid or DASH diet.* *Because of (family history, past concerns, present medical problems), you would benefit from . . .* *Would you like to discuss the possibility of such a change?* *What do you believe will happen if you don't change?*

Resolving Phase (15 minutes) (Continued)	
❍ Discuss importance. ❍ Identify motivating factors.	*How come you picked 3 and not 1 on the ruler?* *What would have to happen for you to move up to the number 8?*
❑ Summarize. ❑ Ask permission to give advice.	*It's up to you—you know best—but small changes can make a difference.*
❑ Express support.	*You probably need some time to think about this. . . .* *Do not hesitate to call me if you have any questions.*
Level 2—Unsure (4–8 on the ruler)	
❑ Raise awareness.	*Summarize benefits of following the pyramid or DASH diet.*
❑ Ask key open-ended questions; promote change talk. ❍ Explore confidence. ❍ Explore barriers.	*How come you rated your confidence as a 6 instead of a 1 on the ruler?* *What would you need to get to 10 on the ruler?* *What is preventing you from making changes?*
❑ Examine pros and cons. ❍ Summarize.	*What do you like about your present diet? Dislike?* *Advantages of changing? Disadvantages?*
❑ Imagine the future.	*What would your life be like if . . . ? What is the first thing you notice? How do you feel?*
❑ Explore past successes. ❑ Explore social supports. ❑ Summarize. ❑ Ask about next step—go to goal setting in Level 3, if appropriate	*Were you ever able to . . . ?* *Do you have someone who could support you?*
Level 3—Ready (8–12 on the ruler)	
❑ Praise positive behaviors.	*It is so good that you . . .*
❑ Explore change options to develop a broadly stated goal. ❍ Elicit client's thoughts. ❍ Make an options tool, if appropriate.	*Do you have an idea of what will work for you?* *This is an options tool. Let's brainstorm ideas, and we'll write them in the circles.*
❍ Probe for concerns about the selected option.	*This seems to be the best choice, but will it work for you?*
❑ Explain goal-setting basics. ❑ Identify a specific goal. ❍ Small talk ❍ Look to the past. ❍ Build on past. The stated goal is ❍ achievable. ❍ measurable. ❍ totally under client's control. ❍ stated positively.	*What is the smallest goal you believe is worth pursing?* *When have you eaten . . . before?* *Explain how it happened that you ate . . . before.* *Would this work for someone else?*

Resolving Phase (15 minutes) (Continued)	
❑ Develop an action plan.	
❍ Investigate physical environment.	*Do you have everything you need?*
❍ Examine social support.	*Is there anyone who could help you achieve your goal?*
❍ Review cognitive environment. Explain positive coping talk, if necessary.	*What are you saying to yourself right now about this goal?*
❑ Select tracking technique—chart, journal, etc.	
❑ Ask your client to verbalize the goal.	*Just to be sure we are both clear about your goal, could you please state the goal?*
❑ Write down goal on a card and give it to your client.	

Closing Phase (5 minutes)	
❑ Support self-efficacy.	*I think we did a good job selecting a goal that will work for you.*
❑ Review issues and strengths.	
❑ Use a relationship-building response (respect).	*I am very impressed with . . .*
❑ Restate food goal.	
❑ Give Physical Activity Forms, LMFs 7.4 and 7.5.	
❑ Review next meeting time.	
❑ Set date/time to call with reminder.	
❑ Shake hands.	
❑ Express appreciation for participation.	*Thank you so much for your participation in this program. I look forward to working with you to implement some of the options we've identified.*
❍ Use support and partnership statement.	

POSTSESSION 1

Congratulations! You have just finished your first nutrition counseling session. It is now time to reflect, evaluate, document, and plan for the next session.

EVALUATE AND DOCUMENT SESSION 1

- ❑ Complete the following forms:
  - ❍ LMF 4.7, Client Concerns and Strengths Log
  - ❍ LMF 4.8, Client Progress Log
  - ❍ LMF 8.1, Interview Checklist
- ❑ Document the session by writing a SOAP.

PLAN FOR SESSION 2

- ❑ Prepare a preliminary copy of your client's food monitoring tool as indicated in the Client Assessment Questionnaire.
- ❑ Calculate your client's BMI (Chapter 4).
- ❑ Calculate your client's exercise target zone (Chapter 7).
- ❑ Prepare an education intervention according to the needs and desires of your client as indicated in the Client Assessment Questionnaire, LMF 4.1, and your discussions.

PREPARE FOR SESSION 2

- ❑ Review the following procedures/ guidelines:
  - ❍ Session 2 guidelines and checklist
  - ❍ Physical Activity Algorithm, Figure 7.2, and protocols (Chapter 7)
  - ❍ Goal-setting process (Chapter 4)
  - ❍ Food diary guidelines (Chapter 4)
  - ❍ SOAP and Client Progress Report notes from session 1
- ❑ Bring copies of the following forms:
  - ❍ LMF 3.1, Assessment Ruler
  - ❍ LMF 7.1, Benefits of Regular Moderate Physical Activity
  - ❍ LMF 7.2, Physical Activity Log
  - ❍ LMF 7.3, Physical Activity Options
  - ❍ LMF 7.4, Physical Activity Medical Readiness Form—in case your client does not bring
  - ❍ LMF 7.5, Physical Activity Status—in case your client does not bring
  - ❍ LMF 7.7, Physical Activity Feedback Form, 2 copies
  - ❍ LMF 4.2, Food Record
  - ❍ Preliminary copy of your client's food-monitoring tool
  - ❍ Calculated BMI
- ❑ Bring visuals to estimate portion size.
- ❑ Minimize distractions.
- ❑ Remind yourself of the six relationship-building responses: attending, reflection, legitimation, support, partnership, and respect.

SESSION 2 CHECKLIST	
TASK	**POSSIBLE DIALOGUE**
Involving Phase (5 minutes)	
❑ Greeting ❍ Extend verbal greeting. ❍ Shake hands.	*Good morning. It is nice to see you again.*
❑ Set agenda. ❍ Review week. ❍ Address requested educational issue. ❍ Review selected diet modification method. ❍ Review food diary procedure. ❍ Assess physical activity readiness.	*In the session today I thought we would review how your goal worked out this week, go over [client's requested educational need], discuss the food management tool you indicated that you desired, review how to keep a food diary, and then discuss physical activity. How does this sound?*
Transition ❑ Transition statement	*So let's move on. First I'd like to address last week's goal.*
Exploration-Education Phase (25 minutes)	
❑ Investigate client thoughts about the week. *If appropriate, use* ❍ reflection statements. ❍ legitimation statements. ❍ respect statements.	*How did your week go?* *Let's look at the adherence ruler. What number would you pick to describe how closely you have been following your plan?* *So you feel good about your diet but you are not sure if you can keep it going.*
❑ Evaluate effectiveness of plan. ❍ Identify barriers to goal achievement. ❍ Clarify client strengths and weaknesses. ❑ Summarize—keep present goal or modify.	*So overall this week you . . .*
❑ Tailor educational experience. ❑ Determine appropriate food management tool. ❍ Show preliminary food pattern based on indicated pattern. ❍ If appropriate, review each tool (detailed plan, exchanges, food groups, only goal setting). ❍ Identify advantages and disadvantages, if appropriate.	*On your Client Assessment Questionnaire, you indicated that you wanted some structure but freedom to select foods from food groups. I prepared a preliminary food management tool based on the DASH Food Plan. I thought we could discuss the plan today to see whether this is truly what you think would be useful to you and adjust it according to what you believe would work for you. We could use this plan in two ways: (1) One would be a defined pattern that you would try to implement immediately. (2) It could also be used as a tool that we periodically review to identify what we are working toward.*
❑ Instructions on use of tool (detailed plan, exchanges, food groups, only goal setting), if appropriate	

Exploration-Education Phase (25 minutes) (Continued)	
❑ Food Diary/Food Record ❍ Give instructions, LMF 4.2. ❍ Select length of time (need 3 days for computer analysis). ❍ State purpose of computer analysis.	*Keeping a food diary is an excellent way to monitor and influence your eating habits. I really encourage you to keep records periodically. For the purpose of the computer analysis, I hope I can get you to do the analysis for three days. What do you think?*
❑ Provide physical activity feedback. ❍ Discuss BMI and health. ❍ Collect LMFs 7.4 and 7.5. ❍ Complete LMF 7.7 in duplicate Give one to the client. ❑ Ask opinion of comparison. ❑ Give your opinion, if appropriate. ❑ Summarize.	*What do you think about this comparison?* *Does this information surprise you?*
Transition to Resolving Phase ❑ Transition Statement	*Now I'd like to talk about any changes you would like to make.*
Resolving Phase (25 minutes)	
Level 1—Not Ready (Score 2 on LMF 7.5)	
❑ Summarize benefits of physical activity; use LMF 7.7. ❑ Personalize benefits to health status. ❑ Request permission to discuss possibility of change. ❑ Ask key open-ended questions. ❍ Discuss importance/reasons to be physically active. ❍ Elicit barriers to physical activity. ❑ Summarize. ❑ Ask permission to give advice, if appropriate. ❑ Give advice. ❑ Support self-efficacy.	*Would you be willing to discuss the possibility of a change in your physical activity patterns?* *What benefits of physical activity do you believe most likely apply to you?* *How come you picked 3 and not 1 on the ruler?* *What would have to happen for you to move up to the number 8?* *It's up to you—you know best—but small changes can make a difference.* *I really admire how you. . . . Look how capable you are when you. . . . You have the resources to be physically active. When you are ready, you will be able to increase your physical activity.*
Level 2—Unsure (Score 2 on LMF 7.5)	
❑ Raise awareness. ❑ Ask key open-ended questions to explore ambivalence. ❍ Identify disadvantages of changing. ❍ Explore consequences of inactivity. ❍ Identify hoped-for benefits.	*Summarize benefits of physical activity.* *What are some reasons you would like things to stay just like they are?* *What concerns do you have about not increasing your activity?* *What are some good things that would happen if you were more physically active?*

Resolving Phase (25 minutes) (Continued)	
❑ Explore past successes.	*Have you ever been physically active?* *Have you ever enjoyed a physical activity?*
❑ Imagine the future.	*What would your life be like if . . . ? What is the first thing you notice? How do you feel?*
❑ Summarize ambivalence—use LMF 7.3.	
❑ Ask about next step—go to goal setting in Level 3, if appropriate	
Level 3—Ready (Score 3 or 4 on LMF 7.5)	
❑ Praise positive behaviors.	
❑ Review current activity program.	
❑ Explore change options to develop a broadly stated goal.	
❍ Elicit client's thoughts.	*Do you have an idea of what will work for you?*
❍ Look to the past.	*What activities have you enjoyed in the past?*
❍ Go over list of possibilities. (See LMF 7.3.)	*Here are some options.*
❍ Validate physician approval.	*This seems to be the best choice, but will it work for you?*
❑ Client selects an appropriate activity goal.	
❑ Explain goal-setting basics, if appropriate.	
❑ Identify a specific physical activity goal.	
❍ Small talk.	*What is the smallest physical activity goal you believe is worth pursing?*
❍ Look to the past.	*When have you exercised before?*
❍ Goal is achievable.	*Would this goal work for someone else?*
❍ Goal is measurable.	
❍ Client has total control over achieving the goal?	
❍ Goal is stated positively.	
❑ Develop an action plan.	
❍ Discuss target heart rate zone.	
❍ Investigate physical environment.	*Do you have everything you need? Do you have walking shoes?*
❍ Examine social support.	*Is there anyone who could help you achieve your goal?*
❍ Review cognitive environment.	*What will you say to yourself if you miss a day that you planned to walk?*
❍ Explain positive coping talk, if necessary.	
❍ Discuss target heart rate zone.	
❑ Select tracking technique—chart, journal, etc.	
❑ Ask your client to verbalize the goal.	*Just to be sure we are both clear about your goal, could you please state the goal?*
❑ Write down goal on a card and give it to your client.	

Resolving Phase (25 minutes) (Continued)	
Level 4—Active (Score 5 on LMF 7.5)	
❑ Praise positive behaviors. ❑ Review current activity program. ❑ Review sport-specific nutrient needs. ❑ Relapse prevention. ❍ Explain relapse prevention. ❍ Use Assessment Ruler, LMF 3.1, to assess confidence to continue. ❍ Identify barriers. ❍ Explore solutions to barriers. ❍ Explain that setbacks are common. ❍ Identify social supporters. ❑ Identify a specific goal and action plan, if appropriate. ❑ Go to goal setting in Level 3.	*It is wonderful that you have such a physically active lifestyle.* *Let's look at how confident you are that you can maintain your level of activity.* *How come you chose 10 instead of 12?* *Do you have any ideas of how to overcome difficulties with . . . ?* *Setbacks are to be expected. It is important to just start up again.*
Closing Phase (5 minutes)	
❑ Support self-efficacy. ❑ Review issues and strengths. ❑ Use a relationship-building response (respect). ❑ Restate goals: ❍ Food goal ❍ Food diary ❍ Exercise goal ❑ Review next meeting time. ❑ Discuss audiotaping session 3 or 4. ❑ Set date/time to call with reminder. ❑ Shake hands. ❑ Express appreciation for participation. ❍ Use support and partnership statement.	*I am very impressed with . . .* *I have really enjoyed working with you. If you have any questions or concerns, do not hesitate to call me.*

POSTSESSION 2

EVALUATE AND DOCUMENT SESSION 2

- ❑ Complete the following forms:
  - ❍ LMF 4.8, Client Progress Report—Make additions to the form you started after session 1.
  - ❍ LMF 8.1, Interview Checklist—Use a copy of this form for a self-assessment after each session.

PLAN FOR SESSION 3

- ❑ Make adjustments to your client's food monitoring tool, if necessary.
- ❑ Prepare an education intervention as indicated in the Client Assessment Questionnaire and your discussions.

PREPARE FOR SESSION 3

- ❑ Review the following procedures/guidelines:
  - ❍ Analysis and flow of a counseling interview/counseling session (Chapter 3)
  - ❍ Nutrition Counseling Motivational Algorithm (Figure 3.2 in Chapter 3)
  - ❍ Goal-setting process (Chapter 4)
  - ❍ Session 3 guidelines and checklist
  - ❍ Cue management, countering, and barriers counseling (Chapter 5)
  - ❍ Social support and social disclosure (Chapter 6)
- ❑ Review your notes from session 1:
  - ❍ LMF 4.8, Client Progress Report
- ❑ Bring copies of the following forms:
  - ❍ LMF 3.1, Assessment Ruler
  - ❍ LMF 4.3, 24-Hour Recall/Usual Diet Form—This form will be used to complete a twenty-four-hour recall if your client does not bring three-day food records.
- ❑ Bring visuals to estimate portion size.
- ❑ Bring visuals for tailored education intervention.
- ❑ Bring audio- or videotape recorder if taping session 3.
- ❑ Minimize distractions.
- ❑ Remind yourself of the six relationship-building responses: attending, reflection, legitimation, support, partnership, and respect.

SESSION 3 CHECKLIST	
TASK	**POSSIBLE DIALOGUE**
Involving Phase (5 minutes)	
❑ Greeting ❍ Verbal greeting ❍ Shake hands. ❑ Set agenda. ❍ Review week. ❍ Tailor education intervention. ❍ Describe pending computer analysis. ❍ Discuss role of behavior in food choices. ❍ Identify support system.	*Good morning. It is nice to see you again.*
Transition ❑ Transition statement	*I'd like to start by asking about your week.*
Exploration-Education Phase (25 minutes)	
❑ Investigate client thoughts about the week *If appropriate, use:* ❍ reflection statements. ❍ legitimation statements. ❍ respect statements.	*How did your week go?*
❑ Assess adherence to goals using LMF 3.1. ❍ Food goal ❍ Exercise goal ❍ Identify barriers to goal achievement. ❍ Clarify client strengths and weaknesses. ❍ Summarize—keep present goals or modify. ❑ Review food diary: ❍ Portion sizes ❍ Preparation methods ❍ Condiments ❑ Tailor education intervention. ❑ Review food management tool, if appropriate. ❑ Discuss role of behavior in food choices: ❍ cue management ❍ countering ❍ barriers management ❑ Discuss importance of support system and possible supports: ❍ support buddies ❍ organizations	*Let's look at the adherence ruler again. What number would you pick to describe how closely you have been following your plan?* *So you feel good about your diet but you are not sure if you can keep it going.* *So overall this week you . . .* *A variety of behavior change methods have been used successfully to help people change their habits. I thought we could talk about some of them and choose one to help you with your desired changes.* *Another topic I'd like to address today is the importance of a support system and the different ways to build one.*

Exploration-Education Phase (25 minutes) (Continued)	
❍ self-help groups ❍ classes ❑ Social disclosure	
Transition to Resolving Phase ❑ Transition statement.	*Now, I would like to talk about any specific changes you would like to make.*
Resolving Phase (25 minutes)	
❑ Identify a specific goal. ❍ Select a new or modify a previous goal. ❍ Review options tool, if appropriate. ❍ Small talk ❍ Look to the past. ❍ Build on past. The stated goal is: ❍ achievable. ❍ measurable. ❍ totally under client's control. ❍ stated positively.	*What is the smallest goal you believe is worth pursing?* *When have you eaten . . . before?* *Explain how it happened that you ate . . . before. Would this work for someone else?*
❑ Develop an action plan. ❍ Investigate physical environment. ❍ Examine social support. ❍ Review cognitive environment. Explain positive coping talk, if necessary. ❑ Select a tracking technique—chart, journal, etc. ❑ Do a microanalysis of scenario; review the plan step by step. ❑ Ask your client to verbalize the goal. ❑ Write down goal on a card and give it to your client.	
Closing Phase (5 minutes)	
❑ Support self-efficacy. ❑ Review issues and strengths. ❑ Use a relationship-building response (respect). ❑ Restate food, exercise and behavior goals ❑ Review next meeting time ❑ Set date/time to call with reminder ❑ Shake hands ❑ Express appreciation for participation ❍ Use support and partnership statement	*I am very impressed with . . .*

POSTSESSION 3

EVALUATE AND DOCUMENT SESSION 3

❑ Complete the following forms:
   ❍ LMF 4.8, Client Progress Report—Make additions to the form you have been completing after each session.
   ❍ LMF 8.1, Interview Checklist—Use a copy of this form for a self-assessment after each session.
❑ Complete a computer analysis of the food diary data.

PLAN FOR SESSION 4

❑ Make adjustments on your client's food monitoring tool, if necessary
❑ Prepare an education intervention as indicated in the Client Assessment Questionnaire and your discussions.

PREPARE FOR SESSION 4

❑ Review the following procedures/guidelines:
   ❍ Analysis and flow of a counseling interview/counseling session (Chapter 3)
   ❍ Nutrition Counseling Motivational Algorithm (Figure 3.2 in Chapter 3)
   ❍ Session 4 guidelines and checklist
   ❍ Modifying Cognitions (Chapter 6)
   ❍ Relapse Prevention (Chapter 6)
   ❍ Ending the Counseling Relationship (Chapter 8)
   ❍ Goal-setting process (Chapter 4)
   ❍ Client Progress Report (LMF 4.8)
❑ Bring copies of the following forms:
   ❍ LMF 6.4, Prochaska's and DiClemente's Spiral of Change
   ❍ LMF 6.5, Frequent Cognitive Pitfalls
❑ Bring visuals to estimate portion size.
❑ Bring tape recorder if audiotaping session 4.
❑ Minimize distractions.
❑ Remind yourself of the six relationship-building responses: attending, reflection, legitimation, support, partnership, and respect.

SESSION 4 CHECKLIST	
TASK	**POSSIBLE DIALOGUE**
Involving Phase (5 minutes)	
❑ Greeting ❍ Verbal greeting ❍ Shake hands.	*Good morning. It is nice to see you again.*
❑ Set agenda. ❍ Discuss goal achievement. ❍ Review computerized analysis and comparison to standards. ❍ Tailor education intervention. ❍ Discuss relapse prevention. ❍ Discuss cognitive restructuring.	*Let's discuss your concerns about this past week, how the goals and action plan worked out, your computerized diet analysis, relapse prevention and cognitive restructuring.*
Transition ❑ Transition statement	*So let's move on and take a look at your week.*
Exploration-Education Phase (5 minutes)	
❑ Investigate client thoughts about the week ❑ Summarize. ❑ Assess present regimen: ❍ Food goals ❍ Physical activity ❍ Behavior strategy ❍ Continue/change ❑ Review computerized analysis: ❍ Compare to standard. ❍ Highlight areas of concern. ❍ Provide feedback, education. ❍ Ask opinion of comparison. ❍ Give your opinion, if appropriate. ❑ Summarize. ❑ Tailor education intervention. ❑ Relapse prevention. ❍ Describe behavior change and relapse—show LMF 6.4. ❍ Identify high-risk situations. ❍ Determine apparent irrelevant decisions.	*So overall this week you . . .* *What do you think about the analysis?* *Behavior change can be compared to a journey. . . .* *When are you likely to have the greatest difficulty keeping your goals?* *Sometimes problems occur because of mini decisions that seem harmless on the surface.*
Transition to Resolving Phase ❑ Transition statement	

Resolving Phase (25 minutes)	
❑ Coping strategies: ❍ Urge management: contract, countering, urge surfing ❍ STOP ❑ Cognitive restructuring counseling: ❍ Educate about the process ❍ Investigate dysfunctional thinking patterns; show LMF 6.5. ❍ Explore validity of self-destructive statements. ❍ Explain thought stopping. ❍ Prepare constructive responses. ❍ Use imagery. **Ending the Counseling Relationship** ❑ Review the issues that brought your client to you in the first place. ❑ Identify goals and progress in meeting them. ❑ Emphasize success. ❑ Summarize current status. ❑ Explore the future. ❑ Provide and elicit feedback regarding the significance of the relationship. ❑ Summarize.	*Some people are surprised to learn that what they are thinking can influence their ability to make lifestyle changes. Many people are not aware that thinking patterns can be changed. We are not obligated to keep a thought in our head.*
Closing Phase (5 minutes)	
❑ Support self-efficacy. ❑ Use a relationship-building response (respect). ❑ Shake hands. ❑ Express appreciation for participation. ❑ Give a certificate of appreciation.	

POSTSESSION 4

Now that you have completed the four-session program, take the time to reflect and evaluate the experience.

EVALUATE AND DOCUMENT SESSION 4

- ❑ Complete the following forms:
  - ❍ LMF 4.8, Client Progress Report—Make additions to the form you have been completing after each session.
  - ❍ LMF 8.1, Interview Checklist—Use a copy of this form for a self-assessment after each session.
  - ❍ LMF 8.2, Complete Counseling Responses Competency Assessment

EVALUATE THE TOTAL COUNSELING INTEREVENTION

- ❑ Complete the Chapter 8 assignment, "Evaluate Your Counseling Effectiveness."
- ❑ Reflect on the total four-session program and answer the following questions:

1. Describe your best counseling experience. What did you specifically say or do to facilitate the positive interaction? Give examples.
2. Describe any behavioral strategies you used during the intervention. Were they effective? Explain.
3. Describe any difficulties you encountered. What do you believe caused them? Is there anything you could have done differently to prevent or alleviate the impact of the difficulties?
4. What did you learn from this experience?

APPENDICES

Appendix

DAILY FOOD GUIDE

Breads, Cereals, and Other Grain Products: 6 to 11 servings per day.

These foods are notable for their contributions of complex carbohydrates, riboflavin, thiamin, niacin, iron, protein, magnesium, and fiber.
Serving = 1 slice bread; ½ c cooked cereal, rice, or pasta; 1 oz ready-to-eat cereal; ½ bun, bagel, or English muffin; 1 small roll, biscuit, or muffin; 3 to 4 small or 2 large crackers.

▼ Whole grains (wheat, oats, barley, millet, rye, bulgur, couscous, polenta), enriched breads, rolls, tortillas, cereals, bagels, rice, pastas (macaroni, spaghetti), air-popped corn.
♦ Pancakes, muffins, cornbread, crackers, cookies, biscuits, presweetened cereals, granola, taco shells, waffles, french toast.
■ Croissants, fried rice, doughnuts, pastries, cakes, pies.

Vegetables: 3 to 5 servings per day (use dark green, leafy vegetables and legumes several times a week).

These foods are notable for their contributions of vitamin A, vitamin C, folate, potassium, magnesium, and fiber, and for their lack of fat and cholesterol.
Serving = ½ c cooked or raw vegetables; 1 c leafy raw vegetables; ½ c cooked legumes; ¾ c vegetable juice.

▼ Bamboo shoots, bok choy, bean sprouts, broccoli, brussels sprouts, cabbage, carrots, cauliflower, corn, cucumbers, eggplant, green beans, green peas, leafy greens (spinach, mustard, and collard greens), legumes, lettuce, mushrooms, okra, onions, peppers, potatoes, pumpkin, scallions, seaweed, snow peas, soybeans, tomatoes, water chestnuts, winter squash.
♦ Candied sweet potatoes.
■ French fries, tempura vegetables, scalloped potatoes, potato salad.

Fruits: 2 to 4 servings per day.

These foods are notable for their contributions of vitamin A, vitamin C, potassium, and fiber, and for their lack of sodium, fat, and cholesterol.
Serving = typical portion (such as 1 medium apple, banana, or orange, ½ grapefruit, 1 melon wedge); ¾ c juice; ½ c berries; ½ c diced, cooked, or canned fruit; ¼ c dried fruit.

▼ Apricots, blueberries, cantaloupe, grapefruit, guava, oranges, orange juice, papaya, peaches, strawberries, plums, apples, bananas, pears, watermelon; unsweetened juices.
♦ Canned or frozen fruit (in syrup); sweetened juices.
■ Dried fruit, coconut, avocados, olives.

Key:

▼ Foods generally highest in nutrient density (good first choice).
♦ Foods moderate in nutrient density (reasonable second choice).
■ Foods lowest in nutrient density (limit selections).

Meat, Poultry, Fish, and Alternates: 2 to 3 servings per day.

Meat, poultry, and fish are notable for their contributions of protein, phosphorus, vitamin B_6, vitamin B_{12}, zinc, iron, niacin, and thiamin; legumes are notable for their protein, fiber, thiamin, folate, vitamin E, potassium, magnesium, iron, and zinc, and for their lack of fat and cholesterol.

Servings = 2 to 3 oz lean, cooked meat, poultry, or fish (total 5 to 7 oz per day); count 1 egg, ½ c cooked legumes, 4 oz tofu, or 2 tbs nuts, seeds, or peanut butter as 1 oz meat (or about ⅓ serving).

- ▼ Poultry (light meat, no skin), fish, shellfish, legumes, egg whites.
- ♦ Lean meat (fat-trimmed beef, lamb, pork); poultry (dark meat, no skin); ham; refried beans; whole eggs, tofu, tempeh.
- ■ Hot dogs, luncheon meats, ground beef, peanut butter, nuts, sausage, bacon, fried fish or poultry, duck.

Milk, Cheese, and Yogurt: 2 servings per day; 3 servings per day for teenagers and young adults, pregnant/lactating women, women past menopause; 4 servings per day for pregnant/lactating teenagers.

These foods are notable for their contributions of calcium, riboflavin, protein, vitamin B_{12}, and, when fortified, vitamin D and vitamin A.

Serving = 1 c milk or yogurt; 2 oz process cheese food; 1½ oz cheese.

- ▼ Nonfat and 1% low-fat milk (and nonfat products such as buttermilk, cottage cheese, cheese, yogurt); fortified soy milk.
- ♦ 2% reduced-fat milk (and low-fat products such as yogurt, cheese, cottage cheese); chocolate milk; sherbet; ice milk.
- ■ Whole milk (and whole-milk products such as cheese, yogurt); custard; milk shakes; ice cream.

Note: These serving recommendations were established before the 1997 DRI, which raised the recommended intake for calcium; meeting the calcium recommendation may require an additional serving from the milk, cheese, and yogurt group.

Fats, Sweets, and Alcoholic Beverages: Use sparingly.

These foods contribute sugar, fat, alcohol, and food energy (kcalories). They should be used sparingly because they provide food energy while contributing few nutrients. Miscellaneous foods not high in kcalories, such as spices, herbs, coffee, tea, and diet soft drinks, can be used freely.

- ■ Foods high in fat include margarine, salad dressing, oils, lard, mayonnaise, sour cream, cream cheese, butter, gravy, sauces, potato chips, chocolate bars.
- ■ Foods high in sugar include cakes, pies, cookies, doughnuts, sweet rolls, candy, soft drinks, fruit drinks, jelly, syrup, gelatin, desserts, sugar, and honey.
- ■ Alcoholic beverages include wine, beer, and liquor.

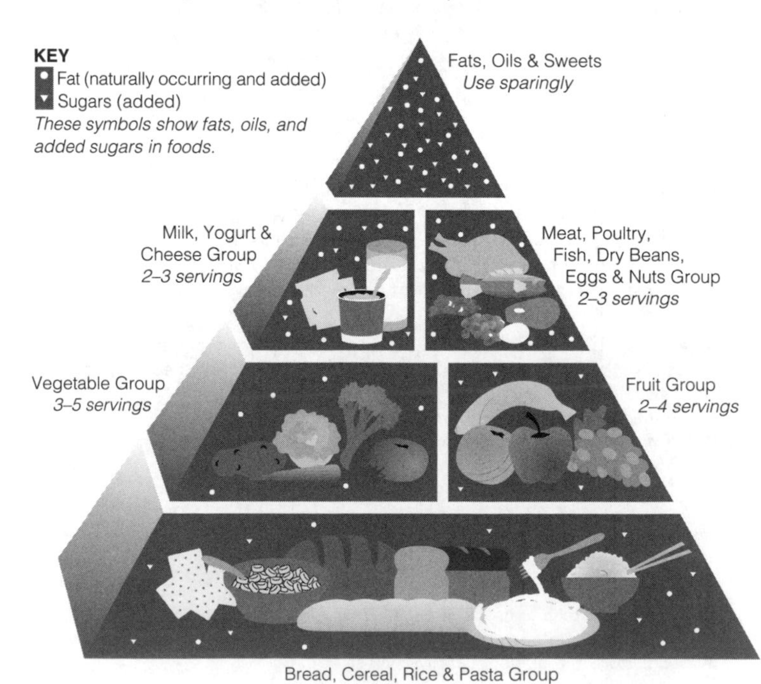

Food Guide Pyramid
A Guide to Daily Food Choices
The breadth of the base shows that grains (breads, cereals, rice, and pasta) deserve most emphasis in the diet. The tip is smallest: use fats, oils, and sweets sparingly.

Appendix

B

Following the DASH Diet

The DASH eating plan shown below is based on 2,000 calories a day. The number of daily servings in a food group may vary from those listed depending on your caloric needs.

Use this chart to help you plan your menus or take it with you when you go to the store.

FOOD GROUP	DAILY SERVINGS (except as noted)	SERVING SIZES	EXAMPLES AND NOTES	SIGNIFICANCE OF EACH FOOD GROUP TO THE DASH EATING PLAN
Grains & grain products	7–8	1 slice bread 1 cup dry cereal* ½ cup cooked rice, pasta, or cereal	whole wheat bread, English muffin, pita bread, bagel, cereals, grits, oatmeal, crackers, unsalted pretzels and popcorn	major sources of energy and fiber
Vegetables	4–5	1 cup raw leafy vegetable ½ cup cooked vegetable 6 oz vegetable juice	tomatoes, potatoes, carrots, green peas, squash, broccoli, turnip greens, collards, kale, spinach, artichokes, green beans, lima beans, sweet potatoes	rich sources of potassium, magnesium, and fiber
Fruits	4–5	6 oz fruit juice 1 medium fruit ¼ cup dried fruit ½ cup fresh, frozen, or canned fruit	apricots, bananas, dates, grapes, oranges, orange juice, grapefruit, grapefruit juice, mangoes, melons, peaches, pineapples, prunes, raisins, strawberries, tangerines	important sources of potassium, magnesium, and fiber
Lowfat or fat free dairy foods	2–3	8 oz milk 1 cup yogurt 1½ oz cheese	fat free (skim) or lowfat (1%) milk, fat free or lowfat buttermilk, fat free or lowfat regular or frozen yogurt, lowfat and fat free cheese	major sources of calcium and protein
Meats, poultry, and fish	2 or less	3 oz cooked meats, poultry, or fish	select only lean; trim away visible fats; broil, roast, or boil, instead of frying; remove skin from poultry	rich sources of protein and magnesium
Nuts, seeds, and dry beans	4–5 per week	⅓ cup or 1½ oz nuts 2 Tbsp or ½ oz seeds ½ cup cooked dry beans	almonds, filberts, mixed nuts, peanuts, walnuts, sunflower seeds, kidney beans, lentils and peas	rich sources of energy, magnesium, potassium, protein, and fiber
Fats & oils**	2–3	1 tsp soft margarine 1 Tbsp lowfat mayonnaise 2 Tbsp light salad dressing 1 tsp vegetable oil	soft margarine, lowfat mayonnaise, light salad dressing, vegetable oil (such as olive, corn, canola, or safflower)	Besides fats added to foods, remember to choose foods that contain less fats
Sweets	5 per week	1 Tbsp sugar 1 Tbsp jelly or jam ½ oz jelly beans 8 oz lemonade	maple syrup, sugar, jelly, jam, fruit-flavored gelatin, jelly beans, hard candy, fruit punch, sorbet, ices	Sweets should be low in fat

*Serving sizes vary between ½–1¼ cups. Check the product's nutrition label.

**Fat content changes serving counts for fats and oils: For example, 1 Tbsp of regular salad dressing equals 1 serving; 1 Tbsp of a lowfat dressing equals ½ serving; 1 Tbsp of a fat free dressing equals 0 servings.

What's On Your Plate?

Use this form to track your food habits before you start on the DASH eating plan or to see how you're doing after a few weeks. To record more than 1 day, just copy the form. Total each day's food groups and compare what you ate with the DASH plan.

FOOD	AMOUNT (serving size)	NUMBER OF SERVINGS BY DASH FOOD GROUP							
		GRAINS	VEGETABLES	FRUITS	DAIRY FOODS	MEAT, POULTRY, & FISH	NUTS, SEEDS, & DRY BEANS	FATS AND OILS	SWEETS
Breakfast									
Example: whole wheat bread & soft margarine	*2 slices* *2 tsp*	*2*						*2*	
Lunch									
Dinner									
Snacks									
DAY'S TOTAL									
Compare yours with the DASH plan		7–8	4–5	4–5	2–3	2 or less	4–5 a week	2–3	5 a week

Appendix

ANTHROPOMETRIC MEASUREMENT PROTOCOLS

Height measurement:

Periodic height measurements (every three to five years and biannually in persons over fifty) are recommended for accuracy of calculations and to establish a baseline to assess possible shortening due to skeletal degeneration.[1] Measure clients without shoes or socks in a standing position using a stadiometer or portable anthropometer. Clients should stand erect with eyes facing forward and with feet, knees, buttocks and shoulder blades making contact with the vertical surface (e.g., a wall). To record height ask your client to take a deep breath, relax shoulders, and let arms hang loose at the sides with palms facing the thighs. Gently lower the moveable headboard until it touches the crown of the head.[2]

Weight measurement:

A platform balance, calibrated periodically with a set of weights, should be used to take weight recordings. Place the balance on a hard flat surface (never a carpet), and check for zero balance before each measurement. Do not take a weight reading after a large meal. Ask your client to wear light clothing and no shoes or socks. Your client should stand still and unattended in the center of the platform. If a client is taking his or her own weight measurements, then the readings should be taken upon arising in the morning after voiding.

Waist circumference:[2]

Clients should not have eaten a large meal within one hour of taking measurements and clothing should be removed from the waist area for proper positioning of the measuring tape. Mark the natural waist with a cosmetic pencil midway between the palpated (felt) iliac crest (hip bone) and the palpated lowest rib margin in the left and right mid-axially lines (each side of body). Use a non-stretchable measuring tape and place evenly around the natural waist covering each natural-waist mark. The tape should be snug but not compressing the skin. Ask your client to breathe minimally but normally.

Hip circumference:[2]

Ask your client to stand erect with arms at side and feet together. The non-stretchable tape should be placed to provide the maximum circumference over the buttocks. Hold the tape firmly against the skin but do not indent the soft tissue.

Frame size:[3]

Wrist circumference of the right arm can be used to determine frame size. Record height without shoes. Use a narrow tape measure (0.7 cm or less) to measure the smallest part of the wrist. Record to the nearest 0.1 cm and use the following formula:

$$r = \text{height (cm)} \div \text{wrist circumference (cm)}$$

FRAME SIZE	MALE r VALUE	FEMALE r VALUE
Small	>10.4	>11.0
Medium	9.6–10.4	10.1–11.0
Large	<9.6	<10.1

REFERENCES

1. Simko MD, Cowell C, Gilbride JA. *Nutrition Assessment: A Comprehensive Guide for Planning Intervention.* Rockville, MD: An Aspen Publication; 1995.

2. Shape Up America and American Obesity Association. *Guidance for Treatment of Adult Obesity.* Bethesda, MD: Shape Up America; 1997–1998, appendix IV.

3. From Grant A, DeHoog S. *Nutrition Assessment Support and Management.* 5th ed. Seattle, WA: Grant & DeHoog; 1999:94.

Appendix

WEIGHT-FOR-HEIGHT TABLES

Height & Weight Table for Women

HEIGHT FEET INCHES	SMALL FRAME	MEDIUM FRAME	LARGE FRAME
4′10″	102–111	109–121	118–131
4′11″	103–113	111–123	120–134
5′0″	104–115	113–126	122–137
5′1″	106–118	115–129	125–140
5′2″	108–121	118–132	128–143
5′3″	111–124	121–135	131–147
5′4″	114–127	124–138	134–151
5′5″	117–130	127–141	137–155
5′6″	120–133	130–144	140–159
5′7″	123–136	133–147	143–163
5′8″	126–139	136–150	146–167
5′9″	129–142	139–153	149–170
5′10″	132–145	142–156	152–173
5′11″	135–148	145–159	155–176
6′0″	138–151	148–162	158–179

Weights at ages 25–59 based on lowest mortality. Weight in pounds according to frame (in indoor clothing weighing 3 lbs.; shoes with 1″ heels)

Height & Weight Table for Men

HEIGHT FEET INCHES	SMALL FRAME	MEDIUM FRAME	LARGE FRAME
5′2″	128–134	131–141	138–150
5′3″	130–136	133–143	140–153
5′4″	132–138	135–145	142–156
5′5″	134–140	137–148	144–160
5′6″	136–142	139–151	146–164
5′7″	138–145	142–154	149–168
5′8″	140–148	145–157	152–172
5′9″	142–151	148–160	155–176
5′10″	144–154	151–163	158–180
5′11″	146–157	154–166	161–184
6′0″	149–160	157–170	164–188
6′1″	152–164	160–174	168–192
6′2″	155–168	164–178	172–197
6′3″	158–172	167–182	176–202
6′4″	162–176	171–187	181–207

Weights at ages 25–59 based on lowest mortality. Weight in pounds according to frame (in indoor clothing weighing 5 lbs; shoes with 1″ heels)

Source of basic data: *Build Study,* 1979. Society of actuaries and association of Life Insurance Medical Directors of America, 1980.

Appendix E

BODY MASS INDEX CHART

HEIGHT	18	19	20	21	22	23	24	25	26	27	28	29	30	31	32	33	34	35	36	37	38	39	40
	BODY WEIGHT (POUNDS)																						
4′10″	86	91	96	100	105	110	115	119	124	129	134	138	143	148	153	158	162	167	172	177	181	186	191
4′11″	89	94	99	104	109	114	119	124	128	133	138	143	148	153	158	163	168	173	178	183	188	193	198
5′0″	92	97	102	107	112	118	123	128	133	138	143	148	153	158	163	168	174	179	184	189	194	199	204
5′1″	95	100	106	111	116	122	127	132	137	143	148	153	158	164	169	174	180	185	190	195	201	206	211
5′2″	98	104	109	115	120	126	131	136	142	147	153	158	164	169	175	180	186	191	196	202	207	213	218
5′3″	102	107	113	118	124	130	135	141	146	152	158	163	169	175	180	186	191	197	203	208	214	220	225
5′4″	105	110	116	122	128	134	140	145	151	157	163	169	174	180	186	192	197	204	209	215	221	227	232
5′5″	108	114	120	126	132	138	144	150	156	162	168	174	180	186	192	198	204	210	216	222	228	234	240
5′6″	112	118	124	130	136	142	148	155	161	167	173	179	186	192	198	204	210	216	223	229	235	241	247
5′7″	115	121	127	134	140	146	153	159	166	172	178	185	191	198	204	211	217	223	230	236	242	249	255
5′8″	118	125	131	138	144	151	158	164	171	177	184	190	197	203	210	216	223	230	236	243	249	256	262
5′9″	122	128	135	142	149	155	162	169	176	182	189	196	203	209	216	223	230	236	243	250	257	263	270
5′10″	126	132	139	146	153	160	167	174	181	188	195	202	209	216	222	229	236	243	250	257	264	271	278
5′11″	129	136	143	150	157	165	172	179	186	193	200	208	215	222	229	236	243	250	257	265	272	279	286
6′0″	132	140	147	154	162	169	177	184	191	199	206	213	221	228	235	242	250	258	265	272	279	287	294
6′1″	136	144	151	159	166	174	182	189	197	204	212	219	227	235	242	250	257	265	272	280	288	295	302
6′2″	141	148	155	163	171	179	186	194	202	210	218	225	233	241	249	256	264	272	280	287	295	303	311
6′3″	144	152	160	168	176	184	192	200	208	216	224	232	240	248	256	264	272	279	287	295	303	311	319
6′4″	148	156	164	172	180	189	197	205	213	221	230	238	246	254	263	271	279	287	295	304	312	320	328
6′5″	151	160	168	176	185	193	202	210	218	227	235	244	252	261	269	277	286	294	303	311	319	328	336
6′6″	155	164	172	181	190	198	207	216	224	233	241	250	259	267	276	284	293	302	310	319	328	336	345
	Underweight (<18.5)	**Healthy Weight** (18.5–24.9)						**Overweight** (25–29.9)					**Obese** (≥30)										

Find your height along the left-hand column and look across the row until you find the number that is closest to your weight. The number at the top of that column identifies your BMI. The area shaded gray represents healthy weight ranges.

Appendix

EXCHANGE LISTS FOR WEIGHT MANAGEMENT

The Exchange Lists

GROUP/LISTS	TYPICAL ITEM/ PORTION SIZE	CARBOHYDRATE (g)	PROTEIN (g)	FAT (g)	ENERGY[a] (kcal)
Carbohydrate Group					
Starch[b]	1 slice bread	15	3	1 or less	80
Fruit	1 small apple	15	—	—	60
Milk					
Nonfat	1 c nonfat milk	12	8	0–3	90
Reduced-fat	1 c reduced-fat milk	12	8	5	120
Whole	1 c whole milk	12	8	8	150
Other carbohydrates[c]	2 small cookies	15	varies	varies	varies
Vegetable	½ c cooked carrots	5	2	—	25
Meat and Meat Substitute Group[d]					
Meat					
Very lean	1 oz chicken (white meat, no skin)	—	7	0–1	35
Lean	1 oz lean beef	—	7	3	55
Medium-fat	1 oz ground beef	—	7	5	75
High-fat	1 oz pork sausage	—	7	8	100
Fat Group					
Fat	1 tsp butter	—	—	5	45

[a]The energy value for each exchange list represents an approximate average for the group and does not reflect the precise number of grams of carbohydrate, protein, and fat. For example, a slice of bread contains 15 grams of carbohydrate (that's 60 kcalories), 3 grams protein (that's another 12 kcalories), and a little fat—rounded to 80 kcalories for ease in calculating. A half-cup of vegetables (not including starchy vegetables) contains 5 grams carbohydrate (20 kcalories) and 2 grams protein (8 more), which has been rounded down to 25 kcalories.
[b]The starch list includes cereals, grains, breads, crackers, snacks, starchy vegetables (such as corn, peas, and potatoes), and legumes (dried beans, peas, and lentils).
[c]The other carbohydrates list includes foods that contain added sugars and fats such as cakes, cookies, doughnuts, ice cream, potato chips, pudding, syrup, and frozen yogurt.
[d]The meat and meat substitutes list includes legumes, cheeses, and peanut butter.

Development of an Eating Plan Based on Exchange Lists for Weight Management

- Determine desired calorie level. See Chapter 4 for help in determining calorie needs.
- Choose a desired percentage carbohydrate, protein, and fat.
- Calculate the grams of carbohydrate, protein, and fat that the eating plan should contain.
- Calculate carbohydrate exchanges first. Select numbers of servings of milk, fruit, and vegetable exchanges. Calculate total grams of carbohydrate accounted for and then divide the remaining grams of carbohydrate by 15, the number of grams of carbohydrate in one exchange of the starch group. Then total the grams of protein accounted for in the carbohydrate exchanges.
- Determine the number of meat and meat alternate exchanges by dividing the remaining grams of protein not accounted for in the carbohydrates by 7, the grams of protein in one protein exchange.
- Determine the number of fat exchanges by adding up the fat already accounted for in the carbohydrate and meat and meat alternate exchanges. Remaining fat grams should be divided by 5, the number of grams of fat in one fat exchange.

Example of Calculations for Development of an Eating Plan Bbased on Exchange Lists for Weight Management

1,800 kilocalories: roughly 50% carbohydrate, 20% protein, 30% fat

Carbohydrate (CH0): 1,800 kcal × 0.50 = 900 kcal/g ÷ 4 kcal/g = 225
Protein (PRO): 1,800 kcal × 0.20 = 360 kcal/g ÷ 4 kcal/g = 90
Fat (FAT): 1,800 kcal × 0.30 = 540 kcal/g ÷ 9 kcal/g = 60

FOOD GROUP	NUMBER OF SERVINGS	CHO 225 g	PRO 90 g	FAT 60 g	KILOCALORIES 1,800
Skim milk	2	24	16	–	180
Vegetables	3	15	6	–	75
Fruit	3	45	–	–	180
		TOTAL 84 g			

225 g total CHO – 84 g CHO used =141 g CHO left
141 CHO g left ÷ 15 g CHO / starch exchange = 9.4 ~ 9 servings of starch

FOOD GROUP	NUMBER OF SERVINGS	CHO 225 g	PRO 90 g	FAT 60 g	KILOCALORIES 1,800
Starch	9	135	27	–	720
			TOTAL 49 g		

90 g total PRO – 49 g PRO used = 41 g PRO left
41 g PRO left ÷ 7 g PRO / meat exchange = 5.8 ~ 6 meat exchanges

FOOD GROUP	NUMBER OF SERVINGS	CHO 225 g	PRO 90 g	FAT 60 g	KILOCALORIES 1,800
Meat (lean)	6	–	42	18	330
				TOTAL 18 g	

60 g total FAT – 18 g FAT used = 42 g FAT left
42 g FAT left ÷ 5 g FAT / FAT exchange = 8 FAT exchanges

FOOD GROUP	NUMBER OF SERVINGS	CHO 225 g	PRO 90 g	FAT 60 g	KILOCALORIES 1,800
Fat	8	–	–	40	360
TOTALS		**219**	**91**	**58**	**1845**

Template for Calculations to Develop an Eating Plan Based on Exchange Lists for Weight Management

____ kcal – roughly ____ % carbohydrate, ____ % protein, ____ % fat

Carbohydrate (CHO): ____ kcal × ____ = ____ kcal/g ÷ 4 kcal/g = ____ g

Protein (PRO): ____ kcal × ____ = ____ kcal/g ÷ 4 kcal/g = ____ g

Fat (FAT): ____ kcal × ____ = ____ kcal/g ÷ 9 kcal/g = ____ g

FOOD GROUP	NUMBER OF SERVINGS	CHO 225 g	PRO 90 g	FAT 60 g	KILOCALORIES 1,800
Milk					
Vegetables					
Fruit					

TOTAL ____ g

____ g total CHO – ____ g CHO used = ____ g CHO left

____ CHO g left ÷ 15 g CHO / starch exchange = ____ servings of starch

FOOD GROUP	NUMBER OF SERVINGS	CHO 225 g	PRO 90 g	FAT 60 g	KILOCALORIES 1,800
Starch					

TOTAL ____ g

____ g total PRO – ____ g PRO used = ____ g PRO left

____ g PRO left ÷ 7 g PRO / meat exchange = ____ meat exchanges

FOOD GROUP	NUMBER OF SERVINGS	CHO 225 g	PRO 90 g	FAT 60 g	KILOCALORIES 1,800
Meat					

TOTAL ____ g

____ g total FAT – ____ g FAT used = ____ g FAT left

____ g FAT left ÷ 5 g FAT / FAT exchange = ____ FAT exchanges

FOOD GROUP	NUMBER OF SERVINGS	CHO 225 g	PRO 90 g	FAT 60 g	KILOCALORIES 1,800
Fat					
TOTALS					

TABLE F-1—U.S. Exchange System: Starch List

1 starch exchange = 15 g carbohydrate, 3 g protein, 0–1 g fat, and 80 kcal

Note: In general, a starch serving is ½ c cereal, grain, pasta, or starchy vegetable; 1 oz of bread; ¾ to 1 oz snack food.

SERVING SIZE	FOOD
Bread	
½ (1 oz)	Bagels
2 slices (1½ oz)	Bread, reduced-kcalorie
1 slice (1 oz)	Bread, white (including French and Italian), whole-wheat, pumpernickel, rye
2 (⅔ oz)	Bread sticks, crisp, 4″ x ½″
½	English muffins
½ (1 oz)	Hot dog or hamburger buns
½	Pita, 6″ across
1 (1 oz)	Plain rolls, small
1 slice (1 oz)	Raisin bread, unfrosted
1	Tortillas, corn, 6″ across
1	Tortillas, flour, 7–8″ across
1	Waffles, 4½″ square, reduced-fat
Cereals and Grains	
½ c	Bran cereals
½ c	Bulgur, cooked
½ c	Cereals, cooked
¾ c	Cereals, unsweetened, ready-to-eat
3 tbs	Cornmeal (dry)
⅓ c	Couscous
3 tbs	Flour (dry)
¼ c	Granola, low-fat
¼ c	Grape nuts
½ c	Grits, cooked
½ c	Kasha
¼ c	Millet
¼ c	Muesli
½ c	Oats
½ c	Pasta, cooked
1½ c	Puffed cereals
½ c	Rice milk
⅓ c	Rice, white or brown, cooked
½ c	Shredded wheat
½ c	Sugar-frosted cereal
3 tbs	Wheat germ
Starchy Vegetables	
⅓ c	Baked beans
½ c	Corn
1 (5 oz)	Corn on cob, medium
1 c	Mixed vegetables with corn, peas, or pasta
½ c	Peas, green
½ c	Plantains
1 small (3 oz)	Potatoes, baked or boiled
½ c	Potatoes, mashed
1 c	Squash, winter (acorn, butternut)
½ c	Yams, sweet potatoes, plain
Crackers and Snacks	
8	Animal crackers
3	Graham crackers, 2½″ square
¾ oz	Matzoh

TABLE F-1—U.S. Exchange System: Starch List (Continued)

SERVING SIZE	FOOD
4 slices	Melba toast
24	Oyster crackers
3 c	Popcorn (popped, no fat added or low-fat microwave)
¾ oz	Pretzels
2	Rice cakes, 4″ across
6	Saltine-type crackers
15–20″ (¾ oz)	Snack chips, fat-free (tortilla, potato)
2–5 (¾ oz)	Whole-wheat crackers, no fat added
Dried Beans, Peas, and Lentils	
½ c	Beans and peas, cooked (garbanzo, lentils, pinto, kidney, white, split, black-eyed)
⅔ c	Lima beans
3 tbs	Miso

SERVING SIZE	FOOD
Starchy Foods Prepared with Fat Count as 1 starch + 1 fat exchange.	
1	Biscuit, 2½″ across
½ c	Chow mein noodles
1 (2 oz)	Corn bread, 2″ cube
6	Crackers, round butter type
1 c	Croutons
16–25 (3 oz)	French-fried potatoes
¼ c	Granola
1 (1½ oz)	Muffin, small
2	Pancake, 4″ across
3 c	Popcorn, microwave
3	Sandwich crackers, cheese or peanut butter filling
⅓ c	Stuffing, bread (prepared)
2	Taco shell, 6″ across
1	Waffle, 4½″ square
4–6 (1 oz)	Whole-wheat crackers, fat added

= 400 mg or more of sodium per serving.

TABLE F-2—U.S. Exchange System: Fruit List

1 fruit exchange = 15 g carbohydrate and 60 kcal
Note: In general, a fruit serving is 1 small to medium fresh fruit; 1/2 c canned or fresh fruit or fruit juice; 1/4 c dried fruit.

SERVING SIZE	FOOD
1 (4 oz)	Apples, unpeeled, small
½ c	Applesauce, unsweetened
4 rings	Apples, dried
4 whole (5½ oz)	Apricots, fresh
8 halves	Apricots, dried
½ c	Apricots, canned
1 (4 oz)	Bananas, small
¾ c	Blackberries
¾ c	Blueberries
⅓ melon (11 oz) or 1 c cubes	Cantaloupe, small
12 (3 oz)	Cherries, sweet, fresh
½ c	Cherries, sweet, canned
3	Dates
1½ large or 2 medium (3½ oz)	Figs, fresh
1½	Figs, dried
½ c	Fruit cocktail
½ (11 oz)	Grapefruit, large
¾ c	Grapefruit sections, canned
17 (3 oz)	Grapes, small
1 slice (10 oz) or 1 c cubes	Honeydew melon
1 (3½ oz)	Kiwi
¾ c	Mandarin oranges, canned
½ (5½ oz) or ½ c	Mangoes, small
1 (5 oz)	Nectarines, small
1 (6½ oz)	Oranges, small
½ (8 oz) or 1 c cubes	Papayas
1 (6 oz)	Peaches, medium, fresh
½ c	Peaches, canned
½ (4 oz)	Pears, large, fresh
½ c	Pears, canned
¾ c	Pineapple, fresh
½ c	Pineapple, canned
2 (5 oz)	Plums, small
½ c	Plums, canned
3	Prunes, dried
2 tbs	Raisins
1 c	Raspberries
1¼ c whole berries	Strawberries
2 (8 oz)	Tangerines, small
1 slice (13½ oz) or 1¼ c cubes	Watermelon

Fruit Juice

SERVING SIZE	FOOD
½ c	Apple juice/cider
⅓ c	Cranberry juice cocktail
1 c	Cranberry juice cocktail, reduced-kcalorie
⅓ c	Fruit juice blends, 100% juice
⅓ c	Grape juice
½ c	Grapefruit juice
½ c	Orange juice
½ c	Pineapple juice
⅓ c	Prune juice

TABLE F-3—U.S. Exchange System: Milk List

SERVING SIZE	FOOD
Nonfat and Low-Fat Milk	
1 nonfat/low-fat milk exchange = 12 g carbohydrate, 8 g protein, 0–3 g fat, 90 kcal	
1 c	Nonfat milk
1 c	½% milk
1 c	1% milk
1 c	Nonfat or low-fat buttermilk
½ c	Evaporated nonfat milk
⅓ c dry	Dry nonfat milk
¾ c	Plain nonfat yogurt
1 c	Nonfat or low-fat fruit-flavored yogurt sweetened with aspartame or with a nonnutritive sweetener
Reduced-Fat Milk	
1 reduced-fat milk exchange = 12 g carbohydrate, 8 g protein, 5 g fat, 120 kcal	
1 c	2% milk
¾ c	Plain low-fat yogurt
1 c	Sweet acidophilus milk

SERVING SIZE	FOOD
Whole Milk	
1 whole milk exchange = 12 g carbohydrate, 8 g protein, 8 g fat, 150 kcal	
1 c	Whole milk
½ c	Evaporated whole milk
1 c	Goat's milk
1 c	Kefir

TABLE F-4—U.S. Exchange System: Other Carbohydrates List

1 other carbohydrate exchange = 15 g carbohydrate, or 1 starch, or 1 fruit, or 1 milk exchange

FOOD	SERVING SIZE	EXCHANGES PER SERVING
Angel food cake, unfrosted	1⁄12 cake	2 carbohydrates
Brownies, small, unfrosted	2″ square	1 carbohydrate, 1 fat
Cake, unfrosted	2″ square	1 carbohydrate, 1 fat
Cake, frosted	2″ square	2 carbohydrates, 1 fat
Cookie, fat-free	2 small	1 carbohydrate
Cookies or sandwich cookies	2 small	1 carbohydrate, 1 fat
Cupcakes, frosted	1 small	2 carbohydrates, 1 fat
Cranberry sauce, jellied	¼ c	2 carbohydrates
Doughnuts, plain cake	1 medium, (1½ oz)	1½ carbohydrates, 2 fats
Doughnuts, glazed	3¾″ across (2 oz)	2 carbohydrates, 2 fats
Fruit juice bars, frozen, 100% juice	1 bar (3 oz)	1 carbohydrate
Fruit snacks, chewy (pureed fruit concentrate)	1 roll (¾ oz)	1 carbohydrate
Fruit spreads, 100% fruit	1 tbs	1 carbohydrate
Gelatin, regular	½ c	1 carbohydrate
Gingersnaps	3	1 carbohydrate
Granola bars	1 bar	1 carbohydrate, 1 fat
Granola bars, fat-free	1 bar	2 carbohydrates
Hummus	⅓ c	1 carbohydrate, 1 fat
Ice cream	½ c	1 carbohydrate, 2 fats
Ice cream, light	½ c	1 carbohydrate, 1 fat
Ice cream, fat-free, no sugar added	½ c	1 carbohydrate
Jam or jelly, regular	1 tbs	1 carbohydrate
Milk, chocolate, whole	1 c	2 carbohydrates, 1 fat
Pie, fruit, 2 crusts	⅙ pie	3 carbohydrates, 2 fats
Pie, pumpkin or custard	⅛ pie	1 carbohydrate, 2 fats
Potato chips	12–18 (1 oz)	1 carbohydrate, 2 fats
Pudding, regular (made with low-fat milk)	½ c	2 carbohydrates
Pudding, sugar-free (made with low-fat milk)	½ c	1 carbohydrate
Salad dressing, fat-free 🖉	¼ c	1 carbohydrate
Sherbet, sorbet	½ c	2 carbohydrates
Spaghetti or pasta sauce, canned 🖉	½ c	1 carbohydrate, 1 fat
Sweet roll or danish	1 (2½ oz)	2½ carbohydrates, 2 fats
Syrup, light	2 tbs	1 carbohydrate
Syrup, regular	1 tbs	1 carbohydrate
Syrup, regular	¼ c	4 carbohydrates
Tortilla chips	6–12 (1 oz)	1 carbohydrate, 2 fats
Yogurt, frozen, low-fat, fat-free	⅓ c	1 carbohydrate, 0–1 fat
Yogurt, frozen, fat-free, no sugar added	½ c	1 carbohydrate
Yogurt, low-fat with fruit	1 c	3 carbohydrates, 0–1 fat
Vanilla wafers	5	1 carbohydrate, 1 fat

🖉 = 400 mg or more of sodium per exchange.

TABLE F-5—U.S. Exchange System: Vegetable List

1 vegetable exchange = 5 g carbohydrate, 2 g protein, 25 kcal

Note: In general, a vegetable serving is ½ c cooked vegetables or vegetable juice; 1 c raw vegetables. Starchy vegetables such as corn, peas, and potatoes are on the starch list.

Artichokes
Artichoke hearts
Asparagus
Beans (green, wax, Italian)
Bean sprouts
Beets
Broccoli
Brussels sprouts
Cabbage
Carrots
Cauliflower
Celery
Cucumbers
Eggplant
Green onions or scallions
Greens (collard, kale, mustard, turnip)
Kohlrabi
Leeks
Mixed vegetables (without corn, peas, or pasta)
Mushrooms
Okra
Onions
Pea pods
Peppers (all varieties)
Radishes
Salad greens (endive, escarole, lettuce, romaine, spinach)
Sauerkraut
Spinach
Summer squash (crookneck)
Tomatoes
Tomatoes, canned
Tomato sauce
Tomato/vegetable juice
Turnips
Water chestnuts
Watercress
Zucchini

 = 400 mg or more of sodium per exchange.

Table F-6—U.S. Exchange System: Meat and Meat Substitutes List

Note: In general, a meat serving is 1 oz meat, poultry, or cheese; ½ c dried beans (weigh meat and poultry and measure beans after cooking).

Very Lean Meat and Substitutes

1 very lean meat exchange = 7 g protein, 0–1 g fat, 35 kcal

SERVING SIZE	FOOD
1 oz	Poultry: Chicken or turkey (white meat, no skin), Cornish hen (no skin)
1 oz	Fish: Fresh or frozen cod, flounder, haddock, halibut, trout; tuna, fresh or canned in water
1 oz	Shellfish: Clams, crab, lobster, scallops, shrimp, imitation shellfish
1 oz	Game: Duck or pheasant (no skin), venison, buffalo, ostrich
	Cheese with ≤ 1 g fat/oz:
¼ c	Nonfat or low-fat cottage cheese
1 oz	Fat-free cheese
	Other:
1 oz	Processed sandwich meats with ≤ 1 g fat/oz (such as deli thin, shaved meats, chipped beef, turkey ham)
2	Egg whites
¼ c	Egg substitutes, plain
1 oz	Hot dogs with ≤ 1 g fat/oz
1 oz	Kidney (high in cholesterol)
1 oz	Sausage with ≤ 1 g fat/oz

Count as one very lean meat and one starch exchange:

SERVING SIZE	FOOD
½ c	Dried beans, peas, lentils (cooked)

Lean Meat and Substitutes

1 lean meat exchange = 7 g protein, 3 g fat, 55 kcal

SERVING SIZE	FOOD
1 oz	Beef: USDA Select or Choice grades of lean beef trimmed of fat (round, sirloin, and flank steak); tenderloin; roast (rib, chuck, rump); steak (T-bone, porterhouse, cubed), ground round
1 oz	Pork: Lean pork (fresh ham); canned, cured, or boiled ham; Canadian bacon; tenderloin, center loin chop
1 oz	Lamb: Roast, chop, leg
1 oz	Veal: Lean chop, roast
1 oz	Poultry: Chicken, turkey (dark meat, no skin), chicken white meat (with skin), domestic duck or goose (well-drained of fat, no skin)
	Fish:
1 oz	Herring (uncreamed or smoked)
6 medium	Oysters
1 oz	Salmon (fresh or canned), catfish
2 medium	Sardines (canned)
1 oz	Tuna (canned in oil, drained)
1 oz	Game: Goose (no skin), rabbit
	Cheese:
¼ c	4.5%-fat cottage cheese
2 tbs	Grated Parmesan
1 oz	Cheeses with ≤ 3 g fat/oz
	Other:
1½ oz	Hot dogs with ≤ 3 g fat/oz
1 oz	Processed sandwich meat with ≤ 3 g fat/oz (turkey pastrami or kielbasa)
1 oz	Liver, heart (high in cholesterol)

Medium-Fat Meat and Substitutes

1 medium-fat meat exchange = 7 g protein, 5 g fat, and 75 kcal

SERVING SIZE	FOOD
1 oz	Beef: Most beef products (ground beef, meatloaf, corned beef, short ribs, Prime grades of meat trimmed of fat, such as prime rib)
1 oz	Pork: Top loin, chop, Boston butt, cutlet
1 oz	Lamb: Rib roast, ground
1 oz	Veal: Cutlet (ground or cubed, unbreaded)
1 oz	Poultry: Chicken dark meat (with skin), ground turkey or ground chicken, fried chicken (with skin)

Table F-6—U.S. Exchange System: Meat and Meat Substitutes List (Continued)

Note: In general, a meat serving is 1 oz meat, poultry, or cheese; ½ c dried beans (weigh meat and poultry and measure beans after cooking).

Medium-Fat Meat and Substitutes

SERVING SIZE	FOOD
1 oz	Fish: Any fried fish product
	Cheese with ≤ 5 g fat/oz:
1 oz	Feta
1 oz	Mozzarella
¼ c (2 oz)	Ricotta
	Other:
1	Egg (high in cholesterol, limit to 3/week)
1 oz	Sausage with ≤ 5 g fat/oz
1 c	Soy milk
¼ c	Tempeh
4 oz or ½ c	Tofu

High-Fat Meat and Substitutes

1 high-fat meat exchange = 7 g protein, 8 g fat, 100 kcal

SERVING SIZE	FOOD
1 oz	Pork: Spareribs, ground pork, pork sausage
1 oz	Cheese: All regular cheeses (American, cheddar, Monterey Jack, swiss)
	Other:
1 oz	Processed sandwich meats with ≤ 8 g fat/oz (bologna, pimento loaf, salami)
1 oz	Sausage (bratwurst, Italian, knockwurst, Polish, smoked)
1 (10/lb)	Hot dog (turkey or chicken)
3 slices (20 slices/lb)	Bacon

Count as one high-fat meat plus one fat exchange:

SERVING SIZE	FOOD
1 (10/lb)	Hot dog (beef, pork, or combination)
2 tbs	Peanut butter (contains unsaturated fat)

= 400 mg or more of sodium per exchange.

TABLE F-7—U.S. Exchange System: Fat List

1 fat exchange = 5 g fat, 45 kcal
Note: In general, a fat serving is 1 tsp regular butter, margarine, or vegetable oil; 1 tbs regular salad dressing. Many fat-free and reduced fat foods are on the Free Foods List.

SERVING SIZE	FOOD
Monounsaturated Fats	
⅛ medium (1 oz)	Avocadoes
1 tsp	Oil (canola, olive, peanut)
8 large	Olives, ripe (black)
10 large	Olives, green, stuffed 🧂
6 nuts	Almonds, cashews
6 nuts	Mixed nuts (50% peanuts)
10 nuts	Peanuts
4 halves	Pecans
2 tsp	Peanut butter, smooth or crunchy
1 tbs	Sesame seeds
2 tsp	Tahini paste
Polyunsaturated Fats	
1 tsp	Margarine, stick, tub, or squeeze
1 tbs	Margarine, lower-fat (30% to 50% vegetable oil)
1 tsp	Mayonnaise, regular
1 tbs	Mayonnaise, reduced-fat
4 halves	Nuts, walnuts, English
1 tsp	Oil (corn, safflower, soybean)
1 tbs	Salad dressing, regular
2 tbs	Salad dressing, reduced-fat
2 tsp	Mayonnaise type salad dressing, regular 🧂
1 tbs	Mayonnaise type salad dressing, reduced-fat
1 tbs	Seeds: pumpkin, sunflower
Saturated Fats*	
1 slice (20 slices/lb)	Bacon, cooked
1 tsp	Bacon, grease
1 tsp	Butter, stick
2 tsp	Butter, whipped
1 tbs	Butter, reduced-fat
2 tbs (½ oz)	Chitterlings, boiled
2 tbs	Coconut, sweetened, shredded
2 tbs	Cream, half and half
1 tbs (½ oz)	Cream cheese, regular
2 tbs (1 oz)	Cream cheese, reduced-fat
	Fatback or salt pork†
1 tsp	Shortening or lard
2 tbs	Sour cream, regular
3 tbs	Sour cream, reduced-fat

🧂 = 400 mg or more sodium per exchange
*Saturated fats can raise blood cholesterol levels.
† Use a piece 1″ × 10″ × ¼″ if you plan to eat the fatback cooked with vegetables. Use a piece 2″ × 1″ × ½″ when eating only the vegetables with the fatback removed.

TABLE F-8—U.S. Exchange System: Free Foods List

Note: A serving of free food contains less than 20 kcalories; those with serving sizes should be limited to 3 servings a day whereas those without serving sizes can be eaten freely.

SERVING SIZE	FOOD
Fat-Free or Reduced-Fat Foods	
1 tbs	Cream cheese, fat-free
1 tbs	Creamers, nondairy, liquid
2 tsp	Creamers, nondairy, powdered
1 tbs	Mayonnaise, fat-free
1 tsp	Mayonnaise, reduced-fat
4 tbs	Margarine, fat-free
1 tsp	Margarine, reduced-fat
1 tbs	Mayonnaise type salad dressing, nonfat
1 tsp	Mayonnaise type salad dressing, reduced-fat
	Nonstick cooking spray
1 tbs	Salad dressing, fat-free
2 tbs	Salad dressing, fat-free, Italian
¼ c	Salsa
1 tbs	Sour cream, fat-free, reduced-fat
2 tbs	Whipped topping, regular or light
Sugar-Free or Low-Sugar Foods	
1 piece	Candy, hard, sugar-free
	Gelatin dessert, sugar-free
	Gelatin, unflavored
	Gum, sugar-free
2 tsp	Jam or jelly, low-sugar or light
	Sugar substitutes
2 tbs	Syrup, sugar-free
Drinks	
	Bouillon, broth, consommé 🧂
	Bouillon or broth, low-sodium
	Carbonated or mineral water
1 tbs	Cocoa powder, unsweetened
	Coffee
	Club soda
	Diet soft drinks, sugar-free
	Drink mixes, sugar-free
	Tea
	Tonic water, sugar-free
Condiments	
1 tbs	Catsup
	Horseradish
	Lemon juice
	Lime juice
	Mustard
1½ large	Pickles, dill 🧂
	Soy sauce, regular or light 🧂
1 tbs	Taco sauce
	Vinegar
Seasonings	
Flavoring extracts	
Garlic	
Herbs, fresh or dried	
Pimento	
Spices	
Hot pepper sauces	
Wine, used in cooking	
Worcestershire sauce	

🧂 = 400 mg or more sodium per exchange.

TABLE F-9—U.S. Exchange System: Combination Foods List

FOOD	SERVING SIZE	EXCHANGES PER SERVING
Entrees		
Tuna noodle casserole, lasagna, spaghetti with meatballs, chili with beans, macaroni and cheese	1 c (8 oz)	2 carbohydrates, 2 medium-fat meats
Chow mein (without noodles or rice)	2 c (16 oz)	1 carbohydrate, 2 lean meats
Pizza, cheese, thin crust	¼ of 100 (5 oz)	2 carbohydrates, 2 medium-fat meats, 1 fat
Pizza, meat topping, thin crust	¼ of 10″ (5 oz)	2 carbohydrates, 2 medium-fat meats, 2 fats
Pot pie	1 (7 oz)	2 carbohydrates, 1 medium-fat meat, 4 fats
Frozen entrees		
Salisbury steak with gravy, mashed potato	1 (11 oz)	2 carbohydrates, 3 medium-fat meats, 3–4 fats
Turkey with gravy, mashed potato, dressing	1 (11 oz)	2 carbohydrates, 2 medium-fat meats, 2 fats
Entree with less than 300 kcalories	1 (8 oz)	2 carbohydrates, 3 lean meats
Soups		
Bean	1 c	1 carbohydrate, 1 very lean meat
Cream (made with water)	1 c (8 oz)	1 carbohydrate, 1 fat
Split pea (made with water)	½ c (4 oz)	1 carbohydrate
Tomato (make with water)	1 c (8 oz)	1 carbohydrate
Vegetable beef, chicken noodle, or other broth-type	1 c (8 oz)	1 carbohydrate
Fast Foods		
Burritos with beef	2	4 carbohydrates, 2 medium-fat meats, 2 fats
Chicken nuggets	6	1 carbohydrate, 2 medium-fat meats, 1 fat
Chicken breast and wing, breaded and fried	1	1 carbohydrate, 4 medium-fat meats, 2 fats
Fish sandwich/tartar sauce	1	3 carbohydrates, 1 medium-fat meat, 3 fats

TABLE F-9—U.S. Exchange System: Combination Foods List (Continued)

FOOD	SERVING SIZE	EXCHANGES PER SERVING
French fries, thin	20–25	2 carbohydrates, 2 fats
Hamburger, regular	1	2 carbohydrates, 2 medium-fat meats
Hamburger, large 🧂	1	2 carbohydrates, 3 medium-fat meats, 1 fat
Hot dog with bun 🧂	1	1 carbohydrate, 1 high-fat meat, 1 fat
Individual pan pizza 🧂	1	5 carbohydrates, 3 medium-fat meats, 3 fats
Soft serve cone	1 medium	2 carbohydrates, 1 fat
Submarine sandwich 🧂	1 (60)	3 carbohydrates, 1 vegetable, 2 medium-fat meats, 1 fat
Taco, hard shell 🧂	1 (6 oz)	2 carbohydrates, 2 medium-fat meats, 2 fats
Taco, soft shell 🧂	1 (3 oz)	1 carbohydrate, 1 medium-fat meat, 1 fat

🧂 = 400 mg or more sodium per exchange.

Meal Plan for ______________________________

Calories: ________
Carbohydrate: ________
Protein: ________
Fat: ________

Total Food Groups:

Milk ______ Bread ______
Fruit ______ Meat ______
Vegetables ______ Fat ______

B/L/D/S*	NUMBER OF FOOD GROUP SERVINGS	MEAL SUGGESTIONS
	______ Milk ______ Fruit ______ Vegetables ______ Bread ______ Meat ______ Fat	
	______ Milk ______ Fruit ______ Vegetables ______ Bread ______ Meat ______ Fat	
	______ Milk ______ Fruit ______ Vegetables ______ Bread ______ Meat ______ Fat	

*Breakfast; lunch; dinner; snack.

Appendix

G

LIFESTYLE MANAGEMENT FORMS

- 3.1 Assessment Ruler
- 3.2 Nutrition Counseling—Lifestyle Management Agreement
- 3.3 Student Nutrition Counseling Assignment—Lifestyle Management Agreement
- 4.1 Client Assessment Questionnaire
- 4.2 Food Record
- 4.3 24-Hour Recall/Usual Diet Form
- 4.4 Food Frequency Questionnaire
- 4.5 Food Group Feedback Form
- 4.6 Anthropometric Feedback Form
- 4.7 Client Concerns and Strengths Log
- 4.8 Client Progress Report
- 5.1 Eating Behavior Journal
- 5.2 Counseling Agreement
- 6.1 Symptoms of Stress
- 6.2 Stress Awareness Journal
- 6.3 Tips to Reduce Stress
- 6.4 Prochaska and DiClemente's Spiral of Change
- 6.5 Frequent Cognitive Pitfalls
- 7.1 Benefits of Regular *Moderate* Physical Activity
- 7.2 Physical Activity Log
- 7.3 Physical Activity Options
- 7.4 Physical Activity Medical Readiness Form
- 7.5 Physical Activity Status
- 7.6 Medical Release
- 7.7 Physical Activity Feedback Form
- 8.1 Interview Checklist
- 8.2 Counseling Responses Competency Assessment
- 9.1 Registration for Nutrition Clinic

Assessment Ruler

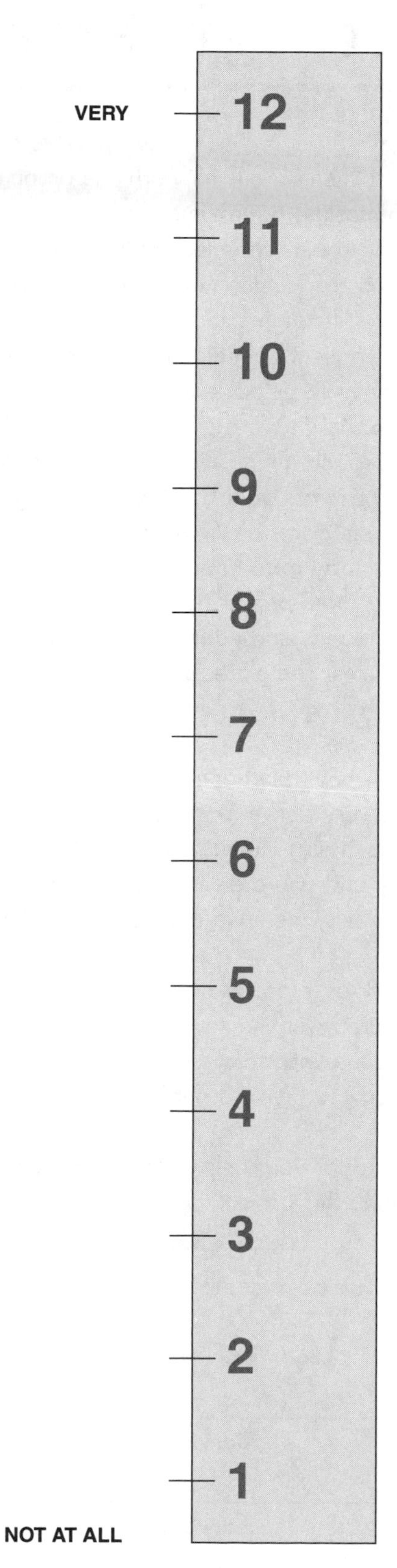

For readiness to change
1 = not at all 12 = very

For adherence to dietary goals
1 = never 12 = always

For confidence in making a lifestyle change
1 = not at all 12 = very

For degree of importance for making a lifestyle change
1 = not at all 12 = very

Nutrition Counseling—Lifestyle Management Agreement

Thank you for your interest in the nutrition counseling clinic offered by __________. It is designed to provide a mutually beneficial experience for both students and volunteer adult clients. You will work one on one with an advanced nutrition counseling student for ____ sessions, each one lasting approximately one hour. During the registration process, clients are assigned a counselor, a counseling room, and meeting times. The counseling sessions provide clients an opportunity to explore and find solutions for nutrition and weight issues. At the same time, students will be working on their nutrition counseling skills. Although students will be following a well-defined counseling guideline, each session will be tailored to their client's needs. Students can only assist clients in achieving weight loss if the client is overweight by National Institutes of Health standards. Normal and underweight clients can still take part in the program with the goal of improving the quality of their diet.

Your student counselor will use a client-centered, motivational approach during his or her sessions with you. This means your counselor will work collaboratively with you to explore your nutrition and weight issues, brainstorm resources and solutions, and help you set achievable goals each week. Students will ask you questions about your health and family history as well as present day food habits. Two of the nutrition assessment forms will be given to you at registration. You can look at them before signing this form. Students will have a variety of tools at their disposal including videos, food models, and educational handouts. Students are encouraged to engage their clients in hands-on experiences. Therefore, at times your counseling session may take place in a grocery store, the student cafeteria, or the gym. Possibly you and your counselor will follow the walk-about map of our campus.

Physical activity is an important part of fitness and weight management. Experience has shown that our clients have a variety of orientations to this topic. If you are already very active in this area, you will be encouraged to continue your program. However, if exercise has not been a joyful experience, you will be invited to explore this issue. As long as you have no medical problem and you are ready to take action, weekly activity goals will be developed with you. For appropriate clients, we have a structured walking protocol that can be followed.

The student may speak occasionally with his or her graduate mentor or instructor about you. The student will write a report about the counseling experience. This report is only shared with the course instructor. Your counselor may give a case study presentation about you to the nutrition counseling class, but at no time in these presentations will your name be used. In all other respects, information you give the student will be held in absolute and strictest confidence.

We thank you very sincerely for your willingness to participate and for your help in the education of future nutrition counselors. If you have any questions or problems during this project, please call the course instructor, ______________________________, at ______________________________.

I, ______________________, have read and understand the above statement and agree to

Print your name here

meet with ______________________ at agreed times and places on the registration form.

______________________________ ______________________

Your signature here *Today's date*

______________________________ ______________________

Counselor signature here *Today's date*

Student Nutrition Counseling Assignment—Lifestyle Management Agreement

Thank you for your willingness to participate in the nutrition counseling clinic offered by ____________________. This interview is designed to provide nutrition counseling students interviewing experience. The objective is for the student to work on counseling skills, gather information about the health problem, and learn something about your health issues. While discussing your situation, you may receive some benefit by clarifying your health problem(s) to yourself, and possibly you will make a resolution to take a new action regarding the problem; however, this experience is not designed to be an intervention.

After this meeting, the student will be required to write a report about his or her findings. This report is only shared with the course instructor. It is possible that information in the report will be shared with other students during classroom discussions; however, at no time will your name be used in those discussions. In all other respects, the information you give will be held in absolute and strictest confidence.

We thank you very sincerely for your willingness to participate and for your help in the education of future nutrition counselors. If you have any questions or problems during this project, please call the course instructor, ________________, at ________________.

I, ____________________________, have read and understand the above statement.

Print your name here

____ *Your signature here*

______________________________ *Today's date*

____ *Counselor's signature here*

______________________________ *Today's date*

Client Assessment Questionnaire

DEMOGRAPHIC DATA

Name ______________________ Date: ______________________
Address ______________________ Home telephone: ______________________
______________________ Office telephone: ______________________
Fax: ______________________ E-mail ______________________
Sex: M F Age: ______ Birth date ______ Height ______ Weight ______

HEALTH HISTORY

1. What medical concerns (e.g., pregnancy), if any, do you have at the present time?

____

2. Indicate whether you have had blood relatives with any of the following problems:

Cancer ☐ yes ☐ no
Diabetes ☐ yes ☐ no
Heart disease ☐ yes ☐ no
High cholesterol ☐ yes ☐ no
High blood pressure ☐ yes ☐ no
Osteoporosis ☐ yes ☐ no
Thyroid disorder ☐ yes ☐ no

3. Do you have complaints about any of the following?

_____ Appetite
_____ Bleeding gums
_____ Bruising
_____ Chewing or swallowing
_____ Constipation
_____ Diarrhea
_____ Edema
_____ Indigestion
_____ Menstrual difficulties
_____ Seeing in dim light
_____ Sudden weight change
_____ Stress

4. Do you use tobacco in any way? ☐ yes ☐ no How much? ______________
Did you recently stop smoking? ☐ yes ☐ no

5. Do you enjoy physical activity? ☐ yes ☐ no Explain: ______________

6. List any food allergies or intolerances.

____
____

DRUG HISTORY

List any prescribed, over-the-counter, herbal, or vitamin/mineral supplements you take.

____

____

____

DIET HISTORY

1. Do you follow a special dietary plan, such as low cholesterol, kosher, or vegetarian? ______________________________

2. Have you ever followed a special diet? _____ Explain: ______________________________

3. Do you have any problems purchasing foods that you want to buy? ______________________________

4. Are there certain foods that you do not eat? ______________________________

5. Do you eat at regular times each day? ☐ yes ☐ no How often? ______________________________

6. Identify any foods you particularly like. ______________________________

7. Do you drink alcohol? ☐ yes ☐ no How often? ______________________________

8. What change would you like to make?
   - ☐ Improve my eating habits
   - ☐ Improve my activity level
   - ☐ Learn to manage my weight
   - ☐ Improve my cholesterol/triglyceride levels
   - ☐ Other ______________

9. Please add any additional information you feel may be relevant to understanding your nutritional health. ______________________________

10. To tailor your counseling experience to your needs, it would be useful to know your expectations. Please check one of the following to indicate the amount of structure you believe meets your needs:
    - ☐ *Just tell me exactly what to eat for all my meals and snacks.* I want a detailed food plan. Example: ¾ cup corn flakes, 1 cup skim milk, 6 oz. orange juice, 1 slice whole wheat toast, 1 teaspoon margarine
    - ☐ *I want a lot of structure but freedom to select foods. I want to use the exchange system.* Example: 1 milk, 2 starch, 1 fruit, and 1 fat exchange
    - ☐ *I want some structure and freedom to select foods. I want to use a food group plan.* Example: 1 serving of dairy foods, fruits, and fat and oil group; 2 servings of grains
    - ☐ *I don't want a diet. I just want to eat better. I will just set food goals each week.*

SOCIOECONOMIC HISTORY

1. Circle the last year of school attended:

   1 2 3 4 5 6 7 8 Grade School   9 10 11 12 High School   1 2 3 4 College   M.A.   Ph.D.

   Other type of school ________________________________

2. Are you employed? _____ Occupation ________________________________

3. How many people in your household? _____ Ages? ________________________

4. Present marital status (circle one):

   Single   Married   Divorced   Widowed   Separated   Engaged

5. Do you have a refrigerator? _____ Stove? _____

6. Who prepares most of the meals in your home? __________ Shopping? __________

7. Do you use convenience foods daily? ☐ yes ☐ no

8. How often do you eat out? __________ Where? ____________________

9. Have you made any food changes in your life you feel good about? ☐ yes ☐ no

10. Who could support and encourage you to make these changes? ______________

EDUCATION INTERESTS

What information would you like from your counselor?

☐ Supermarket shopping tour
☐ Weight management
☐ Healthy food preparation
☐ Fiber
☐ Food labels
☐ Eating out
☐ Portion size
☐ Eating less fat
☐ Walking program
☐ Other ________________________
☐ Exercise
☐ Alcohol calories
☐ Meal planning
☐ Snack foods

Thank you for your willingness to share this information and to take part in the Nutrition Clinic. We look forward to working with you to make lifestyle changes to meet your food and fitness objectives.

Food Record

Name: ______________________________ **Date:** ______________

- Complete this form as accurately as possible, using the examples as a guide.
- Use only one form per day. Do not put anything on this form that pertains to another day.
- Record all foods and beverages, including water, you consumed from the time you wake up to the time you go to bed.

TIME	FOOD / DRINK	TYPE	PREPARATION	AMOUNT
8:00 A.M.	Bagel	Cinnamon raisin	Toasted	Half
8:00 A.M.	Milk	1% fat	Fresh	8 ounces
Noon	Chicken	Leg and thigh	Fried	1 each

24-Hour Recall/ Usual Diet Form

Date:________ **Day of the week:**__________

- Record food and fluid intake from time of awakening until the next morning.

FOOD AND DRINK CONSUMED			NUMBER OF SERVINGS FROM EACH GROUP Milk	Meat	Fruits	Veggies	Breads	Fats, Sweets
Standard for Adults			2-3	2-3	2-4	3-5	6-11	None
Time	Name and Type	Amount						
TOTALS								
***EVALUATION**								

*Evaluation: L = low A = adequate E = excessive

Food Group Serving Sizes

Using the Food Guide Pyramid Serving Sizes

BREADS, CEREALS, AND OTHER GRAIN PRODUCTS

1 slice bread
½ c cooked cereal, rice, or pasta
1 oz. ready-to-eat cereal
½ bun, bagel, or English muffin
1 small roll, biscuit, or muffin
3 to 4 small or 2 large crackers

VEGETABLES

½ c cooked or raw vegetables
1 c leafy raw vegetables
½ c cooked legumes
¾ c vegetable juice

FRUITS

typical portion: 1 medium apple, banana, or orange, ½ grapefruit, or 1 melon wedge
¾ c juice
½ c berries
½ c diced, cooked, or canned fruit
¼ c dried fruit

MEAT, POULTRY, FISH, AND ALTERNATES

2 to 3 oz. lean, cooked meat, poultry, or fish (total 5–7 oz. per day)
Count as 1 oz. meat or ⅓ serving: 1 egg, ½ c cooked legumes, 4 oz. tofu, 2 tbs. nuts, seeds, or peanut butter

MILK, CHEESE, AND YOGURT

1 c milk or yogurt
2 oz. process cheese food
1½ oz. cheese

FATS, SWEETS, AND ALCOHOLIC BEVERAGES

- Foods high in fat include margarine, salad dressing, oils, mayonnaise, sour cream, cream cheese, butter, gravy, sauces, potato chips, and chocolate bars.
- Foods high in sugar include cakes, pies, cookies, doughnuts, sweet rolls, candy, soft drinks, fruit drinks, jelly, syrup, gelatin, desserts, sugar, and honey.
- Alcoholic beverages include wine, beer, and liquor.

Food Frequency Questionnaire

SERVING SIZES	FOOD GROUP	SERVINGS PER DAY	SERVINGS PER WEEK	NEVER or RARELY
1 slice bread 1 c dry cereal ½ c cooked rice, pasta, or cereal ½ bun, bagel, or English muffin 1 small roll, biscuit, or muffin	**Refined Grains**—white bread, pasta, cereals			❑
	Whole Grains—whole-wheat bread, brown rice, oatmeal, bran cereal			❑
1 c raw leafy vegetable ½ c cooked or raw vegetables 6 oz. vegetable juice	**Vegetables**			❑
6 oz. fruit juice 1 medium fruit ¼ c dried fruit ½ c fresh, frozen, or canned fruit	**Fruits**			❑
8 oz. milk 1 c yogurt 1½ oz. cheese 2 oz. process cheese	**Dairy**—low-fat or fat-free ice cream, milk, cheese, yogurt; frozen yogurt			❑
	Dairy—whole milk, regular cheese, regular ice cream			❑
3 oz. cooked meats, poultry, or fish	**Meats, Poultry, Fish**—lean			❑
	Meats, Poultry, Fish—high-fat: sausage, cold cuts, spareribs, hot dogs, eggs, bacon			❑
⅓ c or 1½ oz. nuts 2 tbsp. or ½ oz. seeds ½ c cooked dry beans 4 oz. tofu, 1 c soy milk 2 tbsp. peanut butter	**Nuts, Seeds, and Dry Beans**			❑
1 tbsp. regular dressing 2 tbsp. light salad dressing 1 tsp. oil 1 tbsp. low-fat mayonnaise 1 tsp. margarine, butter	**Fats and Oils**			❑
8 oz. lemonade 1½ oz. candy 8 oz. soda	**Sweets**			❑
12 oz. beer, 4 oz. wine 1 shot hard liquor	**Alcohol**			❑

Food Group Feedback Form

FOOD GROUP	YOUR SERVINGS		RECOMMENDED SERVINGS	
	NUMBER	NEVER OR RARELY	PYRAMID	DASH
Refined Grains—white bread, pasta, cereals		❑		
Whole Grains—whole-wheat bread, brown rice, oatmeal, bran cereal		❑	6–11	7–8
Vegetables		❑	3–5	4–5
Fruits		❑	2–4	4–5
Dairy—low-fat, fat-free, low-fat ice cream or frozen yogurt		❑	2	2–3
Dairy—whole milk, regular ice cream, regular cheese		❑	—	—
Meats, Poultry, Fish—lean: poultry (no skin), egg whites		❑	2–3	2 or less
Meats, Poultry, Fish—high fat: hot dogs, cold cuts, sausage		❑	—	—
Nuts, Seeds, and Dry Beans		❑		4–5 per week
Fats and Oils		❑	Use sparingly	2–3
Sweets		❑	Use sparingly	5 per week
Alcohol		❑	Use sparingly	

Anthropometric Feedback Form

Volunteer's Measurements	Standard
Actual weight =	Insurance table =
	Dietary guidelines =
	Hamwi ideal body weight = BMI desirable weight =
Body mass index =	Desirable = 19–25
Waist circumference =	High risk = males, >102 centimeters (40 inches); females, >88 centimeters (35 inches)
Waist-to-hip ratio =	Increased risk = males ≥1.0; females, ≥ 0.8

Client Concerns and Strengths Log

1. List all concerns expressed by your client or identified by you.

2. Write NC (no control) next to of all concerns over which you or your client have no control.

3. Categorize in the following chart the remaining concerns over which there is some degree of control and as a result could be addressed by a goal:

Nutritional	Behavioral	Exercise

4. List strengths and skills.

5. Categorize the strengths and skills in the following chart:

Nutritional	Behavioral	Exercise

6. What strengths and skills can be used to address the concerns? List them in the following chart.

Strengths and Skills	Concerns	Possible Intervention Strategies

Client Progress Report

Name: ______________________________

Date	Issue	Action	Outcome (Anticipated)	Follow-up*

* O = ongoing; A = achieved; U = unrealistic.

Eating Behavior Journal

Name: ____________________

Day/Date: ____________________ **Physical Activities:**[1] ____________________

Time	Location/ Place	Foods and Beverages Consumed Amounts/Description	Degree of Hunger[2]	Social Situation[3]	Comments[4]

[1]Include type of activities and minutes engaged in the activities.
[2]Use the following rating scale: 0 = not hungry; 1 = hungry; 2 = very hungry.
[3]Indicate activities and who you were with, if anyone.
[4]Record significant thoughts ("I'm doing great"; "I am a loser"); feelings (angry, happy, worried); concerns ("Maybe I should have had the turkey sandwich")
Source: Adapted from Pastors et al., *Facilitating Lifestyle Change: A Resource Manual.* Chicago: American Dietetic Association; © 1996. Reprinted with permission.

Counseling Agreement

Name: ______________________________ **Date:** ____________________

My plan is to do the following:

____

____

This activity will be accomplished by

____

My reward will be (specify when, where, and what)

____

____

______________________________ ____________________

Your signature *Date*

______________________________ ____________________

Counselor's signature *Date*

Symptoms of Stress

Physical Symptoms

- Muscular tension
- Headaches
- Insomnia
- Twitching eyelid
- Fatigue
- Backaches
- Neck/shoulder pain
- Digestive disorders
- Teeth grinding
- Changes in eating/sleep patterns
- Sweaty palms

Emotional Symptoms

- Anxiety
- Frequent crying
- Irritability
- Frustration
- Depression
- Worrying
- Nervousness
- Moodiness
- Anger
- Self-doubt
- Resentment

Mental Symptoms

- Short concentration
- Forgetfulness
- Lethargy
- Pessimism
- Low productivity
- Confusion

Social Symptoms

- Loneliness
- Nagging
- Withdrawal from social contact
- Isolation
- Yelling at others
- Reduced sex drive

Sources: Adapted from Women First Health Care, www.womenfirst.com/ and Goliszek A, 60 *Second Stress Management.* Far Hills, NJ: New Horizon Press; 1992.

Stress Awareness Journal

Name: ______________________________ **Date:** ____________

Time	Symptom of Stress	Activities	Internal Self-Talk

Tips to Reduce Stress

- Learn to say no. Don't overcommit. Delegate tasks at home and work.
- Organize your time. Use a daily planner. Prioritize your tasks. Make a list and a realistic timetable. Check off tasks as they are completed. This gives you a sense of control for overwhelming demands and reduces anxiety.
- Be physically active. Big-muscle activities, such as walking, are the best for relieving tension.
- Develop a positive attitude. Surround yourself with positive quotes, soothing music, and affirming people.
- Relax or meditate. Schedule regular massages, use guided imagery tapes, or just take ten minutes for quiet reflection time in a park.
- Get enough sleep. Small problems can seem overwhelming when you are tired.
- Eat properly. Be sure to eat five servings of fruits and vegetables and three servings of whole grains every day. Limit intake of alcohol and caffeine.
- To err is human. Don't create a catastrophe over a mistake. Ask yourself what will be the worst thing that will happen.
- Work at making friends and being a friend. Close relationships don't just happen. Compliment three people today. Send notes to those who did a good job.
- Accept yourself. Appreciate your talents and your limitations. Everyone has them.
- Laugh. Look at the irony of a difficult situation. Watch movies and plays and read stories that are humorous.
- Take three deep breaths.
- Forgive. Holding onto grudges only causes you more stress and pain.

Prochaska and DiClemente's Spiral of Change

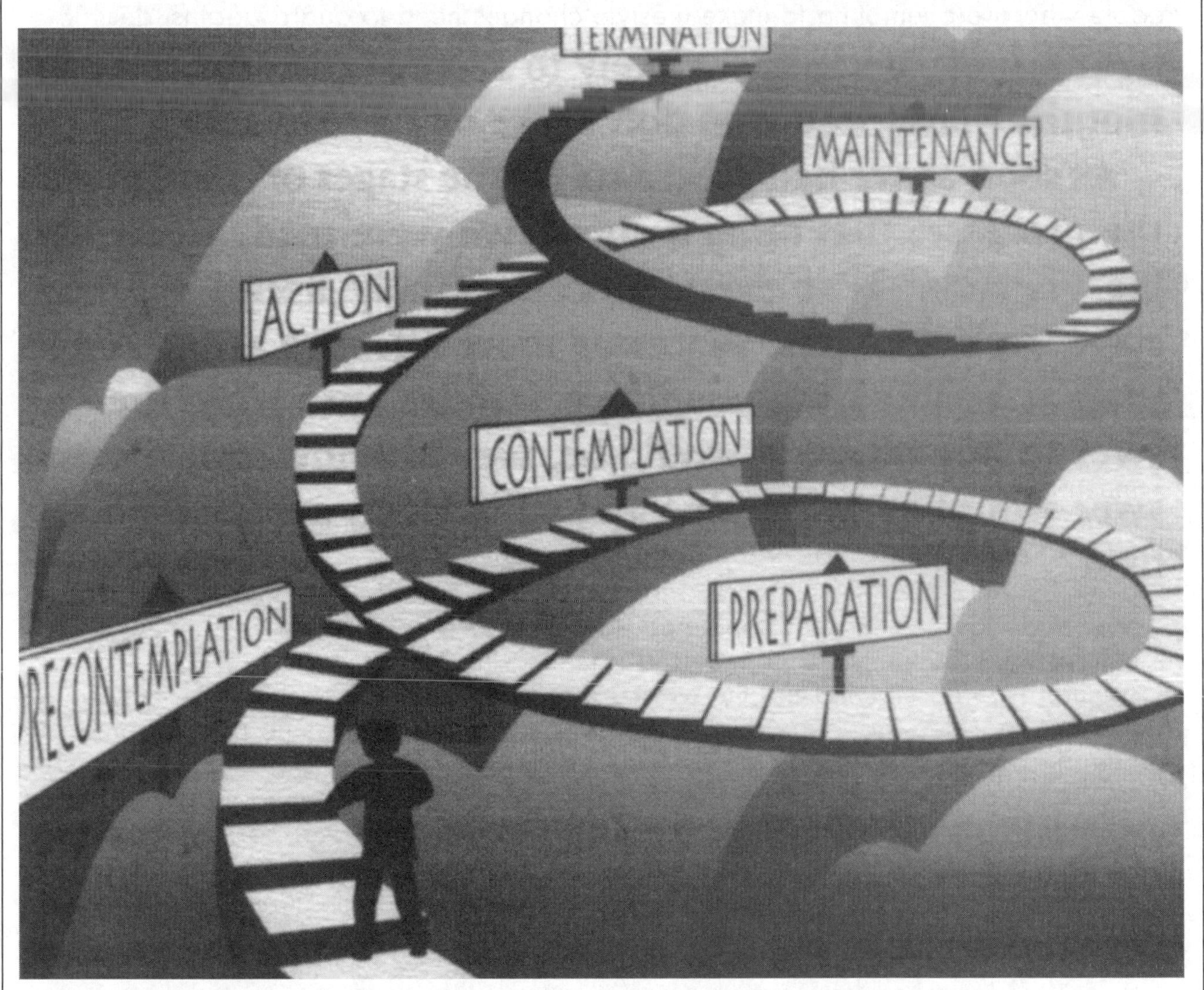

Source: Prochaska JO, Norcross JC, DiClemente CC, *Changing for Good*. New York: Avon; © 1994, p. 49. Used with permission.

Countering Negative Thinking

People who are attempting to make lifestyle changes need to guard against destructive negative thinking. For example, a person who says such statements to himself or herself as the following needs to find substitutions:

- *"A physical activity program is out for me. A woman at the gym said I should be ashamed of myself! She is right! I am going to ignore the people in my support group who say I should accept and love myself. Besides, I tried walking once, but I got a blister. That just goes to show that I wasn't made for exercise! Probably if I walked every day my blood pressure wouldn't come down anyway."* The talk exhibited here will certainly lead to defeat. This woman is focusing on negative feedback, generalizing that a single blister indicates she should stop walking, and assuming the worst outcome will happen. This talk could be transformed into "I am still searching for a way to make physical activity work for me."
- *"I did have an orange for a snack instead of my usual donut, but that didn't mean much because later I ate some potato chips. First, I just ate just one chip, and then I figured, 'this is absolutely awful so the diet is over and I might as well eat the whole bag of chips.' I guess I just don't have willpower. I am such a jerk!"* This person should give herself more credit for eating fruit for a snack, and focus on how and why that success happened. Identifying an episode as "awful" is not helpful since such a label can lead to a feeling that the situation is so bad that a solution is impossible. Using the word "absolutely" compounds the negativity of the phrase. The idea that once she started eating chips, there was no use stopping often happens when certain foods are considered off-limits. Also, blaming an indulgence on lack of willpower is always counterproductive since the characteristic is considered a personal failing so a change in lifestyle could not possibly occur. Instead, this woman should ask herself what she learned from the situation, and tell herself what she will do differently next time. In addition, all clients should be discouraged from using derogatory terms to describe themselves, such as "jerk." Once denigrated, a person is not likely to expect success in future attempts at lifestyle change. Instead, the person should remind himself or herself that he or she is learning so that better choices can be made in the future.
- *"I really do not want to go on the hayride because I must not have the cookies and hot chocolate afterwards. The other people in my group ought to be more considerate of the fact that I have diabetes. They should know better. I can't stand this!"* Sometimes people focus on one small difficulty and distort the total picture. Instead of searching for acceptable options, this obstacle thinker is caught-up criticizing others. This person uses words such as "should," "ought," and "must" to create an impossible standard, resulting in negative feelings and can lead to a relapse.

Benefits of Regular *Moderate* Physical Activity

➤ Reduces risk of dying prematurely

➤ Reduces risk or aids in the management of
- heart disease,
- diabetes,
- high blood pressure,
- colon cancer,
- strong bones, and
- falls.

➤ Improves mood, self-esteem, and self-image

➤ Increases energy

➤ Maintains weight or aids loss of weight

➤ Maintains function and preserves independence in older adults

Physical Activity Log

- Record all physical activity for a week. Remember to include regular daily activities such as climbing stairs, gardening, and walking to the office from a parking lot.
- Include all forms of physical fitness activities including stretching, weight lifting, balancing, and aerobic movement.

Day of the Week	Type of Activity	Amount of Time
Sunday		
Monday		
Tuesday		
Wednesday		
Thursday		
Friday		
Saturday		

Physical Activity Options

➤ Look for Everyday Opportunities

Short bursts of activity throughout the day make a difference.

- Use steps instead of elevators or escalators.
- Park your car in a distant section of the parking lot.
- Leave work five minutes later. Take a walk around the building.
- Get off the train or bus one stop earlier and walk the rest of the way.
- Take a walk during lunch.
- March, stretch, or do squats while brushing your teeth.
- Pace around the house or do arm curls with a can of food while talking on the phone.
- Jump rope, stretch, jog in place, or lift weights while watching TV.
- Be prepared. Keep walking shoes in your car or in your desk.
- Take your bike with you to a conference and explore the local scenery before driving home.

➤ Plan a Daily Routine

Think about cost, convenience, and bad weather options when planning a program. Look for creative ways to keep the activities enjoyable.

- Schedule time for physical activity. Write it in your calendar.
- Vary the physical activities. Plan to bike one day a week, jog two days a week, and go to the gym three days a week.
- Join a walking club, a biking club, and so forth.
- Add variety to the activity. Have several walking trails, ask a friend to join you in your walks, or listen to music or recorded books during your walks.

➤ Plan Physically Active Leisure-Time Events

Look for activities the whole family can enjoy.

- Have a family baseball or soccer game.
- Plan a bike tour, mountain hike, or canoe trip.
- Explore a cave.

Need more ideas? The American Heart Association has an inexpensive paperback with hundreds of simple, affordable, and practical ideas. *Fitting in Fitness* (Times Books–Random House, 1997) is available in bookstores.

Physical Activity Medical Readiness Form

Regular physical activity is fun and healthy and being more active is very safe for most people. Checking with your doctor is always a good idea before starting to become much more physically active. The questions below can help guide you on the necessity of getting a physician's opinion. Your best guide when answering the questions is to use common sense. Please read the questions carefully and check YES or NO.

YES	NO	
❑	❑	1. You have a heart condition, and your doctor recommends only medically supervised physical activity.
❑	❑	2. During or right after you exercise, you frequently have pains or pressure in the left or midchest area, left side of your neck, or left shoulder or arm.
❑	❑	3. You have developed chest pain within the last month.
❑	❑	4. You tend to lose consciousness or fall over because of dizziness.
❑	❑	5. You feel extremely breathless after mild exertion.
❑	❑	6. Your doctor recommended that you take medicine for high blood pressure or a heart condition.
❑	❑	7. You have bone or joint problems.
❑	❑	8. You have a medical condition or other physical reason not mentioned here that might need special attention in an exercise program (such as insulin-dependent diabetes).
❑	❑	9. You are more than 25 to 30 pounds overweight.
❑	❑	10. You are a man over the age of 40 or a woman over the age of 50, have not been physically active, and are planning a vigorous exercise program.

Source: American Heart Association. *Fitting in Fitness.* New York: Times Books; 1997, p.33. Reprinted with permission. The American Heart Association checklist was developed from several sources, particularly the Physical Activity Readiness Questionnaire, British Columbia Ministry of Health, Department of National Health and Welfare, Canada (revised 1992).

If you answered YES to one or more questions:

Talk with your doctor by phone or in person BEFORE you start becoming much more physically active or BEFORE you have a fitness appraisal.

- You may be able to do any activity you want—as long as you start slowly and build up gradually. Or, you may need to restrict your activities to those that are safe for you. Talk with your doctor about the kinds of activities you wish to participate in and follow his or her advice.
- Develop an exercise plan with the aid of an exercise specialist.

If you answered NO honestly to *all* the questions, you can be reasonably sure that you can:

- Start becoming much more physically active—begin slowly and build up gradually. This is the safest and easiest way to go.
- Take part in a fitness appraisal—this is an excellent way to determine your basic fitness so that you can plan the best way for you to live actively.

DELAY BECOMING MUCH MORE ACTIVE:

- If you are not feeling well because of a temporary illness such as a cold or a fever—wait until you feel better; or
- If you are or may be pregnant—talk to your doctor before you start becoming more active.

Please note: If your health changes so that you then answer YES to any of these questions, tell your fitness or health professional. Ask whether you should change your physical activity plan.

Physical Activity Status

Moderate physical activity includes swimming, cycling, dancing, gardening, domestic and occupational activities at an intensity level equivalent to 30 minutes of brisk walking.

Vigorous physical activity refers to activities that make you work as hard as jogging for 20 minutes; generally you sweat and feel out of breath, and your heart rate increases. Activities in this category include running, lap swimming, jumping rope, and cross-country skiing.

Muscular strength activities include weight training using dumbbells or machines or resistance activities using elastic bands.

Flexibility training activities include stretching, yoga, and T'ai Chi Chuan.

Leisure Time Physical Activity Status

Muscular Strength Activity (Check if the statement applies to you.)

❑ I am involved in muscular strength activities consisting of at least one set of 8 to 10 exercises (8–12 repetitions of each) that conditions the major muscle groups at least 2 times per week.

Flexibility Training Activity (Check if the statement applies to you.)

❑ I am involved in flexibility exercises that stretch the major muscle groups at least 2 times per week.

Moderate or Vigorous Activity (Circle one number only.)

1. I do not exercise or walk regularly now, and I do not intend to start in the near future.
2. I do not exercise or walk regularly, but I have been thinking of starting.
3. I am trying to start to exercise or walk. *(or)* During the last month I have started to exercise or walk on occasion *(or on weekends only)*.
4. I am doing vigorous or moderate exercise, less than 3 times per week *(or moderate exercise less than 2 hours per week)*.
5. I have been doing moderate or vigorous exercise, 3 or more times per week *(or more than 2 hours per week)* for the last 1 to 6 months. If this is the case, put a check next to either of the following if they apply to you:

   ❑ I have been doing at least 30 minutes of moderate activity or 20 minutes of vigorous activity most days of the week.

   ❑ I have been doing at least 20 minutes of vigorous activity 3 or more days of the week.

Source: This form is based on What is Your PACE SCORE assessment form. Long BL et al., *Project PACE Physician Manual.* Atlanta, GA: Centers for Disease Control, Cardiovascular Health Branch, 1992. Physical activity standards are from American College of Sports Medicine Position Stand, *Med Sci Sports Exerc.* 1998:30(6):975–991 and U.S. Department of Health and Human Services, *Healthy People 2010.*

Physical Activity Standard†	Standard Met	Standard Not Met
Muscular strength: Engage in strength activities consisting of one set of eight to ten exercises (8–12 repetitions of each) that conditions the major muscle groups at least 2 times per week.	❑	❑
Flexibility: Engage in activities that stretch major muscle groups at least 2 times per week.	❑	❑
Endurance: Engage in at least 30 minutes of moderate activity or 20 minutes of vigorous activity most days of the week.	❑	❑
Endurance: Engage in at least 20 minutes of vigorous activity 3 or more days of the week.	❑	❑

Motivation Level	Implication
Level 1—Not ready	❑ Would you consider learning more about how moderate physical activity could help your health?
Level 2—Unsure	❑ For some reason you are not sure that you are ready to begin a physical activity program. Your counselor will explore your ambivalence with you to see whether you are ready to make plans to increase your physical activity level.
Level 3—Ready	❑ Great—you are ready to begin or increase your activity level. Your counselor can provide you with resources to aid in developing a plan.
Level 4—Active	❑ Congratulations—you are already actively involved in a physical activity program. Your counselor will review with you the standards set by authorities. If you do not meet all of them, you may wish to make some alterations.

Physical Activity Readiness

❑ Talk to your doctor before becoming much more physically active or having a fitness appraisal as indicated by the following:

❍ Medical Readiness Questions ❍ Woman over age 50 ❍ Man over age 40

❑ Delay an increase in physical activity due to pregnancy or illness.

†Standards are based on American College of Sports Medicine Position Standards, 1998 and Healthy People 2010 physical activity goals.
Note: Reevaluate readiness if you experience dizziness, chest pain, undue shortness of breath, difficulty breathing, or unusual discomfort after beginning an exercise program.

Interview Checklist

Interviewer:________________ Observer: ________________ Date:________

Goal of the interview: ________________________________

I. FLOW OF THE INITIAL INTERVIEW

A. Involving Phase

1. Greeting — Yes ❑ No ❑
   a. Verbal greeting — Yes ❑ No ❑
   b. Shakes hands — Yes ❑ No ❑
2. Introduction of self — Yes ❑ No ❑
3. Attention to self-comfort—other obligations finished or planned for a later time, attention focused (self-evaluation only) — Yes ❑ No ❑
4. Attention to client's comfort—physical comfort, noise and visual distractions minimized — Yes ❑ No ❑
5. Small talk, if appropriate — Yes ❑ No ❑
6. Establishes counseling objectives — Yes ❑ No ❑
   a. Opening question—*What brings you here today?* — Yes ❑ No ❑
   b. Establishes client's long-term objectives — Yes ❑ No ❑
   c. Explains counseling process — Yes ❑ No ❑
   d. Discusses weight monitoring, if appropriate — Yes ❑ No ❑
7. Establishes agenda — Yes ❑ No ❑
8. Transition statement—*Now that we have gone over the basics of the program, we can explore your needs in greater detail.* — Yes ❑ No ❑

B. Exploration-Education Phase

1. Reviews completed assessment forms — Yes ❑ No ❑
2. Compares assessment to a standard, point by point, nonjudgmental — Yes ❑ No ❑
3. Asks client thoughts about comparison — Yes ❑ No ❑
4. Segment summary—identifies problems, reiterates self-motivational statement, checks accuracy — Yes ❑ No ❑
5. Asks client whether he or she would like to make changes — Yes ❑ No ❑
6. Assesses motivation—use a ruler to determine readiness to change — Yes ❑ No ❑
7. Tailors educational experiences to client needs — Yes ❑ No ❑

Source: This evaluation form is based on the Brown Interview Checklist, Brown University School of Medicine, Novack, DH, Goldstein, MG, Dubé CE, 1986. Used with permission.

C. Resolving Phase

Level 1 (0–4 on ruler)

1. Raises awareness—discusses benefits of change	Yes ❑	No ❑
2. Raises awareness—personalizes benefits	Yes ❑	No ❑
3. Asks key open-ended questions regarding importance of change	Yes ❑	No ❑
4. Segment summary	Yes ❑	No ❑
5. Offers advice, if appropriate	Yes ❑	No ❑
6. Expresses support	Yes ❑	No ❑

Level 2 (4–8 on ruler)

1. Raises awareness—discusses benefits of change and diet options	Yes ❑	No ❑
2. Asks key open-ended questions regarding confidence in ability to change	Yes ❑	No ❑
3. Asks key open-ended questions to identify barriers	Yes ❑	No ❑
4. Examines pros and cons	Yes ❑	No ❑
5. Imagines the future	Yes ❑	No ❑
6. Explores past successes	Yes ❑	No ❑
7. Explores support networks	Yes ❑	No ❑
8. Summarizes ambivalence	Yes ❑	No ❑

Level 3 (8–12 on ruler)

1. Praises positive behaviors	Yes ❑	No ❑
2. Explores change options	Yes ❑	No ❑
a. Asks client's ideas for change	Yes ❑	No ❑
b. Uses an options tool, if appropriate	Yes ❑	No ❑
c. Explores concerns regarding selected option	Yes ❑	No ❑
3. Explains goal setting process	Yes ❑	No ❑
4. Identifies a specific goal from a broad goal—uses small talk, explores past experiences, builds on past	Yes ❑	No ❑
5. Goal is achievable, measurable, under client control, stated positively	Yes ❑	No ❑
6. Designs a plan of action	Yes ❑	No ❑
a. Investigates physical environment	Yes ❑	No ❑
b. Examines social support	Yes ❑	No ❑
c. Examines cognitive environment; explains coping talk, if needed	Yes ❑	No ❑
d. Defines a tracking technique	Yes ❑	No ❑
e. Client verbalizes goal	Yes ❑	No ❑
7. Writes down goal	Yes ❑	No ❑

D. Closing Phase

1. Supports self-efficacy — Yes ☐ No ☐
2. Reviews issues and strengths — Yes ☐ No ☐
3. Uses relationship-building response—respect — Yes ☐ No ☐
4. Restates food goal — Yes ☐ No ☐
5. Reviews next meeting time — Yes ☐ No ☐
6. Shakes hands — Yes ☐ No ☐
7. Expresses appreciation for participation — Yes ☐ No ☐
8. Uses relationship-building responses—support and partnership — Yes ☐ No ☐

II. INTERPERSONAL SKILLS

A. Facilitation (Attending) Skills

1. Eye contact—appropriate length to enhance client comfort — Yes ☐ No ☐
2. Uses silences to facilitate client's expression of thoughts and feelings — Yes ☐ No ☐
3. Open posture—arms uncrossed, facing client — F ☐ P ☐ No ☐
4. Head nod, "Mm-hm," repeats client's last statement — F ☐ P ☐ No ☐

F = Frequently; P = Partially

B. Relationship Skills (Conveying Empathy)

1. Reflection—restates the client's expressed emotion or inquires about emotions — F ☐ P ☐ No ☐
2. Legitimation—expresses understandability of client's emotions — Yes ☐ No ☐
3. Respect—expresses respect for the client's coping efforts or makes a statement of praise — Yes ☐ No ☐
4. Support—expresses willingness to be helpful to client addressing his or her concerns — Yes ☐ No ☐
5. Partnership—expresses willingness to work together with client — Yes ☐ No ☐

F = Frequently; P = Partially

III. PATIENT RESPONSES

	OFTEN	SOMETIMES	SELDOM
A. Client freely discusses his or her concerns.			
B. Client appears comfortable and relaxed.			
C. Client appears engaged in the counseling session.			
D. Client freely offers information about his or her condition and life context.			

IV. GENERAL COMMENTS

Counseling Responses Competency Assessment

Audio- or videotape a counseling session, and listen to the tape several times to complete the following assessment:

- Track the number of times *you* made each response by placing slash marks next to the name of the response. Note that this is an evaluation of your responses, not your client's responses.
- For each category of responses, give an example from the tape. In cases where the particular response category was not demonstrated on the tape, write an example that may have been effective with your client and then complete the category evaluation.
- Select an intent and focus of the response. You may wish to review a discussion of these topics in Chapter 2.
- Indicate the effectiveness of your particular response, and explain why it was or was not effective. For responses that do not receive the most effective rating, write alternative responses that you believe would have worked better.
- Some of your responses may not fit any of the categories. This assessment covers many basic counseling responses, but it is possible that some of your statements do not appear to fit into any of the categories. If that is the case, such material would not be evaluated. The following is an example of a competency evaluation for one response:

Example

Questions ///

Example *What brings you here? Are you looking to lower your blood pressure?*

Intent (circle one): *To acknowledge* *(To explore)* *To challenge*

Focus (circle one): *information* *(experience)* *feelings* *thoughts* *behaviors*

❑ Effective ❑ Somewhat Effective ❑ Not Effective Explain *I asked two questions at the same time. I made an assumption that the main issue was blood pressure.*

Alternative Response *What brings you here today?*

1. *Attending* ________

Example ________________________________

Intent (circle one): *To acknowledge* *To explore* *To challenge*

Focus (circle one): *information* *experience* *feelings* *thoughts* *behaviors*

❑ Effective ❑ Somewhat Effective ❑ Not Effective Explain ________________

Alternative Response ________________________________

2. *Reflection (Empathizing)* ________

Example ____

Intent (circle one): *To acknowledge* *To explore* *To challenge*

Focus (circle one): *information* *experience* *feelings* *thoughts* *behaviors*

❑ Effective ❑ Somewhat Effective ❑ Not Effective Explain ____________

Alternative Response ________________________________

3. *Legitimation* ________

Example ____

Intent (circle one): *To acknowledge* *To explore* *To challenge*

Focus (circle one): *information* *experience* *feelings* *thoughts* *behaviors*

❑ Effective ❑ Somewhat Effective ❑ Not Effective Explain ____________

Alternative Response ________________________________

4. *Respect* ________

Example ____

Intent (circle one): *To acknowledge* *To explore* *To challenge*

Focus (circle one): *information* *experience* *feelings* *thoughts* *behaviors*

❑ Effective ❑ Somewhat Effective ❑ Not Effective Explain ____________

Alternative Response ________________________________

5. *Personal Support* ________

Example ____

Intent (circle one): *To acknowledge* *To explore* *To challenge*

Focus (circle one): *information* *experience* *feelings* *thoughts* *behaviors*

❑ Effective ❑ Somewhat Effective ❑ Not Effective Explain ____________

Alternative Response ________________________________

6. *Partnership* ________

Example ____

Intent (circle one): *To acknowledge* *To explore* *To challenge*

Focus (circle one): *information* *experience* *feelings* *thoughts* *behaviors*

❑ Effective ❑ Somewhat Effective ❑ Not Effective Explain ____________

Alternative Response ________________________________

7. *Mirroring (Parroting)* ________

Example ____________________

Intent (circle one): *To acknowledge* *To explore* *To challenge*

Focus (circle one): *information* *experience* *feelings* *thoughts* *behaviors*

❑ Effective ❑ Somewhat Effective ❑ Not Effective Explain ____________________

Alternative Response ____________________

8. *Paraphrasing* ________

Example ____________________

Intent (circle one): *To acknowledge* *To explore* *To challenge*

Focus (circle one): *information* *experience* *feelings* *thoughts* *behaviors*

❑ Effective ❑ Somewhat Effective ❑ Not Effective Explain ____________________

Alternative Response ____________________

9. *Giving Feedback (Immediacy)* ________

Example ____________________

Intent (circle one): *To acknowledge* *To explore* *To challenge*

Focus (circle one): *information* *experience* *feelings* *thoughts* *behaviors*

❑ Effective ❑ Somewhat Effective ❑ Not Effective Explain ____________________

Alternative Response ____________________

10. *Questioning* ________

Example ____________________

Intent (circle one): *To acknowledge* *To explore* *To challenge*

Focus (circle one): *information* *experience* *feelings* *thoughts* *behaviors*

❑ Effective ❑ Somewhat Effective ❑ Not Effective Explain ____________________

Alternative Response ____________________

11. *Clarifying (Probing, Prompting)* ________

Example ____________________

Intent (circle one): *To acknowledge* *To explore* *To challenge*

Focus (circle one): *information* *experience* *feelings* *thoughts* *behaviors*

❑ Effective ❑ Somewhat Effective ❑ Not Effective Explain ____________________

Alternative Response ____________________

12. ***Noting a Discrepancy (Confrontation, Challenging)*** ________

Example ____

Intent (circle one): *To acknowledge* *To explore* *To challenge*

Focus (circle one): *information* *experience* *feelings* *thoughts* *behaviors*

❑ Effective ❑ Somewhat Effective ❑ Not Effective Explain ____________

Alternative Response ____

13. ***Directing (Instructions)*** ________

Example ____

Intent (circle one): *To acknowledge* *To explore* *To challenge*

Focus (circle one): *information* *experience* *feelings* *thoughts* *behaviors*

❑ Effective ❑ Somewhat Effective ❑ Not Effective Explain ____________

Alternative Response ____

14. ***Advice*** ________

Example ____

Intent (circle one): *To acknowledge* *To explore* *To challenge*

Focus (circle one): *information* *experience* *feelings* *thoughts* *behaviors*

❑ Effective ❑ Somewhat Effective ❑ Not Effective Explain ____________

Alternative Response ____

15. ***Allowing Silence*** ________

Example ____

Intent (circle one): *To acknowledge* *To explore* *To challenge*

Focus (circle one): *information* *experience* *feelings* *thoughts* *behaviors*

❑ Effective ❑ Somewhat Effective ❑ Not Effective Explain ____________

Alternative Response ____

16. ***Self-Referent*** ________

Example ____

Intent (circle one): *To acknowledge* *To explore* *To challenge*

Focus (circle one): *information* *experience* *feelings* *thoughts* *behaviors*

❑ Effective ❑ Somewhat Effective ❑ Not Effective Explain ____________

Alternative Response ____

Registration for Nutrition Clinic

Counselor

____
Name

____
Business telephone

Best times to call: _______________

____

____
Home telephone

Best times to call: _______________

____

____
E-mail

____
Fax

Participant

____
Name

____
Business telephone

Best times to call: _______________

____

____
Home telephone

Best times to call: _______________

____

____
E-mail

____
Fax

Your meeting day is: _______________

Location of meetings: _______________

Your meeting time is: _______________

Room number: _______________

Length of meetings is approximately one hour. If welcome packet forms have not been completed previous to the first session, the first counseling session may take an extra twenty minutes.

The dates of your four meetings are as follows: _______________ _______________ _______________ _______________

- Please complete two copies of this agreement form. The client copy should be given to the participant, and the clinic copy should be given to the counselor.
- Thank you for your interest in our program. Please note that any cancellations of meetings should be made directly between each participant and counselor.
- If you have any questions about the program, please call the instructor, _______________, at _______________.

Index

Bold numbers indicate major discussions of topic.